Structural Analysis in Microelectronic and Fiber-Optic Systems

Volume I

Basic Principles of Engineering Elasticity and Fundamentals of Structural Analysis

E. SUHIR, Ph.D.

AT&T Bell Laboratories
Murray Hill, New Jersey

VAN NOSTRAND REINHOLD
New York

Library of Congress Catalog Card Number 90-24201
ISBN 0-442-20771-9

Manufactured in the United States of America.

Published by Van Nostrand Reinhold
115 Fifth Avenue
New York, New York 10003

Chapman and Hall
2–6 Boundary Row
London, SE1 8HN, England

Thomas Nelson Australia
102 Dodds Street
South Melbourne 3205
Victoria, Australia

Nelson Canada
1120 Birchmount Road
Scarborough, Ontario MIK 5G4, Canada

16 15 14 13 12 11 10 9 8 7 6 5 4 3 2 1

Library of Congress Cataloging-in-Publication Data

Suhir, Ephraim.
Structural analysis in microelectronic and fiber-optic systems / Ephraim Suhir.
p. cm.
Includes bibliographical references and index.
ISBN 0-442-20771-9 (Van Nostrand Reinhold)
1. Microelectronics. 2. Fiber optics. 3. Structural analysis (Engineering) I. Title.
TK7874.S856 1991 90-24201
621.381—dc20 CIP

Contents

Preface *ix*
List of Basic Symbols *xiii*

Introduction **1**

I.1 Research Models in Mechanical Problems for Microelectronics and Fiber Optics 1
I.2 Theoretical Modeling 3
I.3 Analytical versus Numerical Modeling 6
I.4 Interaction with Experiment 8
I.5 Theoretical Modeling in Structural Analysis 8
I.6 Historical Sketch 11

Part 1. BASIC PRINCIPLES OF ENGINEERING ELASTICITY **13**

1. General Properties of Elastic Bodies **15**

2. Equations and Conceptions **18**

2.1 Stress 18
2.2 Stress Components 19
2.3 Stress Tensor 20
2.4 Equilibrium Conditions for an Elementary Tetrahedron 22
2.5 Equilibrium Conditions for an Elementary Parallelepiped 25
2.6 Principal Stresses and Stress Invariants 26
2.7 Normal Stress Surface and Stress Ellipsoid 29

2.8 Principal Shearing Stresses 30
2.9 Octahedral Stresses 31
2.10 Stress Deviator and Spherical Tensor 31
2.11 Displacements and Strains 33
2.12 Compatibility Equations 35
2.13 Strain Tensor, Principal Strains, and Strain Invariants 37
2.14 Hooke's Law 38
2.15 Young's Modulus and Poisson's Ratio 40
2.16 Hooke's Law with Consideration of Thermoelastic Strains 43

3. A View of Solution Procedures 45

3.1 Direct and Inverse Problems in Elasticity Theory 45
3.2 Semi-Inverse Method and Principle of Saint-Venant 46
3.3 Beltrami-Michell, Lamé, and Duhamel-Neumann Equations 47

4. The Elementary Problems 49

4.1 Pure Bending of a Prismatic Bar 49
4.2 Simple Torsion of a Circular Shaft 53
4.3 All-Round Uniform Compression 55
4.4 Prismatic Bar Subjected to Tension 56
4.5 Prismatic Bar Under Its Own Weight 56
4.6 Pure Bending of Plates 57

5. Strength Theories 60

5.1 Strain Energy 60
5.2 Classical Strength Theories 61
5.3 Some Modern Strength Theories 62

6. Two-Dimensional Problem in Rectangular Coordinates 66

6.1 Plane Strain and Plane Stress 66
6.2 Maurice Lévy Theorem 68
6.3 Stress (Airy) Function 69

6.4 Solution in Polynomials 70
6.5 Solution in Trigonometric Series 73

7. Two-Dimensional Problem in Polar Coordinates 82

7.1 Basic Relationships 82
7.2 Stresses and Displacements in a Circular Ring 86
7.3 A Circular Ring Loaded at Its Inner Boundary 89
7.4 Axisymmetric Problem 91
7.5 Thick-Walled Tube 93
7.6 Thermal Stress in Coaxial Cylinders 94

8. Torsion 98

8.1 Saint-Venant's Torsion Function 98
8.2 Prandtl's Stress Function 100
8.3 Prandtl's Formula 101
8.4 Prandtl's Formula for a Multiply Connected Region 103
8.5 Torsion of an Elliptical Shaft 104
8.6 Torsion of a Rectangular Shaft 107
8.7 Bredt's Theorem 109
8.8 Torsion of a Closed Thin-Walled Section 110
8.9 Torsion of a Multiconnected Section 112

9. Fracture Mechanics 115

9.1 The Griffith Theory 115
9.2 Irwin's Stress Intensity Approach 117
9.3 K_{IC} Tests 121
9.4 Fracture Toughness 121

10. Plasticity 123

10.1 Uniaxial Plastic Deformation 123
10.2 Multiaxial Plastic Deformation: Small Elastoplastic Strains 125
10.3 Elastoplastic Bending of Beams 130
10.4 Theory of Plastic Flow 138

11. Viscoelasticity **143**

11.1 Viscoelastic Materials 143
11.2 Time Effects 144
11.3 Mathematical Models of Prerupture Deformation of Plastics 149
11.4 Theory of Linear Viscoelasticity 150
11.5 Thermodynamics of Viscoelastic Deformation 155

Questions and Problems **160**
Bibliography **166**

PART 2. FUNDAMENTALS OF STRUCTURAL ANALYSIS **167**

12. Bending of Beams **169**

12.1 Basic Definitions, Hypotheses, and Relationships 170
12.2 Solutions Based on Direct Integration of the Equation of Bending 180
12.3 Method of Initial Parameters 183
12.4 Reference Tables for Beam Deflections 186
12.5 Shear Deformations 187
12.6 Beams on Elastic Foundations 190
12.7 Effect of Shear on the Bending of Beams on Elastic Foundations 194
12.8 Beams Under Combined Action of Lateral and Axial Loads 198
12.9 Nonlinear Bending Under Combined Action of Lateral and Axial Loads 201
12.10 Large Deflections of Bent Beams (the Elastica) 205
12.11 Thermal Bending of Beams 209

13. The Variational and Energy Methods, and Some General Principles of Structural Analysis **213**

13.1 Variational Methods 213
13.2 Variational Principles of Structural Analysis 214
13.3 The Rayleigh–Ritz and Bubnov–Galerkin Methods 226
13.4 Finite-Element Method 234

14. Bending of Frames **243**

14.1 Frame Structures, Simplest Frames 243
14.2 Complex Frames 248
14.3 Force Method and Deformation Method 254

15. Bending of Plates **256**

15.1 Major Definitions and Assumptions 256
15.2 Bending of Rectangular Plates 258
15.3 Bending of Circular Plates 307

16. Buckling **324**

16.1 Buckling of Bars 324
16.2 Buckling of Plates 340

17. Numerical Methods **353**

17.1 Finite Differences 353
17.2 Collocation Method 358
17.3 Finite Elements 360

18. Experimental Techniques **374**

18.1 Introductory Remarks 374
18.2 Bonded Strain Gauges 375
18.3 Photoelastic Analysis 380
18.4 Moiré Method 387

Questions and Problems **397**
Bibliography **401**

Appendix : Tables of Beam Deflections **403**

Index **413**

To the memory of my father

Preface

This book contains the fundamentals of a discipline, which could be called Structural Analysis in Microelectronics and Fiber Optics. It deals with mechanical behavior of microelectronic and fiber-optic systems and is written in response to the crucial need for a textbook for a first in-depth course on mechanical problems in microelectronics and fiber optics. The emphasis of this book is on electronic and optical packaging problems, and analytical modeling. This book is apparently the first attempt to select, advance, and present those methods of classical structural mechanics which have been or can be applied in various stress-strain problems encountered in "high technology" engineering and some related areas, such as materials science and solid-state physics.

The following major objectives are pursued in *Structural Analysis in Microelectronic and Fiber-Optic Systems*:

Identify structural elements typical for microelectronic and fiber-optic systems and devices, and introduce the student to the basic concepts of the mechanical behavior of microelectronic and fiber-optic structures, subjected to thermally induced or external loading.

Select, advance, and present methods for analyzing stresses and deflections developed in microelectronic and fiber-optic structures; demonstrate the effectiveness of the methods and approaches of the classical structural analysis in the diverse mechanical problems of microelectronics and fiber optics; and give students of engineering, as well as practicing engineers and designers, a thorough understanding of the main principles involved in the analytical evaluation of the mechanical behavior of microelectronic and fiber-optic systems.

Provide reference information for research and practicing engineers encountering stress-strain related problems in microelectronics and fiber optics.

Discuss, wherever feasible and possible, how structural parameters can be changed to achieve a desired mechanical performance of the structure, and to teach a creative designer how to extend and/or modify the presented analytical model to address other practically important problems.

It should be pointed out that the overwhelming majority of papers dealing with mechanical problems in microelectronics and fiber optics are experimental. Not too many investigations use numerical (mostly finite-element) methods to solve stress-strain problems. There are very few studies devoted to analytical stress modeling in microelectronics and fiber optics. Although experimental techniques and numerical methods are briefly discussed in this book, I focused primarily on the developing of simple and easy-to-use analytical relationships which would clearly indicate the role of the major factors affecting the mechanical behavior of the system. This would enable designers to predict the effect of thermal or external loading on the structure they deal with, and assist them in finding a way to alter this structure so that a particular mechanical performance is achieved.

The book is intended for a rather broad audience which, in one way or another, is interested in stress and mechanical reliability problems in microelectronics and fiber optics: college students and teachers, students and instructors of various continuing education programs, practicing engineers and designers, technical managers, research engineers, and scientists. The book is also well suited as a guide for self-study.

Although directed principally to specialists in the area of microelectronics and fiber optics, this book can also be of help to those involved in other areas of engineering and applied science, where similar structural elements are used or where the same or similar analytical methods and models are applicable.

I have tried to provide a combination of a clear, detailed, and in-depth introductory manual, as well as an up-to-date reference source on some basic stress-strain problems in microelectronics and fiber optics. Many theoretical results and almost all the engineering problems, which will be examined in Volume II, are based on the author's findings, and therefore cannot be found in any other technical source. I believe that the utilization of the solutions and recommendations developed will result in significantly better understanding of the mechanical behavior and performance of microelectronic and fiber-optic structures. Application of these solutions and recommendations in engineering practice will also result in substantial savings of time and expenses, since in many cases they enable the user to minimize the amount of time- and labor-consuming experimental and computer-aided investigations. In more rare cases utilization of the presented analytical solutions and formulas will enable one to eliminate any other investigations at all.

As is known, microelectronic and fiber-optic packaging is an overlapping of different branches of engineering, such as fabrication, electrical, thermal, mechanical, environmental, etc. Our book discusses primarily static and elastic stress-strain mechanical problems which, generally speaking, could occur in and should be referred to any of the above branches, since stresses and strains can arise during manufacturing, testing, operation, shipping, handling, or storage of electronic and fiber-optic equipment and components. The examined applied problems include bi- and multimaterial assemblies, compliant attachments, multilayered structures, die attach designs, mechanical behavior of solder-joints, encapsulation materials, compliant external leads, flexible test probes, coated optical fibers, etc. All the applied topics discussed and particular problems examined are limited to those which have proven to be of genuine practical interest. Although our analyses are based chiefly on static, elastic, and deterministic (nonrandom) approaches, some dynamic, elastoplastic, viscoelastic, and probabilistic problems and approaches are also partially covered, to give the reader a general idea of these problems and approaches.

The reader is, generally speaking, not expected to have any prior knowledge of the mechanics of solids and structural analysis. However, a background in elementary engineering mechanics and strength of materials would certainly be useful. Of course, the book assumes that the reader has a basic knowledge of high school mathematics. Knowledge of the basics of analytical geometry, calculus, and the theory of differential equations is desirable, although not essential. Wherever the methods of these disciplines are used, the solution is discussed in sufficient detail. In all the cases, when additional mathematical or mechanical equipment has been found necessary, it is given with the appropriate, although brief, explanations. The level of this book is such that it could be used by practicing engineers as well as final-year undergraduate and graduate engineering students.

Let me emphasize that this book is written by an engineer for engineers. I tried to select problems of the utmost practical importance and present them in a readily accessible format, combining theoretical analysis and clear physical reasoning. The purpose has been to present practical information in a form useful to a rather diverse engineering audience, but not to sacrifice usefulness for rigor. The solutions presented are simple and easy-to-use. They illustrate the basic concepts, although they may not be the most general and accurate. They are obtained, as a rule, in the form of algebraic formulas for the quantities of interest by using, when necessary, reasonable simplifying assumptions. The mathematical equipment applied is rather modest. The use of simple mathematical models is employed throughout the book. The reader will always find reasonably sophisticated results supported by step-by-step derivations that clearly identify any assumption made. I always tried to secure a clear physical understanding before or, in some cases, during the undertaking of an analysis.

The material is organized into two volumes—theoretical and applied. Such a breakdown is done in order to serve most effectively both the readers who are interested in theoretical fundamentals (students, researchers, instructors), and those seeking primarily practical and applicable information (practicing engineers, designers, technical managers). However, I always indicate the connection between the theory and the applications wherever this connection exists.

Volume I (theoretical) is organized by topic, and deals with fundamentals of engineering elasticity (Part 1) and structural analysis (Part 2). The references and selected bibliography at the end of each part include general books appropriate to the subject matter of the corresponding part, and will be a helpful guide for further study. The major mathematical symbols which are used throughout the book are defined in the list of basic symbols preceding the text. An attempt has been made to use mathematical notation that is standard in other fields of engineering. In some cases, particular symbols have more than one meaning in different parts of the book, although this should not cause any serious ambiguity. Other symbols are also introduced within the limited contexts of particular examples or pieces of analysis. The appendix is a tabulation of beam deflections.

Volume II (applied) will give applied topics, and will not necessarily require that the reader be acquainted with the material of Volume I. Portions of the Volume II were presented at various technical conferences and published in different professional journals. The reference list in Volume II will contain the bibliography on the mechanical behavior of microelectronic and fiber-optic systems. Though not meant to be exhaustive, it will be quite extensive and will not only support all phases of the book with up-to-date background and additional information, but will be useful for further study as well.

List of Basic Symbols

A = area, constant
a, b, c, d = dimensions, coefficients
C = torsional rigidity (stiffness), constant
D = flexural rigidity (plate constant), diameter
E = modulus of elasticity (Young's modulus)
F = resultant external loading, force
f = function
G = shear modulus (modulus of rigidity)
g = acceleration of gravity
h = thickness, height
I = moment of inertia of a cross section about its neutral axis
I_p = centrifugal (polar) moment of inertia
K = spring constant, bulk modulus, stress intensity factor
k = numerical factor, parameter of interfacial stiffness
L = length
l = half-length of a beam (plate) or a crack, span, direction cosine
M = moment (especially bending moment)
N = force (especially shearing force)
m, n = integers, numerical factors, direction cosines
P = force
p = pressure, intensity of distributed transverse load, peeling stress
Q = torque (twisting moment)
q = intensity of distributed load, pressure
R = reactive force, radius
r = radius
r, θ = polar coordinates
S = Saint-Venant's torsion function
SM = section modulus of a cross section
s = distance (especially distance along a boundary)

T = axial force, interfacial force, tensile force, surface tension, temperature
t = time, temperature, thickness
U, V = potential energy, strain energy, volume
W = work, weight, strain energy
u, v, w = displacements in x, y, z directions; radial, tangential, and axial displacements
X, Y, Z = projections of the body force
x, y, z = rectangular coordinates
α = angle, coefficient of thermal expansion
β = angle, numerical factor
Γ = circulation of the shearing stress
γ = shear strain, angle, specific weight, material constant in Irwin–Orowan's theory
Δ = an increment as in Δr, Δx, etc.
δ = deflection, numerical factor
ε = normal strain
θ = angle, dilatation
λ, μ = Lamé constants (elastic properties)
κ = curvature, interfacial compliance
ν = Poisson's ratio, outward normal to the surface
ξ, η, ζ = rectangular coordinates
ρ = density, radius of curvature
σ = normal stress
τ = shearing stress
Φ = Prandtl's stress function
ϕ = angle, stress function
ω = rotation, area
$\nabla^2 = \Delta$ = harmonic (Laplacean) differential operator
$\nabla^4 = \Delta^2$ = biharmonic differential operator

Introduction

I will quote our famous mathematician P. L. Chebyshev. He said: "In ancient times the mathematical problems were set by the Gods, as, for example, the problem of the doubling of a cube, in connection with the alteration in the dimensions of the Delphic altar. Then, in a later period, the problems were set by the Demigods: Newton, Euler, Lagrange. Today we have a third period, when problems are set by practice." I think that these words are completely true, with just one correction: in my opinion, only the third period has ever existed.

—D. Grave, *Encyclopedia of Mathematics*,
Saint Petersburg, Russia, 1912

I.1 RESEARCH MODELS IN MECHANICAL PROBLEMS FOR MICROELECTRONICS AND FIBER OPTICS

I.1.1 Experimental and Theoretical Models

Modeling is the basic approach of any science. It is used to mimic those properties of a real object or phenomenon that are of interest in the given research or engineering problem.

Research models can be experimental or theoretical. Experimental models are usually of the same physical nature as the actual object. They reproduce this object in a simplified way and, quite often, on a different scale. Therefore various similitude criteria are often needed to predict the behavior of the actual object on the basis of the model experiment. In some cases, experimental models are based on rather frequently encountered similarities in the mathematical description of different objects. These are called analogous models.

Theoretical models represent real objects or phenomena by using abstract notions and usually employ more or less sophisticated mathematical equipment. A theoretical model can be either analytical or numerical (computer-aided).

I.1.2 Experimental Modeling: Why It Is the Most Widespread Approach in Microelectronics and Fiber Optics

Experimental and theoretical models are equally important and equally indispensable for the design of a viable, reliable, and low-cost product or structure. In aerospace, civil, ocean, and many other areas of engineering, experimental and theoretical models are indeed equal partners, successfully supplementing each other in any significant engineering effort. The situation is different, however, in microelectronics and fiber optics. The overwhelming majority of studies dealing with stress and reliability of electronic and fiber-optic components and systems are experimental. For instance, many investigators in the area of electronic packaging rely heavily on the results of thermal cycling tests to evaluate materials and structures.

There are several reasons why the experimental approach has become the most important one in microelectronics and fiber optics. First of all, the experimental approach is the only one that can be carried out with, so to speak, full autonomy, without necessarily requiring theoretical support. Unlike theory, testing can be, and is, used for final proof of the workability and reliability of a product. Therefore, some testing procedures are essential requirements of many military and commercial specifications in high technology.

Another reason why experimental modeling has become so popular is that experiments in microelectronics and fiber optics, as expensive as they may be, are considerably less costly than, say, those in naval architecture in which the "specimens"—ships—cost millions of dollars and in which destructive tests are, in effect, impossible. In addition, experimental investigations in microelectronics and fiber optics are much easier to plan, organize, and perform than in most other branches of engineering. The incentive for the development and application of theoretical models is consequently not as great in microelectronics and fiber optics as it is in aerospace, civil, and ocean engineering. Furthermore, lack of information about the mechanical properties of microelectronic and fiber-optic materials is often an obstacle to implementing theoretical methods.

Finally, we would like to mention that many of the active and leading workers in microelectronics and fiber optics, such as materials scientists, chemists, and experimental physicists, traditionally use experimental methods as their major research tools. Mechanical engineers, on the other hand, resort to theoretical models much more often.

I.1.3 Experimental Modeling: Shortcomings

Although experimental modeling is a reasonable approach and even a requirement of many specifications, its application, as a rule, requires considerable time and often significant expense, especially if testing is carried out as part of a research effort. What is more important, though, is that testing alone is often insufficient to understand the behavior of materials and structures. This is because experimental data inevitably reflect the combined effect of a variety of factors, whereas what is needed for the prediction of stresses and optimal structural design is knowledge of the role of each particular parameter affecting the mechanical behavior and reliability of the system. Such a lack of insight frequently leads to trial-and-error experimental procedures, the results of which often cannot be extended to new designs that are different from those already tested.

Note that it is always easy to recognize purely empirical relationships obtained as a result of "blind" processing of experimental data and not based on any rational (theoretical or experimental) considerations underlying the physical nature of the phenomena. These relationships contain, as a rule, fractional exponents and coefficients, "odd" units, etc. Although such relationships may have a certain practical value, the very fact of their existence should be attributed to lack of knowledge in the given area of applied science.

We would also like to point out that although testing can reveal insufficiently reliable joints and structures, it is often unable to detect superfluously reliable ones that may have excessive weight or elevated safety margins. This is definitely a shortcoming of the experimental approach, especially in mass-producing high-cost and high-quality products when superfluous reliability may entail substantial additional expense.

I.2 THEORETICAL MODELING

I.2.1 The Major Requirement for a Theoretical Model Is Simplicity

In view of the foregoing, it is becoming clear what can be expected from, and gained by, employing theoretical models.

Of course, a crucial requirement for a theoretical model, especially in engineering applications, is simplicity. In the mid-1950s, K. K. Darrow, who was secretary of the Physical Society and a theoretical physicist on the staff of Bell Laboratories, noted in jest that "theory discerns the underlying simplicity of phenomena. The nontheorist sees a crazy welter of phenomena; when he becomes a theorist, they fuse into a simple and dignified structure."

Having in mind engineering design theories, it should be emphasized that the "crazy welter of phenomena" can "fuse into a dignified theory" only if the theory is sufficiently simple. It is quite easy and, in many cases, does not take even much creativity to concoct a "theory" loaded with series, non-linearities, etc., and, indeed, those trapped by mathematics often generate such engineering "theories." Evidently, the welter of the theory is even worse than the welter of the phenomena.

It goes without saying that a good theoretical model should provide easy-to-use formulas, clearly indicating the role of the major factors affecting the given object or phenomenon. The simpler the model, the better. One authority in applied mechanics remarked, perhaps only partly in jest, that the degree of understanding of a phenomenon is inversely proportional to the number of variables used for its description, and that any equation longer than three inches is most likely wrong.

I.2.2 How Analytical Models Are Developed

"Applied mechanicians" must have, first of all, good physical intuition and be able to see the underlying qualitative relationships between the major parameters of the object to be investigated. Then, proceeding from intuitively obvious and/or available experimentally established relationships, they try to "translate" the technical problem into an adequate mathematical language, i.e., to formulate it in terms of mathematical equations with relevant boundary conditions. This is, of course, the most important step. It is never straightforward and, in addition to knowledge of engineering and applied mathematics, requires creativity, imagination, good taste, and a sense of proportion. Therefore, the ability to create adequate and effective analytical models is on the border between science and art.

The next step is to choose suitable mathematical equipment to solve the equations. Of course, the choice of one or another analytical model depends primarily on the goals of the study and the state of the art in the given area. However, the availability of a suitable mathematical means often plays a decisive role in the final selection of a theoretical model. It is the mathematical skills that are important at this step.

After the mathematical problem is solved, the solution, containing quantitative relationships between the parameters of interest, must be thoroughly analyzed and, if possible, simplified. Usually, the first thing to do is to ensure that the already known, or self-evident, relationships can be obtained from the obtained solution as extreme or special cases. Then, the researcher can hope that the theory is at least not manifestly erroneous. It is always advisable to carry out numerical examples for several typical situations to ensure that the developed formulas produce reasonable figures. "I do not get satisfaction from the formulas, until I have a feeling for the numerical values of the results," said Lord Kelvin.

Now, researchers can try to employ their theory by applying the derived relationships to predict the behavior of the object, i.e., to try to interpret the mathematical solutions in terms of the real world, carrying out a "back translation" from the mathematical into the physical language. "One should always be able to say 'tables, chairs, glasses of beer,' instead of 'points, straight lines, and planes'" (D. Gilbert).

Note that the ultimate goal of any theoretical modeling is to reveal the nonobvious and relatively complicated relationships which exist, but are hidden in the initial information. Therefore no theoretical model can provide any "new" results, i.e., those not contained in the input data, assumptions, and hypotheses.

I.2.3 Role of Mathematics

Engineers dealing with analytical modeling use a wide range of mathematical methods. Although these methods are, in many cases, rather deep, sophisticated, and subtle, obtaining new results in mathematics is by no means an ultimate goal for engineers. Mathematics for them is simply an effective tool for getting a better understanding of an actual physical or technical phenomenon, simply a means for seeking a quantitative or qualitative result, which is important primarily from the viewpoint of practical applications.

It should be emphasized that time pressure often plays an important role in applied research. Of course, it is better to solve a problem approximately but in time rather than to find an exact, complete, and rigorous solution, taking so much time that the results are rendered obsolete.

It is probably not an exaggeration to say that if the advancement of the applied sciences had depended from the very beginning solely on the subtle and rigorous minds of some modern mathematicians, these sciences would not have progressed at all. The requirement for "complete" mathematical rigor in engineering theories would not only result in restraining technical progress today but would also lead to the conclusion that almost all the existing results in the engineering sciences are wrong since many are based on "insufficiently rigorous" mathematical considerations. A good example is the famous Galerkin method, suggested about 70 years ago in an engineering theory of plates and, since then, successfully used in many areas of applied science. Despite enormous mathematical effort, this method still remains unproven and, hence, "nonrigorous."

There is an anecdote about nineteenth-century balloonists who lost their way in foggy weather and descended to find someone to direct them. After a long search, they noticed a man who was not too far away and shouted to him, "Where are we?" The man thought for a while and, when the travelers were already rather far from him, he shouted back, "You are on a balloon!" After some time, one of the balloonists said to his co-travelers, "There are three reasons why I think that the man we met was a mathematician. First, he answered after thinking for a minute. Secondly, his answer was quite precise. And, thirdly, the information that we received from him was absolutely useless...."

Perhaps because such "rigorous mathematicians" exist, some practicing engineers mistakenly think that mathematics is a science that tries to prove obvious things by using nonobvious methods and that a technical college graduate becomes a good practicing engineer at the very moment the last mathematical formula flies out of his mind.

I.2.4 What Can Be Gained by Using Theoretical Modeling?

In many cases, theory can predict the result of an experiment in less time and at smaller expense than it would take to perform the actual experiment. In other cases, theory serves to discourage wasting time on useless experiments. For instance, many attempts to build impossible heat engines have been prevented by a study of the laws of thermodynamics. This is, of course, a classical and outstanding example of the triumph of a theory. There are also numerous, though less famous, examples, when plenty of time and effort were saved because of prior theoretical analysis of a technical problem.

But theory is, of course, much more than a device to substitute for testing or prevent useless experiments. In the majority of serious engineering endeavors, a preliminary theoretical analysis provides valuable information about the phenomenon to be investigated. This information enables an experimentalist to decide how and what should be tested and measured and in what directions success might be expected. In fact, theory often serves to suggest new experiments. It can also be used to interpret empirical results, to bridge the gap between experiments, and to extend existing experience on new materials and structures. It can also be utilized in experimental planning.

One definitely cannot do without a good theory when developing a rational design. The idea of optimization of structures, materials, functions, and costs, although new in microelectronics and fiber optics, is rapidly penetrating many areas of modern engineering. Clearly, no progress in this direction could be achieved without broad application of theoretical methods.

I.3 ANALYTICAL VERSUS NUMERICAL MODELING

Since the mid-1950s, numerical, especially finite-element, modeling has become the major research tool for theoretical evaluations in mechanical and structural engineering, including the area of microelectronics and fiber optics. This can be attributed primarily to the availability of powerful and flexible computer programs, which enable one to obtain solutions to almost any stress-strain–related problem within a reasonable time.

Broad application of computers, however, has, by no means, made analytical solutions unnecessary, whether exact, approximate, or asymptotic. Analytical solutions, especially if they are simple enough, have invaluable advantages because of the clarity and "compactness" of the obtained information. The advantages of analytical solutions are especially great when the parameter under investigation depends on more than one variable. As to the asymptotic formulas, they can be successful in those cases in which there are difficulties in the application of numerical methods, e.g., in problems containing singularities.

But, even when application of numerical methods encounters no difficulties, it is always advisable to investigate the problem analytically before carrying out computer-aided analyses. Such preliminary analytical investigation may help reduce computer time and expense, develop the most feasible preprocessing model and, in some cases, avoid fundamental errors.

It is noteworthy that the finite-element method was originally developed for structures with complicated geometry and/or complicated boundary conditions, when it might be difficult to apply analytical approaches. As a consequence, this method is widely used in those areas of engineering where structures of complex configuration are typical. In contrast, microelectronic and fiber-optic structures are usually characterized by relatively simple configurations. Therefore, for these structures, there is an obvious incentive for a broad application of analytical modeling.

Adjacent structural elements in microelectronic and fiber-optic systems often have dimensions that differ by orders of magnitude. Examples are thin-film structures fabricated on thick substrates and adhesively bonded assemblies in which the adhesive layer is, as a rule, significantly thinner than the adherends. Since the mesh elements in a finite-element model must be compatible, finite-element analysis of such structures often becomes a problem in itself, especially in regions of high stress concentration. This awkward situation does not occur, however, with an analytical approach.

At the same time, there is often an illusion of simplicity in applying finite-element procedures. Some users of finite-element programs believe that they are not even supposed to have any prior knowledge of structural mechanics and that the "black box" they deal with will automatically provide the right answer. At times, a hasty, thoughtless, and incompetent application of computers results in more harm than good by creating an impression that a well-founded solution was obtained when, actually, this "solution" is simply wrong. Based on the author's own experience, as well as on the experience of many other users of finite-element computer programs, it is usually quite easy to obtain a finite-element solution. It might be very difficult, though, to obtain the right solution. In effect, we have to have a good background in structural analysis, as well as a good intuition about the mechanical behavior of the given structure, in order to develop an adequate, feasible, and economic preprocessing model. Paraphrasing a French proverb, we can say that the finite-element method belongs to those research tools that can be of use only to those who, generally speaking, could manage very well without them.

I.4 INTERACTION WITH EXPERIMENT

Although an experimental approach unsupported by theory is "blind," a theory not supported by an experiment is "dead." It is the experiment that forms a basis for a theory, provides the input data, and determines its viability, accuracy, and limits of application. Limitations of a theoretical model are different in different problems and, in the majority of cases, are not known beforehand. Therefore, the experiment is the supreme and ultimate judge of a theory.

A physical experiment can often be rationally included in a theoretical solution of an applied problem. Even when some relationships and structural characteristics do lend themselves to theoretical evaluation, in principle, it is sometimes simpler and more accurate to find these relationships empirically. For instance, spring constants in compliant lead designs might be most easily determined empirically by loading the leads and measuring the displacements.

I.5 THEORETICAL MODELING IN STRUCTURAL ANALYSIS

1.5.1 Three Steps in Structural Analysis

Rapid advances in integrated electronics and fiber optics during the past two decades have increased the need for a more thorough investigation of the mechanical behavior and rational physical design of electronic and fiber-optic systems and components. Such a design is a process by which the structural components must be proportioned to have ultimate and fatigue strength sufficient to resist the forces and conditions to which they may be subjected. This process involves the consideration of the mechanical properties of the materials and structural analysis. Such an analysis consists of three major steps:

1. Idealization of the structure and development of a theoretical model.
2. Mathematical analysis of the developed model.
3. Application of the obtained results to the actual structure.

In the first step, researchers pick out the most essential features of the structure and develop a stress analysis model. Next, in the second step, they use mathematical methods to discover the hidden relationships inherent in the model. Finally, in the third step, they utilize the results of the analysis to predict the mechanical performance of the structure, establish the margins of safety, and assign the proportions of the structural elements.

The first and the third steps are closely related and involve a rather wide spectrum of problems associated with operational conditions, applied loads, potential failure modes, consequences of failures, etc. Even personnel qualification and production efficiency are sometimes taken into consideration when a stress analysis model is being developed and safety margins are being chosen. Although some of these problems will be partially addressed in this book, the emphasis will be on the second step, i.e., on the mathematical analyses of stress models. One of the reasons for this is that this step can be examined, to a great extent, in isolation from the two other steps. This is true primarily because the methods of mathematical analysis are more or less the same for different structures and structural elements encountered in various areas of engineering. This attracts a great number of researchers, including mathematicians, to the advancement of the methods of mathematical analysis in structural design and makes the results obtained in this field rather general and fundamental. It takes, however, a good engineer to understand the level of accuracy as well as the limitations and drawbacks of the available mathematical solutions. These can be corrected, if necessary, in an engineering design by an appropriate selection of safety margins and, in some cases, by empirical evaluation of the bearing capacity of a structural element or joint.

Since, in most branches of engineering, the analysis of theoretical models is developed considerably farther than the problems associated with structural reliability, and usually occupies more space in text and reference books, the term "structural analysis" is often applied to the second step only. Hereafter, we too will use this term, having in mind just the analysis of stress models. We should, however, never forget that there was "someone" who simplified the actual structure and developed the given stress model and that "someone," perhaps the same engineer, will always be needed to take responsibility for establishing the proportions of the actual structure and assessing its reliability.

I.5.2 Structural Analysis Disciplines

Solutions to structural analysis problems can be obtained by using different methods and approaches. This area of applied science deals with the strength of materials, structural stability, elasticity, plasticity, thermal stresses, theory of plates and shells, fracture mechanics and, of course, various experimental methods and techniques. Obviously, structural analysis widely employs the results from materials science that provides the information concerning material characteristics of the materials.

Strength (mechanics) of materials is a basic engineering discipline used to design all manner of structures, machines, and equipment of suitable strength and stiffness. In this discipline, consideration is generally limited to bodies of simple shapes, such as rods, beams, and shafts. Methods used in strength of materials are also relatively simple, and special assumptions and simplifying hypotheses are introduced for almost every new problem. This enables us to obtain simple and easy-to-use design formulas.

More complicated structural elements, such as plates, shells, thin-walled beams, noncircular shafts, and composite structures, together with more sophisticated, general, and powerful methods of analysis, are examined in a discipline called structural, or engineering, mechanics. This is, in essence, an advanced strength of materials, developed, modified, and extended in application to a particular branch of engineering. For example, the first courses of structural mechanics were developed primarily for various rod systems and reflected the needs of bridge builders and civil engineers. In structural mechanics courses for marine vehicles and ocean structures, the emphasis is mainly on bending and buckling of plates and shells, whereas the structural mechanics of aircraft gives primary attention to bending and torsion of thin-walled and "sandwich" structures. Some typical structural elements in microelectronics are multimaterial and heteroepitaxial systems, adhesively bonded assemblies, solder joint interconnections, and wire bonds. A coaxial cylindrical composite is a typical fiber-optic structure.

Both strength of materials and structural mechanics are in the engineering domain and employ a variety of plausible hypotheses and methods. When possible, however, these disciplines use more rigorous and more accurate solutions based on the methods of elasticity and plasticity, the mathematical theory of plates and shells, and other sciences based on the approaches of mathematical physics. Note that the solutions in these sciences are usually obtained by narrowing the class of the examined problems rather than by introducing additional simplifications. Some stress analysts say that disciplines based on approaches of mathematical physics deal only with problems these disciplines are able to solve. The solutions obtained are, however, such as they "ought to be." The engineering stress analysis disciplines, on the contrary, deal with all the problems, which "ought to be" solved, but solve them only in the way they are able to. Obviously, the role of each of the above-mentioned approaches in a given specific problem depends not only on the nature of the problem and the state of the art in the given area but on personal preference as well.

There are several relatively new disciplines that usually are not included in the traditional courses of structural analysis: viscoelasticity and visco-plasticity, creep, and fracture mechanics. We would like to emphasize that these disciplines are especially important for the analysis and prediction of the mechanical behavior and rational physical design of microelectronic and fiber-optic structures. Fracture mechanics is probably of the utmost importance since it is the only discipline that deals with both parts of the strength condition "actual stress (deformation) $\leq$ design stress (deformation)" and therefore, in many cases, is able to predict the time-to-failure of the given structure for the given operation conditions. All the other disciplines focus mainly on the left-hand part of the strength condition, whereas the right-hand part is simply determined as a portion of the experimental yield or ultimate stress or strain of the material. Clearly, these disciplines can be successfully applied to those materials and conditions for which brittle crack formation and propagation is rather unlikely or for which the critical length of a crack is so small that it is undetectable by available equipment (ultrasonic waves, x-rays, etc.).

I.6 HISTORICAL SKETCH

One can identify three major periods in the development of structural analysis.

The work of the *first period* (the eighteenth century and the beginning of the nineteenth century) concerned primarily strength of materials. During this period, J. Bernoulli (1654–1705) and L. Euler (1707–1783) developed the elementary beam theory; R. Hooke (1635–1703) and T. Young (1773–1829) established the experimental fundamentals of the science of strength of materials; L. Euler and J. Lagrange (1736–1813) suggested theories of buckling of beams; and C. A. Coulomb (1736–1806) developed the theory of torsion of circular shafts.

The beginning of the *second period* (the first half of the nineteenth century) is marked by a series of papers by A. L. Cauchy (1789–1857), H. Navier (1785–1836), and S. D. Poisson (1781–1840), who laid the foundation of the theory of elasticity. At that time, B. de Saint-Venant (1797–1886) suggested his famous "semi-inverse" method, which he used to develop the theory of torsion and bending of noncircular prismatic beams. Saint-Venant also formulated the principle that the stress distribution in the areas sufficiently remote from the areas of application of the external loads can be evaluated with sufficient accuracy by replacing the actual loads with a resultant force and resultant couple. In the second half of the nineteenth century, G. R. Kirchhoff (1824–1887) obtained the basic equations of the thin-walled beam theory. The theory of thin plates was developed as a result of the work of

J. Lagrange, S. Germain (1776–1831), A. Cauchy, S. D. Poisson, and G. R. Kirchhoff. At the end of the nineteenth century, H. Aron (1874) and A. Love (1888) gave the first versions of the modern shell theory, based on the hypothesis of "straight normals."

The *third (modern) period* (the end of the nineteenth century and the twentieth century) is characterized not only by an enormous increase in the number of publications but also by the emergence of several important new disciplines of structural analysis, such as theories of plasticity and creep, theory of viscoelasticity, structural mechanics of complex structures, fracture mechanics, and others. Experimental investigations by F. Schleicher (1926) concerning the effect of three-dimensional pressure on plastic deformation, along with the experiments of A. Nádái (1925), W. Lode (1926), and M. Ros (1926) with thin-walled tubes, enabled R. von Mises (1924), H. Hencky (1928), W. Prager, and A. Nádái to suggest relationships between the stresses and strains during plastic deformations. The work of R. von Mises, D. Drucker, R. Hill, and others resulted in the development of flow theory. Development of the modern theory of viscoelasticity is associated with the names of L. Boltzmann (1844–1906), I. C. Maxwell (1831–1879), W. Thomson (Lord Kelvin) (1824–1907), W. Voigt (1850–1919), and others.

Numerical and, first of all, finite-element methods play an extremely great role in modern structural analysis. Although some basic ideas underlying the finite-element method can be found in papers by G. Kron (1939), R. Courant (1943), and W. Prager and J. L. Synge (1947), it was J. H. Argyris and his colleagues (1954) who advanced this method in application to structural analysis problems and presented its numerical algorithms in a form convenient for computer applications. More recently, this method has been extended into such areas as fluid mechanics and heat transfer. The finite-element method is currently widely used in the structural analysis of microelectronic and fiber-optic systems.

Part 1

Basic Principles of Engineering Elasticity

1

General Properties of Elastic Bodies

We start with the basic principles of the elasticity theory. Utilization of theory-of-elasticity methods and solutions enables us to determine the magnitude and distribution of elastic stresses and displacements in the most substantiated and complete manner, thereby providing a sound basis for a rational design of reliable and lightweight structures. In many cases, an elastic analysis is sufficient. In other cases, it can be successfully used as the first approximation for a further plastic, elastoplastic, viscoelastic, or fracture analysis. At the same time, familiarity with the methods and general relationships of the elasticity theory, and with the general properties of the states of stress and strain, is always a great help when it comes to introducing various simplifying assumptions in those engineering problems that cannot be solved on the basis of theory-of-elasticity methods only.

The theory of elasticity studies the mechanical behavior of elastic bodies. A deformable solid body is called *elastic* if its deformation disappears with the removal of the forces that caused this deformation. When an elastic body is restored to its initial form, it returns completely the work that was expended for the deformation. Therefore, *residual deformations* are not examined in the theory of elasticity. Atomistic structure of actual materials is not considered in this theory either. The material is assumed to be distributed over the body's volume *continuously* without voids or gaps, both in the unloaded and loaded conditions. Stresses, strains, and displacements in continuous bodies are continuous functions of the coordinates. This enables us to utilize the powerful methods of calculus to obtain solutions to theory-of-elasticity problems. Although real materials do not completely satisfy the assumption of continuity, numerous experiments and the entire engineering experience indicate that solutions based on this assumption are generally reliable and are confirmed by practice.

Note that although the theory of elasticity ignores the atomic structure of the matter, treats elastic bodies as continuous media,* and does not consider the forces of the intermolecular interaction, it is these forces that cause the internal stresses in elastic bodies. However, the radius of action for the intermolecular forces is assumed equal to zero in the theory of elasticity. This assumption is equivalent to a hypothesis that the forces acting on the given part of the body from the adjacent parts are transmitted only through the surface of this part. Since the theory of elasticity, as a rule, examines distances significantly exceeding the intermolecular distances and is, in effect, a macroscopic theory, then the disregard of the intermolecular interaction is quite acceptable.

The classical theory of elasticity assumes also that the elastic bodies are *homogeneous* and *isotropic*. Homogenuity of a body means that its properties are the same in all its points. In a homogeneous body, equal stresses result in equal strains. Isotropy of a body means that its properties are the same in different directions.

In the linear theory of elasticity, it is also assumed that the stresses are directly proportional to the strains. These relationships are established by *Hooke's law*. In addition, the elastic body is regarded as sufficiently *rigid*; i.e., elastic displacements are assumed small compared to the linear dimensions of the body. Therefore, the induced strains are small compared to unity. The linear relationship between the stresses and strains, as well as the smallness of strains, enables us to use the *principle of superposition*. This important principle states that the combined effect of a number of loads acting on a body is equal to the sum of the effects of each load applied separately. Problems in which Hooke's law is not fulfilled (*physical nonlinearity*) or in which the displacements cannot be considered small (*geometric nonlinearity*) are examined in the *nonlinear theory of elasticity* if the stresses do not exceed the yield point or in the *theory of plasticity* if the stresses are greater than the yield stress.

Finally, the hypothesis of the *natural unstressed state* of the body is assumed in the theory of elasticity. This means that the initial stresses existing prior to the application of the surface loads are zero.

The *major goal* of the theory of elasticity is to determine the internal forces (stresses), displacements, and strains occurring in an elastic body under the action of the given external forces.

The *major method* of the theory of elasticity is to translate the foregoing assumptions and hypotheses concerning the properties of the elastic body

* In mathematical courses, the theory of elasticity is often examined as a section of a discipline called mechanics of continuous media. Other sections of this discipline are: theories of plasticity, viscoplasticity and creep, hydromechanics, and gas dynamics. Methods of the mechanics of continuous media are also used in electrodynamics, magnetic hydrodynamics, plasma physics, and other areas of engineering and science.

into the form of mathematical relationships. The latter are then analyzed by using various methods of mathematical physics. As mentioned in the Introduction, simplifications in the theory of elasticity are achieved, as a rule, by narrowing the generality of the obtained solution, i.e., by narrowing the circle of problems for which this solution is suitable. This is the major difference between the methods of the theory of elasticity and those of strength of materials and structural mechanics, where special simplifying hypotheses and assumptions are introduced for almost every new problem.

2

Equations and Conceptions

2.1 STRESS

Internal forces are self-equilibrated, form a so-called static zero, and are therefore not examined in courses in statics. The theory of elasticity, on the contrary, sets the determination of the internal forces in a body under external loading as its primary goal since these forces determine the stresses and strains in the body and, ultimately, its strength.

In order to utilize, among other relationships, the equations of equilibrium for the purpose of evaluating the internal forces, we have to transfer the internal forces into the category of the external ones. This is achieved by means of an approach, known as "method of cross-sectioning."

Consider, for instance, a three-dimensional body that is in a state of equilibrium under the action of a system of external forces. We can imagine that the body is cross-sectioned into two parts by slicing it along a plane section of area A (Fig. 2.1). Since each part of the body is in equilibrium, forces must act on the contact areas A of each free body in order to balance the external forces. Let ΔA be an element of the plane area A, and $\bar{v}$ the exterior normal to this element. The distributed forces acting on this element can be reduced to a resultant force $\Delta\bar{P}_v$, applied to point M of the element, and a resultant couple $\Delta\bar{G}_v$ with respect to an axis passing through this point. The ratio

$$\langle \bar{p}_v \rangle = \frac{\Delta\bar{P}_v}{\Delta A}$$

is called the *mean stress* of the internal elastic forces for the element ΔA. Obviously, this stress is a vector acting in the same direction as the force that brought it about. For given external loading, the magnitude of this stress is, in general, different for different plane elements passing through point M, i.e., depends on the orientation of the plane. This means that, in order to define a stress, we have to know not only the magnitude and direction of

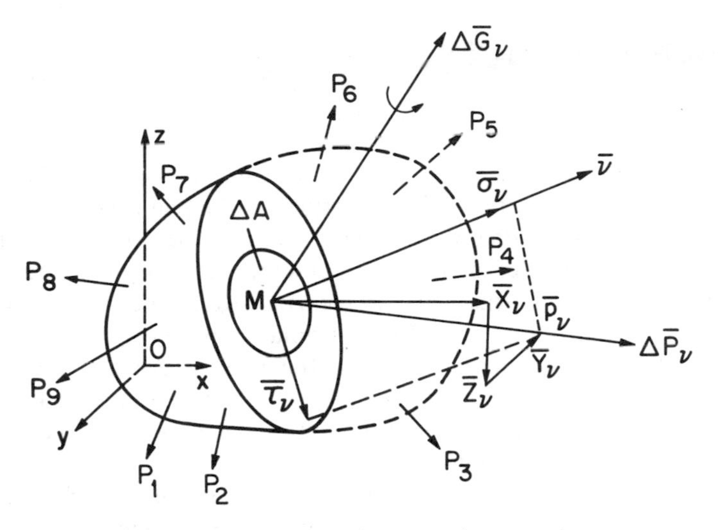

FIGURE 2.1 Components of stress acting on a plane.

the stress but also the orientation of the plane on which it acts. Such vectors are referred to as tensors. The stress tensor will be examined in greater detail in Section 2.3.

In order to be able to use methods of calculus, and assuming the material to be a true continuum, we draw together the contour of the element ΔA to point M. Since all the linear dimensions of the element will, in this case, tend to zero, then,

$$\lim_{\Delta A \to 0} \frac{\Delta \bar{G}_\nu}{\Delta A} = 0$$

The limit

$$\bar{p}_\nu = \lim_{\Delta A \to 0} \frac{\Delta \bar{P}_\nu}{\Delta A}$$

however, has a finite value and is called the *stress* of the internal elastic forces at point M for the element ΔA. The orientation of this element in space is defined by the direction of the normal $\bar{\nu}$. The vector $\bar{p}_\nu$ characterizes the level of the intensity of the internal elastic forces at the point.

2.2 STRESS COMPONENTS

Vector $\bar{p}_\nu$ is oriented, in the general case, under an arbitrary angle with respect to the element ΔA (Fig. 2.1). Evidently, it can be presented as a sum of a normal $\bar{\sigma}_\nu$ and tangential $\bar{\tau}_\nu$ components,

$$\bar{p}_\nu = \bar{\sigma}_\nu + \bar{\tau}_\nu \tag{2.1}$$

which are called *normal* and *shearing stresses*, respectively. The normal stress acts in the direction of the normal-to-the-plane element, and the shearing stress is located in the plane of this element. Note that the mechanical properties of materials are determined to a great extent by the relation between the normal and the shearing stress, since some mechanical processes in solid bodies are affected primarily by the normal stress (e.g., ultimate failure under tensile loading, the final stages of fatigue failure, long-term strength at elevated temperatures), whereas others are due mainly to the shearing stress (e.g., plastic strain, ultimate failure under shear loading, the initial stages of fatigue damage, creep).

The normal stress is assumed positive if it is *tensile*, i.e., acts in the direction of the exterior normal $\bar{\nu}$. *Compressive* stresses are considered negative. The sign of the shearing stress depends on the direction of the normal stress for the given plane element. If the normal stress is directed along a coordinate axis, then the two other axes indicate the positive directions of the components of the shearing stress. If the normal stress acts in the negative direction of the given coordinate axis, then the components of the shearing stress are considered positive if they act in the negative directions of their axes.

Formula (2.1) presents the vector of the total stress as a sum of two components in so-called *natural axes*. In many cases, however, it is necessary to present the total stress as a sum of components acting along the given coordinate axes (Fig. 2.1):

$$\bar{p}_\nu = \bar{X}_\nu + \bar{Y}_\nu + \bar{Z}_\nu \tag{2.2}$$

Obviously, the moduli (absolute values) of the above vectors are related as

$$p_\nu = \sqrt{\sigma_\nu^2 + \tau_\nu^2} = \sqrt{X_\nu^2 + Y_\nu^2 + Z_\nu^2} \tag{2.3}$$

2.3 STRESS TENSOR

As indicated earlier, the magnitude and direction of the vector $\bar{p}_\nu$ of the total stress depends on the orientation of the plane element, containing the given point M, i.e., on the direction of the normal ν. Examining three mutually perpendicular plane elements, with normals parallel to the axes x, y, z, we obtain the following three vectors:

$$\bar{p}_x(\sigma_x, \tau_{xy}, \tau_{xz}) \qquad \bar{p}_y(\tau_{yx}, \sigma_y, \tau_{yz}) \qquad \bar{p}_z(\tau_{zx}, \tau_{zy}, \sigma_z)$$

The matrix

$$T_\sigma = \begin{vmatrix} \sigma_x & \tau_{xy} & \tau_{xz} \\ \tau_{yx} & \sigma_y & \tau_{yz} \\ \tau_{zx} & \tau_{zy} & \sigma_z \end{vmatrix} \tag{2.4}$$

formed by the projections of the vectors $\bar{p}_x$, $\bar{p}_y$, and $\bar{p}_z$ on the coordinate axes is called *stress tensor* at the given point. The values σ are normal, and the values τ are shearing components of the stress tensor, or simply *normal* and *shearing stresses.* Following a standard means of identification, we give normal stress a single subscript to stand for the direction in which they act, and shearing stresses two subscripts, the first indicating the direction of a normal to the plane on which the stress occurs and the second indicating the direction of the stress. Thus, for instance, σ_x is a normal stress acting in the x direction, and τ_{xy} is the shearing stress on a plane normal to the x axis and acting in the y direction (Fig. 2.2). Since the material is continuous, the

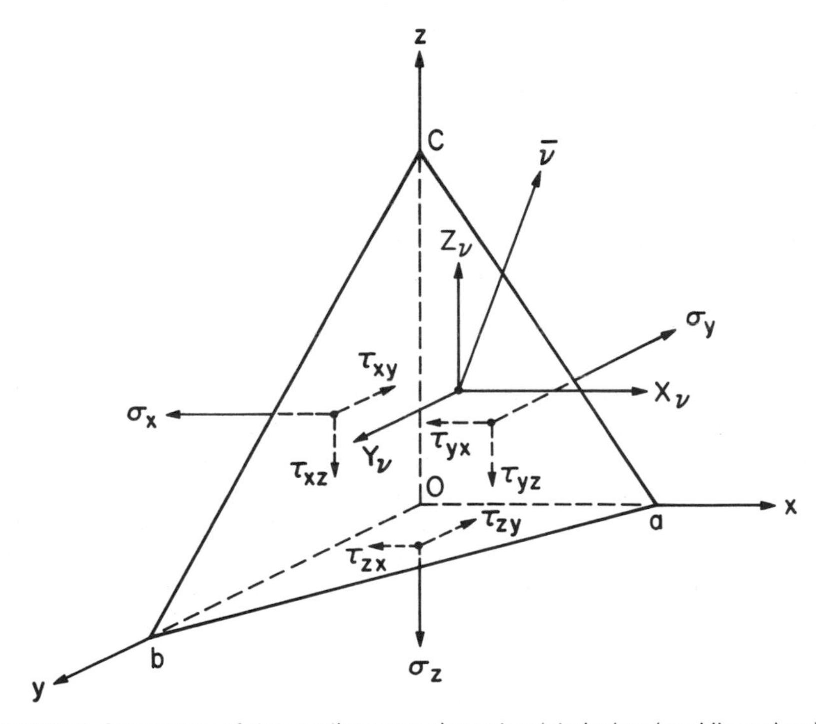

FIGURE 2.2 Components of stress acting on an elementary tetrahedron (an oblique plane).

components of the stress tensor are continuous functions of the coordinates x, y, and z:

$$\sigma_x = \sigma_x(x, y, z) \qquad \tau_{xy} = \tau_{xy}(x, y, z), \ldots \qquad \sigma_z = \sigma_z(x, y, z)$$

Note that although the notion of the stress was introduced here for a body loaded by surface forces, the stress characterizes the level of the intensity of the internal forces, regardless of the cause that brought these forces about. Stresses also can occur under the action of *body forces* (such as gravity, inertia, magnetic) or thermally induced strains or as a result of technological processes leading to the occurrence of residual stresses. However, the latter stresses are not examined in this book (the initial stresses are assumed equal to zero). The stresses we are going to deal with will be those that are due to external surface forces, body forces, and temperature fields.

2.4 EQUILIBRIUM CONDITIONS FOR AN ELEMENTARY TETRAHEDRON

We start with equilibrium conditions for an elementary tetrahedron, which represents a surface element of an elastic body.

Consider an elementary tetrahedron (an infinitely small pyramid), the side faces of which are parallel to the coordinate planes, and the base is an oblique plane with a normal $\bar{\nu}$ (Fig. 2.2). Let the angles that this normal makes with the directions x, y, and z be α, β, and γ, respectively. Then,

$$l = \cos\alpha \qquad m = \cos\beta \qquad n = \cos\gamma$$

are *direction cosines* of the normal $\bar{\nu}$. The tetrahedron will be in equilibrium in the x direction if the following balance of forces is fulfilled:

$$X_\nu\, dA - \sigma_x l\, dA - \tau_{xy} m\, dA - \tau_{zx} n\, dA + X\, dV = 0$$

Here, dA is the area of the base of the tetrahedron abc, dV the volume of the tetrahedron, and X the projection of the body force. The last term in the preceding equation is proportional to the cube of the linear dimensions of the tetrahedron and is of the third order of smallness, whereas the rest of the terms are proportional to the linear dimensions squared and are therefore small quantities of the second order. Then, omitting the last summand, canceling by dA, and forming similar projections for the axes y and z, we have

$$\left.\begin{aligned} X_\nu &= \sigma_x l + \tau_{yx} m + \tau_{zx} n \\ Y_\nu &= \tau_{xy} l + \sigma_y m + \tau_{zy} n \\ Z_\nu &= \tau_{xz} l + \tau_{yz} m + \sigma_z n \end{aligned}\right\} \tag{2.5}$$

These relationships can be interpreted as a law of transformation of a vector $\bar{v}$ with projections l, m, n into the vector $\bar{p}_v$ with projections X_v, Y_v, Z_v by using the matrix T_σ, formula (2.4).

Formulas (2.5) enable us to determine the components of the stress for any oblique plane with the normal $\bar{v}$ if the components of the stress for the planes parallel to the coordinate planes are known. These formulas are also called *static boundary conditions* since the stress components X_v, Y_v, and Z_v are usually given on the body's surface. These conditions are then used to find the stresses within the body.

Formulas (2.5) can also be used to transform the stress components defined in one coordinate system into an equivalent set for some other system.

Let the coordinates x', y', z' be rotated with respect to the coordinates x, y, z in accordance with the following table for the direction cosines:

	x	y	z
x'	l_1	m_1	n_1
y'	l_2	m_2	n_2
z'	l_3	m_3	n_3

From Eqs. (2.5), we have

$$\left.\begin{aligned} X_{x'} &= \sigma_x l_1 + \tau_{yx} m_1 + \tau_{zx} n_1 \\ Y_{x'} &= \tau_{xy} l_1 + \sigma_y m_1 + \tau_{zy} n_1 \\ Z_{x'} &= \tau_{zx} l_1 + \tau_{yz} m_1 + \sigma_z n_1 \end{aligned}\right\} \tag{2.6}$$

In order to obtain an expression for the normal stress σ'_x, we have to project each of the values $X_{x'}$, $Y_{x'}$, and $Z_{x'}$ on the axis x' and to sum up the results:

$$\begin{aligned} \sigma_{x'} &= X_{x'} l_1 + Y_{x'} m_1 + Z_{x'} n_1 \\ &= \sigma_x l_1^2 + \sigma_y m_1^2 + \sigma_z n_1^2 + 2\tau_{xy} l_1 m_1 + 2\tau_{yz} m_1 n_1 + 2\tau_{zx} n_1 l_1 \end{aligned}$$

Similarly, in order to obtain the shearing stress $\tau_{x'y'}$, we have to project each of the values $X_{x'}$, $Y_{x'}$, and $Z_{x'}$ on the axis y' and to sum up the obtained projections

$$\begin{aligned} \tau_{x'y'} = X_{x'} l_2 + Y_{x'} m_2 + Z_{x'} n_2 &= \sigma_x l_1 l_2 + \sigma_y m_1 m_2 + \sigma_2 n_1 n_2 \\ &+ \tau_{xy}(l_1 m_2 + m_1 l_2) + \tau_{yz}(m_1 n_2 + n_1 m_2) + \tau_{zx}(n_1 l_2 + l_1 n_2) \end{aligned}$$

Similar formulas can be obtained for the other components of the stress tensor, so that

$$\left.\begin{aligned}\sigma_{x'} &= \sigma_x l_1^2 + \sigma_y m_1^2 + \sigma_z n_1^2 + 2\tau_{xy} l_1 m_1 + 2\tau_{yz} m_1 n_1 + 2\tau_{zx} n_1 l_1 \\ \sigma_{y'} &= \sigma_x l_2^2 + \sigma_y m_2^2 + \sigma_z n_2^2 + 2\tau_{xy} l_2 m_2 + 2\tau_{yz} m_2 n_2 + 2\tau_{zx} n_2 l_2 \\ \sigma_{z'} &= \sigma_x l_3^2 + \sigma_y m_3^2 + \sigma_z n_3^2 + 2\tau_{xy} l_3 m_3 + 2\tau_{yz} m_3 n_3 + 2\tau_{zx} n_3 l_3\end{aligned}\right\} \quad (2.7)$$

$$\left.\begin{aligned}\tau_{x'y'} &= \sigma_x l_1 l_2 + \sigma_y m_1 m_2 + \sigma_z n_1 n_2 \\ &\quad + \tau_{xy}(l_1 m_2 + m_1 l_2) + \tau_{yz}(m_1 n_2 + n_1 m_2) + \tau_{zx}(n_1 l_2 + l_1 n_2) \\ \tau_{y'z'} &= \sigma_x l_2 l_3 + \sigma_y m_2 m_3 + \sigma_z n_2 n_3 \\ &\quad + \tau_{xy}(l_2 m_3 + m_2 l_3) + \tau_{yz}(m_2 n_3 + n_2 m_3) + \tau_{zx}(n_2 l_3 + l_2 n_3) \\ \tau_{z'x'} &= \sigma_x l_3 l_1 + \sigma_y m_3 m_1 + \sigma_z n_3 n_1 \\ &\quad + \tau_{xy}(l_3 m_1 + m_3 l_1) + \tau_{yz}(m_3 n_1 + n_3 m_1) + \tau_{zx}(n_3 l_1 + l_3 n_1)\end{aligned}\right\} \quad (2.8)$$

Formulas (2.7) and (2.8) determine the components of the stress tensor for any three mutually orthogonal planes passing through the given point if these components are known for certain three mutually orthogonal planes.

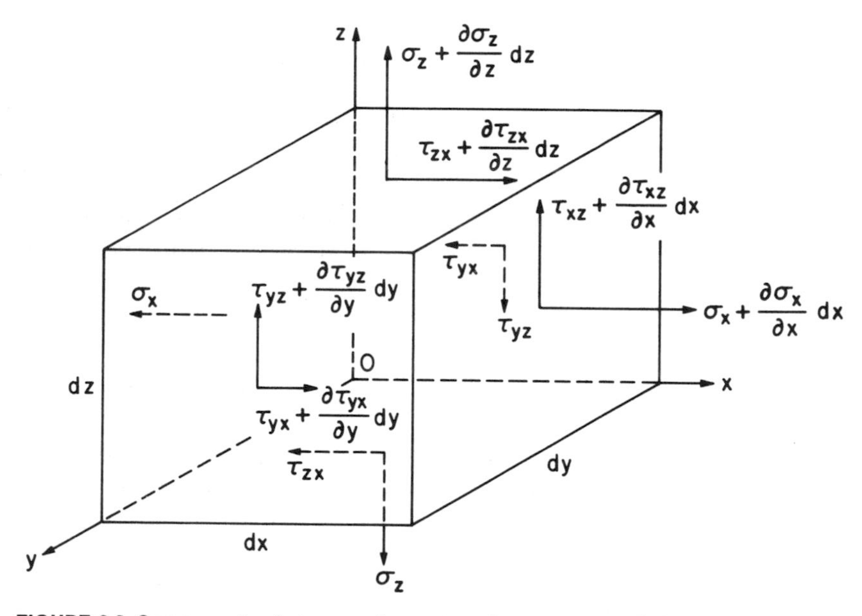

FIGURE 2.3 Components of stress acting on an elementary parallelepiped (at a point).

Thus, the stress condition at the given point is completely defined if the values of the components of the stress tensor are known for any three mutually orthogonal planes.

2.5 EQUILIBRIUM CONDITIONS FOR AN ELEMENTARY PARALLELEPIPED

Equations (2.5) present equilibrium conditions for a surface element of an elastic body. In order to obtain similar conditions for an inner element, we consider an elementary parallelepiped with faces parallel to the coordinate planes and edges dx, dy, and dz (Fig. 2.3). The equations of equilibrium for the forces acting in the direction of the axis x and for the moments about the axis y are as follows (in Fig. 2.3, only those forces are shown that give projections on the axis x or a moment with respect to the axis y):

$$\frac{\partial \sigma_x}{\partial x} dx\, dy\, dz + \frac{\partial \tau_{yx}}{\partial y} dy\, dx\, dz + \frac{\partial \tau_{zx}}{\partial z} dz\, dx\, dy + X\, dx\, dy\, dz = 0 \tag{2.9}$$

$$\begin{aligned}
&\frac{\partial \sigma_x}{\partial x} dx\, dy\, dz \cdot \frac{dz}{2} + \frac{\partial \tau_{yx}}{\partial y} dy\, dx\, dz \cdot \frac{dz}{2} + \left(\tau_{zx} + \frac{\partial \tau_{zx}}{\partial z} dz\right) dx\, dy \cdot dz \\
&\quad - \frac{\partial \sigma_z}{\partial z} dz\, dx\, dy \cdot \frac{dx}{2} - \left(\tau_{xz} + \frac{\partial \tau_{xz}}{\partial x} dx\right) dy\, dz \cdot dx \\
&\quad - \frac{\partial \tau_{yz}}{\partial y} dy\, dx\, dz \cdot \frac{dx}{2} + X\, dx\, dy\, dz \cdot \frac{dz}{2} - Z\, dx\, dy\, dz \cdot \frac{dx}{2} = 0
\end{aligned} \tag{2.10}$$

Here X and Z are projections of the body force on the axes x and z. Dividing (Eq. 2.9) by $dx\, dy\, dz$, we arrive at the equilibrium condition

$$\frac{\partial \sigma_x}{\partial x} + \frac{\partial \tau_{yx}}{\partial y} + \frac{\partial \tau_{zx}}{\partial z} + X = 0$$

Equation (2.10) contains small terms of the third order, which are proportional to the cube of linear dimensions of the parallelepiped, as well as terms of the fourth order, proportional to the fourth power of linear dimensions. After neglecting the latter, we have

$$\tau_{zx} = \tau_{xz}$$

By forming the equations of equilibrium for the forces in the y and z directions, and for the moments about the x and z axes, we obtain, in summary, the following two groups of equilibrium equations:

$$\left.\begin{aligned}\frac{\partial\sigma_x}{\partial x}+\frac{\partial\tau_{yx}}{\partial y}+\frac{\partial\tau_{zx}}{\partial z}+X=0\\ \frac{\partial\tau_{xy}}{\partial x}+\frac{\partial\sigma_y}{\partial y}+\frac{\partial\tau_{zy}}{\partial z}+Y=0\\ \frac{\partial\tau_{xz}}{\partial x}+\frac{\partial\tau_{yz}}{\partial y}+\frac{\partial\sigma_z}{\partial z}+Z=0\end{aligned}\right\} \tag{2.11}$$

$$\tau_{xy}=\tau_{yx}\qquad \tau_{xz}=\tau_{zx}\qquad \tau_{yz}=\tau_{zy} \tag{2.12}$$

Formulas (2.12) express the *theorem of the complementarity of the shearing stresses*: shearing stresses acting in the two mutually perpendicular planes in the direction of the line of their intersection are equal. Hence, of the total of the nine stress components of the stress tensor (2.4), only six are independent and are therefore needed to define the state of stress at a point. These are three normal stresses, $\sigma_x, \sigma_y, \sigma_z$, and three shearing stresses, $\tau_{xy}, \tau_{yz}, \tau_{zx}$. Thus, stress tensor T_σ turns out to be symmetric. Since, for the evaluation of the six unknown stresses, we have only three *equilibrium conditions* (2.11), the problem of determining these stresses is three times *statically indeterminate*. To solve statically indeterminate stress problems, we must supplement the equations of statics (2.11) with additional conditions of deformation. These are derived in Section 2.12.

2.6 PRINCIPAL STRESSES AND STRESS INVARIANTS

In this section, we will demonstrate that if the normal stress for the given plane has a stationary value, the shearing stress for this plane is zero.

Examine, for instance, the normal stress $\sigma_{x'}$, and find the direction cosines for a plane for which $\sigma_{x'}$ has a stationary value

$$\frac{\partial\sigma_{x'}}{\partial l_1}=0\qquad \frac{\partial\sigma_{x'}}{\partial m_1}=0$$

Here the direction cosines l_1 and m_1 are considered as arguments, and the values $\sigma_{x'}$ and n_1 as functions. After substituting the first formula of (2.6) into the above-mentioned conditions, we have

$$\left.\begin{aligned}\sigma_x l_1+\tau_{xy}m_1+\tau_{zx}n_1+(\tau_{zx}l_1+\tau_{yz}m_1+\sigma_z n_1)\frac{\partial n_1}{\partial l_1}=0\\ \tau_{xy}l_1+\sigma_y m_1+\tau_{yz}n_1+(\tau_{zx}l_1+\tau_{yz}m_1+\sigma_z n_1)\frac{\partial n_1}{\partial m_1}=0\end{aligned}\right\} \tag{2.13}$$

As is known from analytical geometry, direction cosines are related by the equation

$$l_1^2 + m_1^2 + n_1^2 = 1 \tag{2.14}$$

From this equation, we find, by differentiation,

$$l_1 + n_1 \frac{\partial n_1}{\partial l_1} = 0 \qquad m_1 + n_1 \frac{\partial n_1}{\partial m_1} = 0 \tag{2.15}$$

Excluding the derivatives $\partial n_1/\partial l_1$ and $\partial n_1/\partial m_1$ from Eqs. (2.13) and (2.15), and using Eqs. (2.6), we obtain

$$\frac{X_{x'}}{l_1} = \frac{Y_{x'}}{m_1} = \frac{Z_{x'}}{n_1}$$

Thus, the components of the total stress for a plane where $\sigma_{x'}$ has a stationary value are proportional to the direction cosines of the normal to this plane. Therefore, the total stress in this plane is directed along the normal to the plane and, hence, the shearing stress in this plane is zero.

The planes on which there are no shearing stresses are called *principal planes*, and the normal stresses acting on these planes are called *principal stresses*.

Let us show that we can draw three principal planes through any point of an elastic body.

Let σ be a principal stress. Then the projections of this stress on the axes x, y, z are, in accordance with Eqs. (2.5),

$$X_\nu = \sigma l \qquad Y_\nu = \sigma m \qquad Z_\nu = \sigma n$$

Introducing these formulas into Eqs. (2.5), we have

$$\left.\begin{aligned}(\sigma_x - \sigma)l + \tau_{yx}m + \tau_{zx}n = 0\\ \tau_{xy}l + (\sigma_y - \sigma)m + \tau_{zy}n = 0\\ \tau_{xz}l + \tau_{yz}m + (\sigma_z - \sigma)n = 0\end{aligned}\right\} \tag{2.16}$$

Since the direction cosines l, m, n must satisfy the condition (2.14), then the solution $l = m = n = 0$ must be excluded. For nonzero solutions to Eqs. (2.16), the determinant of the coefficient matrix must be zero:

$$\begin{vmatrix} \sigma_x - \sigma & \tau_{xy} & \tau_{xz} \\ \tau_{yx} & \sigma_y - \sigma & \tau_{yz} \\ \tau_{zx} & \tau_{xy} & \sigma_z - \sigma \end{vmatrix} = 0$$

Expanding this determinant, we obtain a cubic equation for σ:

$$\sigma^3 - \sigma_I\sigma^2 + \sigma_{II}\sigma - \sigma_{III} = 0 \tag{2.17}$$

where the coefficients are expressed as follows:

$$\left.\begin{aligned} \sigma_I &= \sigma_x + \sigma_y + \sigma_z \\ \sigma_{II} &= \sigma_x\sigma_y + \sigma_y\sigma_z + \sigma_z\sigma_x - \tau_{xy}^2 - \tau_{yz}^2 - \tau_{zx}^2 \\ \sigma_{III} &= \sigma_x\sigma_y\sigma_z + 2\tau_{xy}\tau_{yz}\tau_{zx} - \sigma_x\tau_{zy}^2 - \sigma_y\tau_{zx}^2 - \sigma_z\tau_{xy}^2 \end{aligned}\right\} \tag{2.18}$$

Equation (2.17) is called a *characteristic equation of the stress tensor.* Three roots, σ_1, σ_2, σ_3, of this equation are real and give the values of the three principal stresses. In order to satisfy ourselves in that, we should introduce the value of each of the roots $\sigma_1, \sigma_2, \sigma_3$ in Eqs. (2.16) and, considering the normalizing condition (2.14), determine the three groups of the direction cosines for the three planes on which the stresses $\sigma_1, \sigma_2, \sigma_3$ act. The values of the direction cosines will indicate that all the three planes at each point of the deformed body are mutually orthogonal and that the shearing stresses on these planes are zero. Therefore, the planes on which the stresses $\sigma_1, \sigma_2, \sigma_3$ act are the principal planes, and the stresses themselves are the principal stresses.

In general, the stresses σ_1, σ_2, and σ_3 are different. It is convenient to designate these stresses in declining order of magnitude:

$$\sigma_1 > \sigma_2 > \sigma_3$$

Note that, if $\sigma_2 = \sigma_3$, the state of stress is axisymmetric and any plane in which the vector $\bar{\sigma}_1$ is located is a principal plane. If $\sigma_1 = \sigma_2 = \sigma_3$, then the state of stress is characterized by a center of symmetry, so that any plane drawn through the given point is a principal plane. This is the case of all-round tension or compression.

It is clear that the values $\sigma_1, \sigma_2, \sigma_3$ reflecting the actual state of stress in the elastic body must be independent of the orientation of the coordinate axes. Then the coefficients $\sigma_I, \sigma_{II}, \sigma_{III}$ of this equation must also remain constant during transformation of the coordinate axes, i.e., must be *invariants.* In paticular, when the axes x, y, z are perpendicular to the principal planes, then, as follows from Eqs. (2.18),

$$\left.\begin{aligned} \sigma_I &= \sigma_1 + \sigma_2 + \sigma_3 \\ \sigma_{II} &= \sigma_1\sigma_2 + \sigma_2\sigma_3 + \sigma_3\sigma_1 \\ \sigma_{III} &= \sigma_1\sigma_2\sigma_3 \end{aligned}\right\} \tag{2.19}$$

The values σ_{I}, σ_{II}, and σ_{III} are called *linear*, *quadratic*, and *cubic invariants* of the stress tensor, respectively. It can be shown that, although many other parameters of the state of stress may also be invariants, they are all functions of σ_{I}, σ_{II}, and σ_{III}. In other words, there are only three independent stress invariants. There are some physical phenomena controlled by the state of stress that are described in terms of stress invariants. One of the most important examples is plastic yielding of materials.

2.7 NORMAL STRESS SURFACE AND STRESS ELLIPSOID

A clear geometrical interpretation can be given to some of the foregoing results concerning principal stresses.

The first of the formulas (2.7), written for a normal stress σ_v, yields

$$\sigma_v = \sigma_x l^2 + \sigma_y m^2 + \sigma_z n^2 + 2\tau_{xy} lm + 2\tau_{yz} mn + 2\tau_{zx} nl \tag{2.20}$$

Introducing the notation

$$\xi = \frac{l}{\sqrt{\sigma_v}} \qquad \eta = \frac{m}{\sqrt{\sigma_v}} \qquad \zeta = \frac{n}{\sqrt{\sigma_v}}$$

we can rewrite Eq. (2.20) in the form

$$\sigma_x \xi^2 + \sigma_y \eta^2 + \sigma_z \zeta^2 + 2\tau_{xy}\xi\eta + 2\tau_{yz}\eta\zeta + 2\tau_{zx}\zeta\xi = \pm 1 \tag{2.21}$$

This is an equation of a surface of the second order. It is called the *normal stress surface*, or Cauchy surface. The plus sign on the right side of Eq. (2.21) corresponds to those planes in which σ_v is positive (tensile), and the minus sign corresponds to the planes in which σ_v is negative (compressive).

As we know from analytical geometry, a surface of the second order can be transformed by a proper rotation of the coordinate axes to a so-called *canonical form*, which does not contain terms with the products of the coordinates. As one can see from Eq. (2.21), such a transformation can be achieved if all the three shearing stresses in the new axes are zero. Therefore, the very possibility that the normal stress surface can be transformed to a canonical form

$$\sigma_1 \xi^2 + \sigma_2 \eta^2 + \sigma_3 \zeta^2 = \pm 1 \tag{2.22}$$

is an indication of the fact that, at each point of a body, we can always find three mutually perpendicular planes on which the shearing stresses are zero.

If all the principal stresses are of the same sign, then the surface defined by Eq. (2.22) is an ellipsoid. This ellipsoid is called a *stress ellipsoid* or *Lame' ellipsoid*. Its semiaxes, as we can see from Eq. (2.22), give the values of the three principal stresses. If the principal stresses have different signs, however, the surface (2.22) consists of one- and two-sheet hyperboloids.

2.8 PRINCIPAL SHEARING STRESSES

We have shown that, in general, three independent shearing stresses must be known as part of the definition of the state of stress at a point and that it is always possible to choose such a set of rectangular coordinate axes where all the shearing stresses simultaneously disappear on planes that are normal to these axes. These, then, are the principal axes. We have also shown that the principal stresses can be obtained as stationary values of the normal stresses. In the same way, we can determine the stationary values and the directions of the shearing stresses.

From Eq. (2.3), we have

$$\tau_\nu^2 = X_\nu^2 + Y_\nu^2 + Z_\nu^2 - \sigma_\nu^2 \tag{2.23}$$

Evidently, if the axes x, y, z are directed along the normals to the principal planes, then the projections on the coordinate axes of the total stress acting on the plane with the normal ν are

$$X_\nu = \sigma_1 l \qquad Y_\nu = \sigma_2 m \qquad Z_\nu = \sigma_3 n$$

The normal stress on this plane, in accordance with Eq. (2.20), is

$$\sigma_\nu = \sigma_1 l^2 + \sigma_2 m^2 + \sigma_3 n^2 \tag{2.24}$$

Then Eq. (2.23) yields

$$\tau_\nu^2 = (\sigma_1^2 l^2 + \sigma_2^2 m^2 + \sigma_3^2 n^2) - (\sigma_1 l^2 + \sigma_2 m^2 + \sigma_3 n^2)^2 \tag{2.25}$$

Considering this expression as a function of the direction cosines l, m, n and evaluating the stationary values of the function $\tau_\nu^2(l, m, n)$, we conclude that the shearing stresses take stationary values

$$\tau_1 = \pm\frac{\sigma_1 - \sigma_3}{2} \qquad \tau_2 = \pm\frac{\sigma_1 - \sigma_2}{2} \qquad \tau_3 = \pm\frac{\sigma_2 - \sigma_3}{2} \tag{2.26}$$

for planes that lie in the bisector planes of the dihedral angles formed by the principal planes, i.e., that are inclined 45° from the principal planes. The stresses τ_1, τ_2, τ_3 are called *principal shearing stresses.* Note that, as follows from Eqs. (2.26),

$$\tau_1 + \tau_2 + \tau_3 = 0 \tag{2.27}$$

2.9 OCTAHEDRAL STRESSES

There is another set of planes that is of importance in particular applications such as strength theories and plasticity. These are the eight planes whose normals are equally inclined to the principal axes, so that

$$l = m = n = \pm \frac{1}{\sqrt{3}}$$

They are known as the *octahedral planes.* After introducing these direction cosine values into Eqs. (2.24) and (2.25), we find that the stresses acting in the octahedral planes are expressed by the formulas

$$\sigma_0 = \frac{1}{3}(\sigma_1 + \sigma_2 + \sigma_3) \tag{2.28}$$

$$\tau_0 = \frac{1}{3}\sqrt{(\sigma_1 - \sigma_2)^2 + (\sigma_2 - \sigma_3)^2 + (\sigma_3 - \sigma_1)^2} = \frac{2}{3}\sqrt{\tau_1^2 + \tau_2^2 + \tau_3^2} \tag{2.29}$$

These stresses are called *octahedral stresses.* They were introduced into elasticity theory by Nádái.

2.10 STRESS DEVIATOR AND SPHERICAL TENSOR

Stress tensor T_σ given by the formulas (2.4) can be presented as a sum of two tensors as follows:

$$T_\sigma = D_\sigma + T_s \tag{2.30}$$

where the tensor

$$D_\sigma = \begin{vmatrix} \sigma_x - \sigma_0 & \tau_{xy} & \tau_{yz} \\ \tau_{yx} & \sigma_y - \sigma_0 & \tau_{yz} \\ \tau_{zx} & \tau_{zy} & \sigma_z - \sigma_0 \end{vmatrix} \tag{2.31}$$

is called the *stress deviator*, and the tensor

$$T_s = \begin{vmatrix} \sigma_0 & 0 & 0 \\ 0 & \sigma_0 & 0 \\ 0 & 0 & \sigma_0 \end{vmatrix} \tag{2.32}$$

is called the *spherical tensor*. In these formulas, σ_0 are normal octahedral stresses defined by the formula (2.28). The stress deviator characterizes the state of stress that is due to the distortion in the form of the solid body and is associated with shearing deformations. The spherical tensor characterizes the state of stress taking place during uniform all-round ("hydrostatic") tension or compression, when only the volume of the body element is changing while its form remains unchanged. An example of a state of stress resulting in form distortion without any change in volume is the torsion of a prismatic rod. An example of a state of stress in which only the volume is changed is all-round ("hydrostatic") compression.

In effect, for all the planes passing through the given point, the stress deviator plays the same role as the shearing stresses for a certain elementary plane, whereas the role of the spherical tensor is similar to the role of the normal stress.

The breakdown of the total state of stress and the corresponding tensors into two components is feasible because the body volume changes elastically and linearly up to the very failure of the material. This takes place under very high σ_0 values, well above the yield point, especially in compression. At the same time, the deviation from linearity in the distortion of the body form and the occurrence of plastic strains start with deviatoric stresses not exceeding the yield stress. Therefore, it is the relationships between the deviatoric stresses and strains, and not the Hooke's law equations, that are employed in the plasticity theory. As far as the elasticity theory is concerned, the deviatoric stresses are used to determine the ultimate elastic state of stress, i.e., the limitations of the elastic approach (see Section 5.3).

The invariants of the stress deviator can be formed by analogy with the invariants of the stress tensor and are as follows:

$$\left.\begin{aligned} d_{\mathrm{I}} &= (\sigma_1 - \sigma_0) + (\sigma_2 - \sigma_0) + (\sigma_3 - \sigma_0) = \sigma_1 + \sigma_2 + \sigma_3 - 3\sigma_0 = 0 \\ d_{\mathrm{II}} &= (\sigma_1 - \sigma_0)(\sigma_2 - \sigma_0) + (\sigma_2 - \sigma_0)(\sigma_3 - \sigma_0) + (\sigma_1 - \sigma_0)(\sigma_3 - \sigma_0) \\ &= -\tfrac{1}{6}[(\sigma_1 - \sigma_2)^2 + (\sigma_2 - \sigma_3)^2 + (\sigma_3 - \sigma_1)^2] = -\tfrac{3}{2}\tau_0^2 \\ d_{\mathrm{III}} &= (\sigma_1 - \sigma_0)(\sigma_2 - \sigma_0)(\sigma_3 - \sigma_0) \\ &= \tfrac{1}{3}[(\sigma_1 - \sigma_0)^3 + (\sigma_2 - \sigma_0)^3 + (\sigma_3 - \sigma_0)^3] \end{aligned}\right\} \tag{2.33}$$

The linear invariant d_{I} of the deviator is zero, which is an indication that, on average, there is neither tension nor compression. The quadratic d_{II} and cubic d_{III} invariants characterize the mean quadratic and the mean cubic deviations of the state of stress $\sigma_1, \sigma_2, \sigma_3$ from the average "hydrostatic" stress.

2.11 DISPLACEMENTS AND STRAINS

Suppose that the body is deformed so that one of its particles is displaced from its original position A in the nondeformed body to position A' after deformation. The vector connecting points A and A' is called *displacement.* The projections

$$u = u(x, y, z) \qquad v = v(x, y, z) \qquad w = w(x, y, z)$$

of this vector on the coordinate axes are called components of the displacement vector or simply *displacements.* Obviously, the functions u, v, and w are, in general, different for different points of the body. The relative increment in the distance between the two points of the body due to its deformation is called *linear (normal) strain,* or linear elongation in the given direction. The decrease in the initially right angle between the two directions in the body due to its deformation is called *angular (shearing) strain,* or relative shear at the given point. The angle of rotation of the bisector of the initially right angle between the two directions is called *rotation* at the given point.

In order to obtain the relationships between the components of the strain and rotation, on the one hand, and the displacements, on the other, we consider a simple case of a two-dimensional deformation in which all the particles of the body originally located in the same plane remain in this plane after deformation (Fig. 2.4). Let the coordinates of a particle before the deformation be x, y, z and, after the deformation, become $x + u, y + v, z + w$. Now, examine a small rectangular element $ABCD$ of a nondeformed body. After deformation, this element will occupy a position $A'B'C'D'$. As one can see from Fig. 2.4,

$$(A'B')^2 = \left(dx + \frac{\partial u}{\partial x}\,dx\right)^2 + \left(\frac{\partial v}{\partial x}\,dx\right)^2 \cong \left(1 + \frac{\partial u}{\partial x}\right)^2 (AB)^2$$

Then the longitudinal strain in the x direction is

$$\varepsilon_x = \frac{A'B' - AB}{AB} = \frac{\partial u}{\partial x}$$

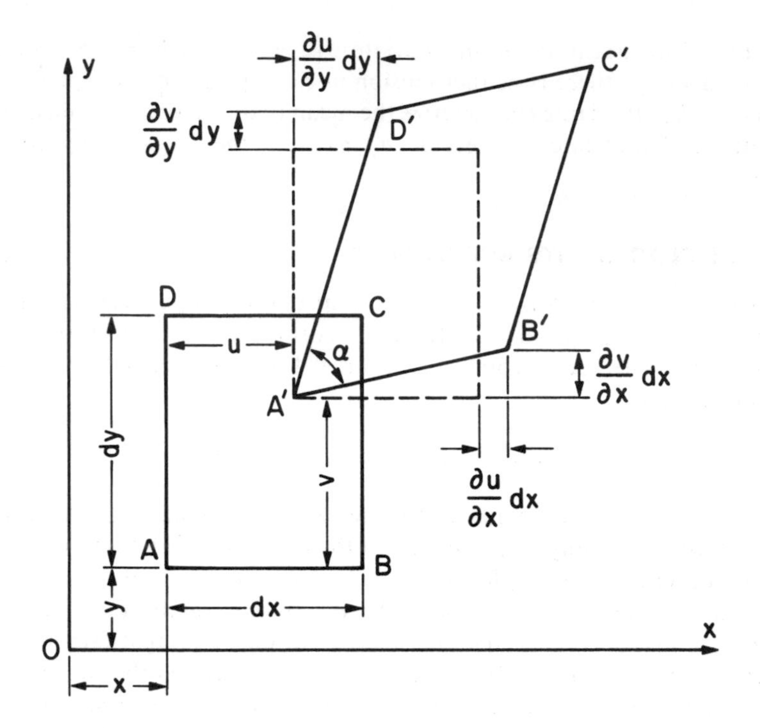

FIGURE 2.4 A plane element in the unstrained (*ABCD*) and strained (*A′B′C′D′*) state.

In a similar manner, we find

$$\varepsilon_y = \frac{\partial v}{\partial y}$$

The shearing strain at point *A*, as evident from Fig. 2.4, is

$$\gamma_{xy} = \frac{\pi}{2} - \alpha = \frac{\partial v}{\partial x} + \frac{\partial u}{\partial y}$$

and the rotation component is

$$\omega_z = \frac{1}{2}\left(\frac{\partial v}{\partial x} - \frac{\partial u}{\partial y}\right)$$

In a three-dimensional case, the strains and rotations are related to the displacements by the formulas

$$\left.\begin{aligned} \varepsilon_x &= \frac{\partial u}{\partial x} \qquad \varepsilon_y = \frac{\partial v}{\partial y} \qquad \varepsilon_z = \frac{\partial w}{\partial z} \\ \gamma_{xy} &= \frac{\partial v}{\partial x} + \frac{\partial u}{\partial y} \qquad \gamma_{yz} = \frac{\partial v}{\partial z} + \frac{\partial w}{\partial y} \qquad \gamma_{zx} = \frac{\partial w}{\partial x} + \frac{\partial u}{\partial z} \end{aligned}\right\} \tag{2.34}$$

$$\omega_x = \frac{1}{2}\left(\frac{\partial w}{\partial y} - \frac{\partial v}{\partial z}\right) \qquad \omega_y = \frac{1}{2}\left(\frac{\partial u}{\partial z} - \frac{\partial w}{\partial x}\right) \qquad \omega_z = \frac{1}{2}\left(\frac{\partial v}{\partial x} - \frac{\partial u}{\partial y}\right) \tag{2.35}$$

Equations (2.34) are called *Cauchy's formulas.* Obviously,

$$\gamma_{xy} = \gamma_{yx} \qquad \gamma_{yz} = \gamma_{zy} \qquad \gamma_{zx} = \gamma_{xz} \tag{2.36}$$

Normal strains are positive if they correspond to elongation in the given direction. Shearing strains are positive if they are accompanied by reduction in the angles between the positive directions of the coordinate axes.

From Eqs. (2.34) and (2.35), we can see that the strains and rotations will not change if the quantities

$$\bar{u} = u_0 + \beta z - \gamma y \qquad \bar{v} = v_0 + \gamma x - \alpha z \qquad \bar{w} = w_0 + \alpha y - \beta x$$

are added to the displacement components. These quantities characterize the displacements of the body as a nondeformable whole. In the preceding formulas, u_0, v_0, w_0 are progressive displacements, and α, β, γ are rotation angles about the coordinate axes.

2.12 COMPATIBILITY EQUATIONS

We will now derive compatibility equations for the strains. The physical rationale behind these equations is as follows.

Imagine that an elastic body, prior to deformation, consisted of numerous elementary cubes. If each of these cubes, taken separately, is subjected to an arbitrary deformation, then it may appear that the parallelepipeds or even more complicated small volumes brought about by this deformation will not form a continuous body anymore. Therefore, certain conditions must be imposed on the components of the strain, so that the continuity of the body is not violated. These conditions are called compatibility equations. Formally, these equations follow from Cauchy's formulas (2.34), which state that six components of the strain are determined by only three independent functions, u, v, w, and, therefore, must be somehow related among themselves.

Having differentiated the first formula in Eqs. (2.34) twice with respect to y and the second formula with respect to x, and summing up the results, we find

$$\frac{\partial^2 \varepsilon_x}{\partial y^2} + \frac{\partial^2 \varepsilon_y}{\partial x^2} = \frac{\partial^3 u}{\partial x \partial y^2} + \frac{\partial^3 v}{\partial x^2 \partial y} = \frac{\partial^2}{\partial x \partial y}\left(\frac{\partial u}{\partial y} + \frac{\partial v}{\partial x}\right) = \frac{\partial^2 \gamma_{xy}}{\partial x \partial y}$$

Then, using Eqs. (2.34), we determine the quantity

$$\begin{aligned}
&\frac{\partial}{\partial x}\left(-\frac{\partial \gamma_{yz}}{\partial x} + \frac{\partial \gamma_{zx}}{\partial y} + \frac{\partial \gamma_{xy}}{\partial z}\right) \\
&\quad = \frac{\partial}{\partial x}\left(-\frac{\partial^2 v}{\partial x \partial z} - \frac{\partial^2 w}{\partial x \partial y} + \frac{\partial^2 w}{\partial x \partial y} + \frac{\partial^2 u}{\partial y \partial z} + \frac{\partial^2 v}{\partial x \partial z} + \frac{\partial^2 u}{\partial y \partial z}\right) \\
&\quad = 2\frac{\partial^3 u}{\partial x \partial y \partial z} = 2\frac{\partial^2 \varepsilon_z}{\partial x \partial y}
\end{aligned}$$

By analogy with the preceding expressions, the rest of the equations can be obtained. In the final form, these *compatibility equations* are as follows:

$$\left.\begin{aligned}
\frac{\partial^2 \varepsilon_x}{\partial y^2} + \frac{\partial^2 \varepsilon_y}{\partial x^2} &= \frac{\partial^2 \gamma_{xy}}{\partial x \partial y} \\
\frac{\partial^2 \varepsilon_y}{\partial z^2} + \frac{\partial^2 \varepsilon_z}{\partial y^2} &= \frac{\partial^2 \gamma_{yz}}{\partial y \partial z} \\
\frac{\partial^2 \varepsilon_z}{\partial x^2} + \frac{\partial^2 \varepsilon_x}{\partial z^2} &= \frac{\partial^2 \gamma_{zx}}{\partial z \partial x}
\end{aligned}\right\} \tag{2.37}$$

$$\left.\begin{aligned}
\frac{\partial}{\partial x}\left(-\frac{\partial \gamma_{yz}}{\partial x} + \frac{\partial \gamma_{zx}}{\partial y} + \frac{\partial \gamma_{xy}}{\partial z}\right) &= 2\frac{\partial^2 \varepsilon_x}{\partial y \partial z} \\
\frac{\partial}{\partial y}\left(\frac{\partial \gamma_{yz}}{\partial x} - \frac{\partial \gamma_{zx}}{\partial y} + \frac{\partial \gamma_{xy}}{\partial z}\right) &= 2\frac{\partial^2 \varepsilon_y}{\partial z \partial x} \\
\frac{\partial}{\partial z}\left(\frac{\partial \gamma_{yz}}{\partial x} + \frac{\partial \gamma_{zx}}{\partial y} - \frac{\partial \gamma_{xy}}{\partial z}\right) &= 2\frac{\partial^2 \varepsilon_z}{\partial x \partial y}
\end{aligned}\right\} \tag{2.38}$$

The relationships (2.37) and (2.38) are also known as *differential relationships of Saint-Venant*. They are the very additional conditions that enable us to "uncover" the statistical indeterminacy of the theory-of-elasticity problems (see Section 2.5).

The formulas (2.37) and (2.38) are the necessary conditions of the continuity of the body. They are also sufficient conditions if the body is limited by a *simply connected region.* This is a region within which any closed curve can be contracted in a continuous manner into a point without intersecting the boundaries of the region. For regions not simply connected, however, the preceding compatibility relationships are insufficient for the continuity of the body and do not guarantee the uniqueness of displacements. In this case, some additional conditions have to be fulfilled. We will discuss these conditions later when examining torsion of hollow sections.

2.13 STRAIN TENSOR, PRINCIPAL STRAINS, AND STRAIN INVARIANTS

The theory of strains can be structured by analogy with the theory of stresses. Thus, the matrix

$$T_\varepsilon = \begin{vmatrix} \varepsilon_x & \gamma_{xy} & \gamma_{xz} \\ \gamma_{yx} & \varepsilon_y & \gamma_{yz} \\ \gamma_{zx} & \gamma_{zy} & \varepsilon_z \end{vmatrix} \tag{2.39}$$

is called the *strain tensor.* The *strain surface,* which is analogous to the stress surface, is a surface of the second order, with the center at the given point. This surface has three mutually orthogonal principal axes, indicating the *directions of the principal strains.* These coincide with the directions of the principal stresses. The shearing strains are zero for these directions, the normal strains $\varepsilon_1, \varepsilon_2, \varepsilon_3$ are called *principal strains* at this point, and $\varepsilon_1 > \varepsilon_2 > \varepsilon_3$.

In order to determine the principal strains, we have to solve the equation

$$\begin{vmatrix} \varepsilon_x - \varepsilon & \gamma_{xy} & \gamma_{xz} \\ \gamma_{xy} & \varepsilon_y - \varepsilon & \gamma_{yz} \\ \gamma_{xz} & \gamma_{yz} & \varepsilon_z - \varepsilon \end{vmatrix} = 0$$

which leads to the following *characteristic equation for the strain tensor:*

$$\varepsilon^3 - \varepsilon_{\rm I}\varepsilon^2 + \varepsilon_{\rm II}\varepsilon - \varepsilon_{\rm III} = 0$$

Here,

$$\left.\begin{aligned} \varepsilon_{\rm I} &= \varepsilon_x + \varepsilon_y + \varepsilon_z = \varepsilon_1 + \varepsilon_2 + \varepsilon_3 \\ \varepsilon_{\rm II} &= \varepsilon_x\varepsilon_y + \varepsilon_y\varepsilon_z + \varepsilon_z\varepsilon_x - \gamma_{xy}^2 - \gamma_{yz}^2 - \gamma_{zx}^2 = \varepsilon_1\varepsilon_2 + \varepsilon_2\varepsilon_3 + \varepsilon_3\varepsilon_1 \\ \varepsilon_{\rm III} &= \varepsilon_x\varepsilon_y\varepsilon_z + 2\gamma_{xy}\gamma_{yz}\gamma_{zx} - \varepsilon_x\gamma_{yz}^2 - \varepsilon_y\gamma_{zx}^2 - \varepsilon_z\gamma_{xy}^2 = \varepsilon_1\varepsilon_2\varepsilon_3 \end{aligned}\right\} \tag{2.40}$$

are *invariants* of the strain tensor.

The first, linear, invariant characterizes the relative increase in the body's volume at the given point. Indeed, let us consider an elementary parallelepiped with edges dx, dy, dz, which coincide with the directions of the principal strains. The volume of this parallelepiped after deformation is

$$V' = (1 + \varepsilon_1)(1 + \varepsilon_2)(1 + \varepsilon_3)\, dx\, dy\, dz$$

so that the relative increase in the volume is

$$\theta = \frac{V' - dx\, dy\, dz}{dx\, dy\, dz} = (1 + \varepsilon_1)(1 + \varepsilon_2)(1 + \varepsilon_3) - 1 \cong \varepsilon_1 + \varepsilon_2 + \varepsilon_3 \quad (2.41)$$

Therefore, the value $\varepsilon_{\mathrm{I}} = \theta$ is called *dilatation.*

2.14 HOOKE'S LAW

In order to be able to utilize the equilibrium equations (2.11), containing stresses, simultaneously with the compatibility equations (2.37) and (2.38), containing strains, we have to have at our disposal some relationships linking stresses and strains. It should be emphasized that, whereas the equations of equilibrium and conditions of compatibility are applicable to any static problem in solid mechanics involving small strains, the form of the equations expressing stresses in terms of strains depends on the nature of the material under consideration.

The simplest relationship between the stresses and strains can be presented in the form of *Hooke's law*, which simply states that the stresses are directly proportional to the strains:

$$\sigma_x = C_{11}\varepsilon_x + C_{12}\varepsilon_y + C_{13}\varepsilon_z + C_{14}\gamma_{xy} + C_{15}\gamma_{yz} + C_{16}\gamma_{zx}$$

$$\tau_{xy} = C_{21}\varepsilon_x + C_{22}\varepsilon_y + C_{23}\varepsilon_z + C_{24}\gamma_{xy} + C_{25}\gamma_{yz} + C_{26}\gamma_{zx}$$

$$\ldots$$

The directions of the principal strains coincide with the directions of the principal stresses. Therefore, without making any difference between these directions, we will refer the preceding equations to the principal directions:

$$\left.\begin{aligned} \sigma_1 &= C_{11}\varepsilon_1 + C_{12}\varepsilon_2 + C_{13}\varepsilon_3 \\ \sigma_2 &= C_{21}\varepsilon_1 + C_{22}\varepsilon_2 + C_{23}\varepsilon_3 \\ \sigma_3 &= C_{31}\varepsilon_1 + C_{32}\varepsilon_2 + C_{33}\varepsilon_3 \end{aligned}\right\} \quad (2.42)$$

For an isotropic body, the effect of the strain ε_1 on the stress σ_1 must be the same as the effect of the strain ε_2 on the stress σ_2 or the strain ε_3 on the stress σ_3, so that

$$C_{11} = C_{22} = C_{33} \tag{2.43}$$

From the same property of isotropy, it follows that the effect of the strains ε_2 and ε_3 on the stress σ_1 must also be the same, so that $C_{12} = C_{13}$. Similarly, we conclude that $C_{21} = C_{23}$ and $C_{31} = C_{32}$. Finally, the effect of the strain ε_2 or ε_3 on σ_1 must be the same as the effect of the strain ε_1 or ε_3 on σ_2, and the effect of ε_1 or ε_3 on σ_1 must be the same as the effect of ε_1 and ε_2 on σ_3. Therefore, $C_{21} = C_{12}$, $C_{31} = C_{13}$, and $C_{23} = C_{32}$. Thus,

$$C_{12} = C_{21} = C_{13} = C_{31} = C_{23} = C_{32} \tag{2.44}$$

On the basis of the relationships (2.43) and (2.44), we conclude that, when referring the stresses and strains to the principal axes, it is sufficient to introduce only two independent elastic constraints, say, a and b. Then, the relationships (2.42) can be presented as follows:

$$\left.\begin{aligned} \sigma_1 &= a\varepsilon_1 + b(\varepsilon_2 + \varepsilon_3) \\ \sigma_2 &= a\varepsilon_2 + b(\varepsilon_1 + \varepsilon_3) \\ \sigma_3 &= a\varepsilon_3 + b(\varepsilon_1 + \varepsilon_2) \end{aligned}\right\}$$

These equations can be rewritten in a more convenient form,

$$\left.\begin{aligned} \sigma_1 &= \lambda\theta + 2\mu\varepsilon_1 \\ \sigma_2 &= \lambda\theta + 2\mu\varepsilon_2 \\ \sigma_3 &= \lambda\theta + 2\mu\varepsilon_3 \end{aligned}\right\} \tag{2.45}$$

where new constants $\mu = (a - b)/2$ and $\lambda = b$ are introduced, and the notation (2.41) is used. The constants μ and λ are called *Lamé constants.*

Equations (2.45) present the relationships between the principal stresses and principal strains. In order to obtain the relationships between the stresses and strains in arbitrary rectangular coordinates, we should use the formulas (2.7) and (2.8) for the transformation of the stress tensor and similar formulas for the transformation of the strain tensor as, for instance,

$$\sigma_x = \sigma_1 l_1^2 + \sigma_2 m_1^2 + \sigma_3 n_1^2 \qquad \tau_{xy} = \sigma_1 l_1 l_2 + \sigma_2 m_1 m_2 + \sigma_3 n_1 n_2 \tag{2.46}$$

$$\varepsilon_x = \varepsilon_1 l_1^2 + \varepsilon_2 m_1^2 + \varepsilon_3 n_1^2 \qquad \gamma_{xy} = 2(\varepsilon_1 l_1 l_2 + \varepsilon_2 m_1 m_2 + \varepsilon_3 n_1 n_2) \tag{2.47}$$

After substituting Eqs. (2.45) into Eqs. (2.46), we obtain

$$\left.\begin{aligned}\sigma_x &= (l_1^2 + m_1^2 + n_1^2)\lambda\theta + 2\mu(l_1^2\varepsilon_1 + m_1^2\varepsilon_2 + n_1^2\varepsilon_3)\\ \tau_{xy} &= (l_1 l_2 + m_1 m_2 + n_1 n_2)\lambda\theta + 2\mu(l_1 l_2\varepsilon_1 + m_1 m_2\varepsilon_2 + n_1 n_2\varepsilon_3)\end{aligned}\right\} \tag{2.48}$$

Considering the formulas (2.47) and the identities

$$l_1^2 + m_1^2 + n_1^2 = 1 \qquad l_1 l_2 + m_1 m_2 + n_1 n_2 = 0$$

we rewrite Eqs. (2.48) in the form

$$\sigma_x = \lambda\theta + 2\mu\varepsilon_x \qquad \tau_{xy} = \mu\gamma_{xy}$$

where

$$\theta = \varepsilon_1 + \varepsilon_2 + \varepsilon_3 = \varepsilon_x + \varepsilon_y + \varepsilon_z \tag{2.49}$$

After obtaining, by a similar manner, other relationships, we have

$$\left.\begin{aligned}&\sigma_x = \lambda\theta + 2\mu\varepsilon_x \qquad \sigma_y = \lambda\theta + 2\mu\varepsilon_y \qquad \sigma_z = \lambda\theta + 2\mu\varepsilon_z\\ &\tau_{xy} = \mu\gamma_{xy} \qquad \tau_{yz} = \mu\gamma_{yz} \qquad \tau_{zx} = \mu\gamma_{zx}\end{aligned}\right\} \tag{2.50}$$

These formulas express *Hooke's law.*

2.15 YOUNG'S MODULUS AND POISSON'S RATIO

In engineering applications, however, the elastic constants

$$E = \frac{\mu(3\lambda + 2\mu)}{\lambda + \mu} \qquad \nu = \frac{\lambda}{2(\lambda + \mu)} \tag{2.51}$$

are used instead of the Lamé constants. Then, the formulas (2.50) yield

$$\left.\begin{aligned}\sigma_x &= \frac{E}{1+\nu}\left(\varepsilon_x + \frac{\nu}{1-2\nu}\theta\right) \qquad \tau_{xy} = G\gamma_{xy}\\ \sigma_y &= \frac{E}{1+\nu}\left(\varepsilon_y + \frac{\nu}{1-2\nu}\theta\right) \qquad \tau_{yz} = G\gamma_{yz}\\ \sigma_z &= \frac{E}{1+\nu}\left(\varepsilon_z + \frac{\nu}{1-2\nu}\theta\right) \qquad \tau_{zx} = G\gamma_{zx}\end{aligned}\right\} \tag{2.52}$$

where

$$G = \frac{E}{2(1 + \nu)} \tag{2.53}$$

Solving Eqs. (2.52) for the strains, we have

$$\left.\begin{aligned} \varepsilon_x &= \frac{1}{E}[\sigma_x - \nu(\sigma_y + \sigma_z)] & \gamma_{xy} &= \frac{1}{G}\tau_{xy} \\ \varepsilon_y &= \frac{1}{E}[\sigma_y - \nu(\sigma_x + \sigma_z)] & \gamma_{yz} &= \frac{1}{G}\tau_{yz} \\ \varepsilon_z &= \frac{1}{E}[\sigma_z - \nu(\sigma_x + \sigma_y)] & \gamma_{zx} &= \frac{1}{G}\tau_{zx} \end{aligned}\right\} \tag{2.54}$$

In order to determine the physical meaning of the elastic constants E, ν, and G, we assume that a rectangular element whose edges are parallel to the coordinate planes is subjected to normal stresses σ_x, evenly distributed over the two opposite faces, whereas all the other faces are stress-free. Then the first three formulas in Eqs. (2.54) yield

$$\varepsilon_x = \frac{\sigma_x}{E} \qquad \varepsilon_y = \varepsilon_z = -\frac{\nu}{E}\sigma_x = -\nu\varepsilon_x \tag{2.55}$$

As evident from the first of Eqs. (2.55), the ratio

$$E = \frac{\sigma_x}{\varepsilon_x} \tag{2.56}$$

is constant for the given material and is the same in tension and compression, providing the stress is not very great, so that the yield point is not exceeded. This ratio is called *Young's modulus* or the *modulus of elasticity* of the material. Historically, Young's modulus was determined experimentally from the simplest tests on tension and compression of bars. Young's modulus can be geometrically interpreted as the tangent of the angle of the slope of the linear part of the stress-strain diagram for a material. A typical stress-strain diagram for carbon steel is shown schematically in Fig. 2.5.

As long as the stress is small, the material behaves elastically, and the original size of the specimen is regained on removal of the applied load. Point A represents the *proportional limit* at which the linear relationship between the stress and the strain ceases to hold. The elastic range generally

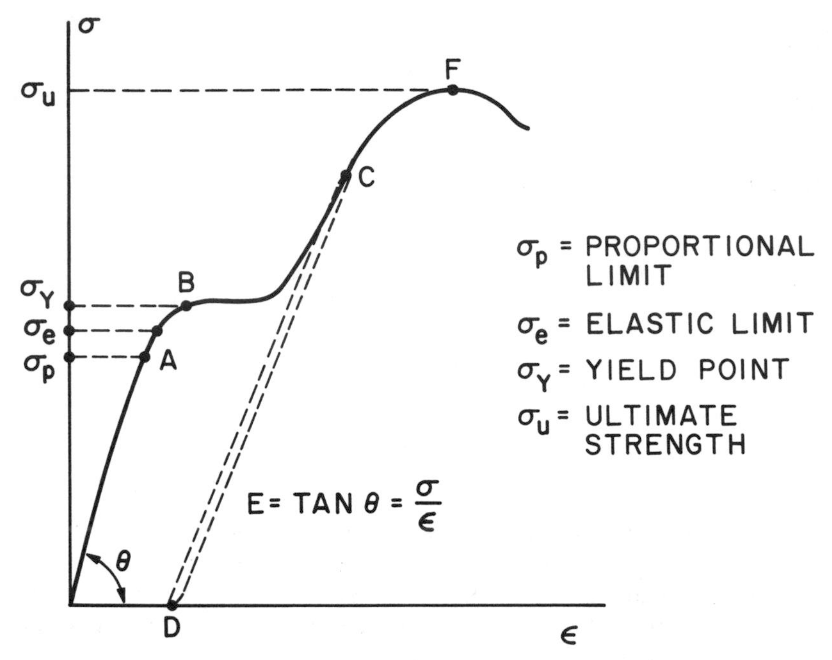

FIGURE 2.5 Typical stress-strain diagram for a metal.

extends slightly beyond the proportional limit, and the corresponding stress is called the *elastic limit*. For most metals, the transition from elastic to plastic behavior is gradual and, therefore, the location of yield point B is largely a matter of convention. The corresponding stress σ_y is known as *yield stress*. It is generally defined as that for which a specified small amount of permanent deformation is observed. Beyond the yield point, the stress at first remains constant with the further increase in strain and then continually increases, while the slope of the stress-strain curve, representing the rate of strain *hardening*, steadily decreases with increasing stress. If the specimen is stressed to some point C in the plastic range and the load is subsequently released, there is an elastic recovery following the path CD, which is very nearly a straight line of slope E. The permanent (residual) strain that remains on complete unloading is equal to OD. On reapplication of the load, neglecting the hysteresis loop of narrow width formed during the unloading and loading, we bring the stress-strain diagram back to point C. On further loading, the curve proceeds virtually as a continuation of the previous portion of the diagram until the specimen fails at point F, which defines the *ultimate strength* of the material.

The second of Eqs. (2.55) indicates that the elongation of an element in one direction is accompanied by its shortening in two other directions and that the ratio

$$\nu = -\frac{\varepsilon_y}{\varepsilon_x} = -\frac{\varepsilon_z}{\varepsilon_x}$$

of the lateral compression to the longitudinal tension is also constant for the given material as long as its deformation is within the range of linear elasticity. This constant is called *Poisson's ratio*. Although Young's modulus has the dimensions of stress, Poisson's ratio is dimensionless.

As to the constant G, its physical meaning can be seen from the formulas in Eqs. (2.52) or (2.54). The ratio

$$G = \frac{\tau_{xy}}{\gamma_{xy}} = \frac{\tau_{yz}}{\gamma_{yz}} = \frac{\tau_{zx}}{\gamma_{zx}}$$

is constant for the given material and is called *shear modulus*, or the modulus of rigidity. Historically, this constant was determined experimentally from tests on the torsion of circular bars. By substituting the formulas (2.51) into Eq. (2.53), we can see that $G = \mu$.

From consideration of the dilatation resulting from a "hydrostatic" state of stress, one more elastic constant

$$K = \frac{\sigma_I}{\theta} = \frac{\sigma_I}{\varepsilon_I} = \frac{E}{1 - 2\nu} \tag{2.57}$$

can be introduced. This constant is the *bulk* or *compressibility modulus*. The relationship (2.57) between K and elastic constants E, ν can be obtained by summing up the first three equations in Eqs. (2.52). As follows from the relationship (2.57), the K value becomes infinitely great for a noncompressible body, and Poisson's ratio, in this case, is $\nu = 0.5$. Obviously, this is the maximum Poisson's ratio possible. Silicone gels, widely used in electronic packaging as encapsulation materials and in fiber optics as primary coating materials, have Poisson's ratios very close to 0.5.

2.16 HOOKE'S LAW WITH CONSIDERATION OF THERMOELASTIC STRAINS

Changes in temperature can also produce strains although, if these changes are uniform throughout the body, which is free to expand or contract, no additional internal stresses are involved. However, in the presence of either thermal gradients or external constraints, stresses are produced.

Provided that the material remains linearly elastic, the stress pattern resulting from the temperature distribution can be superimposed on the stress pattern due to the applied external force. Alternatively, both effects can be incorporated in a single set of constitutive equations. Let α be the linear coefficient of thermal expansion of the material and ΔT the change in temperature, which is, in general, a function of the coordinates x, y, z. In a thermally isotropic body, the thermally induced normal strains are

$$\varepsilon_x = \varepsilon_y = \varepsilon_z = \alpha \Delta T \tag{2.58}$$

and the shearing strains are zero. According to the *Duhamel–Neumann hypothesis*, these strains can simply be added to the strains due to the external forces. Then the Hooke's law equations (2.54) can be presented in the form

$$\left.\begin{aligned}
\varepsilon_x &= \frac{1}{E}[\sigma_x - \nu(\sigma_y + \sigma_z)] + \alpha\Delta T \qquad & \gamma_{xy} &= \frac{1}{G}\tau_{xy} \\
\varepsilon_y &= \frac{1}{E}[\sigma_y - \nu(\sigma_x + \sigma_z)] + \alpha\Delta T \qquad & \gamma_{yz} &= \frac{1}{G}\tau_{yz} \\
\varepsilon_z &= \frac{1}{E}[\sigma_z - \nu(\sigma_x + \sigma_y)] + \alpha\Delta T \qquad & \gamma_{zx} &= \frac{1}{G}\tau_{zx}
\end{aligned}\right\} \tag{2.59}$$

Solving these equations for the stresses, we obtain

$$\left.\begin{aligned}
\sigma_x &= \frac{E}{1+\nu}\left(\varepsilon_x + \frac{\nu}{1-2\nu}\theta - \frac{1+\nu}{1-2\nu}\alpha\Delta T\right) \qquad & \tau_{xy} &= G\gamma_{xy} \\
\sigma_y &= \frac{E}{1+\nu}\left(\varepsilon_y + \frac{\nu}{1-2\nu}\theta - \frac{1+\nu}{1-2\nu}\alpha\Delta T\right) \qquad & \tau_{yz} &= G\gamma_{yz} \\
\sigma_z &= \frac{E}{1+\nu}\left(\varepsilon_z + \frac{\nu}{1-2\nu}\theta - \frac{1+\nu}{1-2\nu}\alpha\Delta T\right) \qquad & \tau_{zx} &= G\gamma_{zx}
\end{aligned}\right\} \tag{2.60}$$

These formulas express Hooke's law with consideration of thermoelastic strains. Note that only the normal stresses include contributions from the temperature change.

3

A View of Solution Procedures

3.1 DIRECT AND INVERSE PROBLEMS IN ELASTICITY THEORY

The differential equations of equilibrium of an elementary tetrahedron (2.5) and elementary parallelepiped (2.11), Cauchy's formulas (2.34), strain compatibility conditions (2.37) and (2.38), and the formulas of Hooke's law, (2.50), (2.52), or (2.54), form a system of the major equations of the theory of elasticity. These equations are sufficient to determine the states of stress and strain in an elastic body. Evidently, solutions to particular problems must satisfy, in addition to the basic equations, the appropriate boundary conditions.

A direct and an inverse problem of the theory of elasticity can be distinguished. In the *direct problem*, the external forces or the temperature fields are given, and the stresses, strains, and displacements are sought. As to the *inverse problem*, it exists in the following three versions:

1. The stresses are given, and the displacements, strains, and external forces are sought.
2. The displacements are given, and the stresses, strains, and external forces are sought.
3. The strains are given, and the stresses, displacements, and external forces are sought.

Solutions to the inverse problem can be obtained, as a rule, without any difficulty, especially when the displacement functions u, v, w are known. Then the strain components can be determined by Cauchy's formulas, the stresses can be evaluated using Hooke's law relationships, and the surface and the body forces can be obtained on the basis of the equilibrium conditions for the elementary tetrahedron and parallelepiped. If the stresses (strains) are given, then the strains (stresses) can be determined from Hooke's law, and the surface and the body forces from the equilibrium equations. Then, in order to find the displacements, one has to integrate Cauchy's formulas. This, however, is not associated with any essential difficulties. In Section 4, we will show how it can be done for some elementary problems of the theory of elasticity.

There is no single general method though for the solution of the direct problem.

Obviously, in some practical cases, a combination of the direct and the inverse problems may be encountered when the external forces are known for one portion of the body surface and the displacements are given for other portions. This is a so-called *mixed-boundary-value problem* of the theory of elasticity.

3.2 SEMI-INVERSE METHOD AND PRINCIPLE OF SAINT-VENANT

The majority of solutions to various special problems in the elasticity theory were obtained on the basis of the so-called *semi-inverse method of Saint-Venant*. In accordance with this method, we should try to guess the form of the unknown solution and then check to see if it satisfies the major equations of the theory of elasticity and the boundary conditions of the given problem. During this procedure, some relationships are fulfilled automatically and identically, whereas the others serve to determine the unknown functions and constants entering the sought solution. This approach will be widely utilized throughout the book.

Solutions to many theory-of-elasticity problems can be substantially facilitated if other, statically equivalent, systems are substituted for the given actual loading systems. According to the *Saint-Venant principle*, if a system of forces acting on a small region of a body is replaced by a different, but statically equivalent system of forces acting on the same region, then such replacement does not cause essential changes in the state of stress and strain at points sufficiently remote from this region. Evidently, this principle is of great value to researchers since it offers an effective means to simplify many theory-of-elasticity problems. Static equivalence merely requires that the two systems of applied forces have the same resultant force and couple. The solutions obtained will be identical everywhere except at areas close to the region where the forces are applied. Of course, the Saint-Venant principle is not an exact mathematical theorem but rather a commonsense proposition.

3.3 BELTRAMI–MICHELL, LAMÉ, AND DUHAMEL–NEUMANN EQUATIONS

There are two major schemes to obtain solutions to the direct problem of the theory of elasticity. In one of them, suggested by E. Beltrami and generalized by A. G. M. Michell, the stress components are sought first and, in the other, suggested by Lamé, it is the displacements that are chosen as the principal unknowns.

After excluding from the system of the major equations of the theory of elasticity all the variables but stresses, we can obtain the following equations:

$$\left.\begin{aligned}
\Delta\sigma_x + \frac{1}{1+\nu}\frac{\partial^2\sigma_{\mathrm{I}}}{\partial x^2} &= -2\frac{\partial X}{\partial x} - \frac{\nu}{1-\nu}\left(\frac{\partial X}{\partial x} + \frac{\partial Y}{\partial y} + \frac{\partial Z}{\partial z}\right) \\
\Delta\sigma_y + \frac{1}{1+\nu}\frac{\partial^2\sigma_{\mathrm{I}}}{\partial y^2} &= -2\frac{\partial Y}{\partial y} - \frac{\nu}{1-\nu}\left(\frac{\partial X}{\partial x} + \frac{\partial Y}{\partial y} + \frac{\partial Z}{\partial z}\right) \\
\Delta\sigma_z + \frac{1}{1+\nu}\frac{\partial^2\sigma_{\mathrm{I}}}{\partial z^2} &= -2\frac{\partial Z}{\partial z} - \frac{\nu}{1-\nu}\left(\frac{\partial X}{\partial x} + \frac{\partial Y}{\partial y} + \frac{\partial Z}{\partial z}\right) \\
\Delta\tau_{xy} + \frac{1}{1+\nu}\frac{\partial^2\sigma_{\mathrm{I}}}{\partial x\partial y} &= -\frac{\partial X}{\partial y} - \frac{\partial Y}{\partial x} \\
\Delta\tau_{yz} + \frac{1}{1+\nu}\frac{\partial^2\sigma_{\mathrm{I}}}{\partial y\partial z} &= -\frac{\partial Y}{\partial z} - \frac{\partial Z}{\partial y} \\
\Delta\tau_{zx} + \frac{1}{1+\nu}\frac{\partial^2\sigma_{\mathrm{I}}}{\partial z\partial x} &= -\frac{\partial Z}{\partial x} - \frac{\partial X}{\partial z}
\end{aligned}\right\} \tag{3.1}$$

where

$$\Delta = \frac{\partial^2}{\partial x^2} + \frac{\partial^2}{\partial y^2} + \frac{\partial^2}{\partial z^2} \tag{3.2}$$

is the *harmonic (Laplacean) operator*. Equations (3.1) were obtained for the case of absence of the body forces ($X = Y = Z = 0$) by Beltrami and then generalized by Michell for the case of nonzero body forces. They are called *Beltrami–Michell's equations*. In these equations, the stresses are linked with the external forces in the simplest way and should be used when external surface forces are given.

If all the variables but the displacements are excluded from the major equations of the theory of elasticity, the following equations can be obtained:

$$\left.\begin{aligned} \Delta u + \frac{1}{1-2\nu}\frac{\partial\theta}{\partial x} &= -\frac{X}{G} \\ \Delta v + \frac{1}{1+2\nu}\frac{\partial\theta}{\partial y} &= -\frac{Y}{G} \\ \Delta w + \frac{1}{1-2\nu}\frac{\partial\theta}{\partial z} &= -\frac{Z}{G} \end{aligned}\right\} \tag{3.3}$$

These equations are called *Lamé equations.* The major merit of Lamé equations is that the number of the unknown functions in them is smaller than in Beltrami–Michell's equations. These Lamé equations should be applied when displacements at the body surface are given.

When the thermally induced strains are considered, the Lamé equations are generalized as follows:

$$\left.\begin{aligned} \Delta u + \frac{1}{1-2\nu}\frac{\partial\theta}{\partial x} &= -\frac{X}{G} + 2\alpha\frac{1+\nu}{1-2\nu}\frac{\partial\Delta T}{\partial x} \\ \Delta v + \frac{1}{1-2\nu}\frac{\partial\theta}{\partial y} &= -\frac{Y}{G} + 2\alpha\frac{1+\nu}{1-2\nu}\frac{\partial\Delta T}{\partial y} \\ \Delta w + \frac{1}{1-2\nu}\frac{\partial\theta}{\partial z} &= -\frac{Z}{G} + 2\alpha\frac{1+\nu}{1-2\nu}\frac{\partial\Delta T}{\partial z} \end{aligned}\right\} \tag{3.4}$$

These are known as *Duhamel–Neumann equations.* In these equations, α is the coefficient of linear thermal expansion (CTE), and ΔT is the change in temperature. Obviously, the temperature-dependent terms should be considered if the temperature is distributed nonuniformly over the body's volume. Therefore, in complex problems of thermoelasticity, an analysis of the temperature distribution should precede an elastic analysis.

4

The Elementary Problems

Inverse problems, in which the stresses and strains are linear functions of the coordinates, are referred to as the elementary problems of the theory of elasticity. In these problems, the compatibility conditions are fulfilled identically, and the solutions are sought in simple forms and predicted by methods used in strength of materials. We examine several such problems.

4.1 PURE BENDING OF A PRISMATIC BAR

As is known from principles in strength of materials, the deformation of pure bending of a prismatic bar is due to two mutually equilibrating couples M, acting in the x, z plane. Assume that (Fig. 4.1)

$$\sigma_x = \frac{E}{\rho} z \qquad \sigma_y = \sigma_z = \tau_{xy} = \tau_{yz} = \tau_{zx} = 0 \tag{4.1}$$

where ρ is the radius of curvature of the deflected bar.

Let us check to see whether these relationships satisfy the theory-of-elasticity equations. After substituting them into the equations of equilibrium of the elementary tetrahedron (2.5) and elementary parallelepiped (2.11), we find that the latter are satisfied identically and that the first yield $X_v = Y_v = Z_v = 0$. Hence, the lateral surface of the beam is free of normal stresses. Substitution of Eq. (4.1) into the Hooke's law formulas (2.54) results in the relationships

$$\varepsilon_x = -\frac{z}{\rho} \qquad \varepsilon_y = \varepsilon_z = \frac{v}{\rho} z \qquad \gamma_{xy} = \gamma_{yz} = \gamma_{zx} = 0 \tag{4.2}$$

The compatibility conditions (2.37) and (2.38) are fulfilled identically.

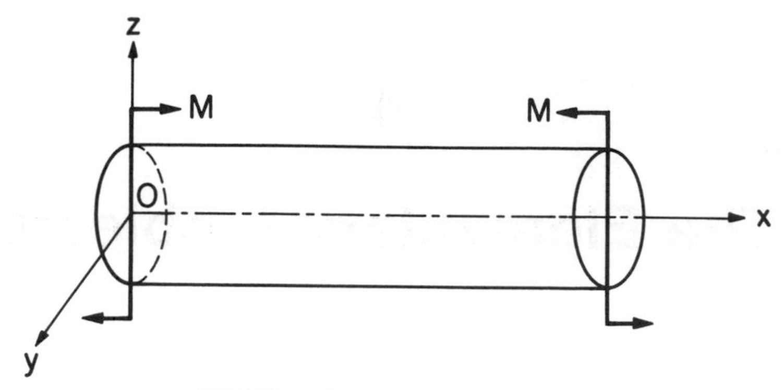

FIGURE 4.1 Pure bending of a beam.

Determine the displacements of the points of the beam. Cauchy's formulas (2.34) yield

$$\frac{\partial u}{\partial x} = -\frac{z}{\rho} \qquad \frac{\partial v}{\partial y} = \frac{\partial w}{\partial z} = \frac{\nu}{\rho} z \tag{4.3}$$

$$\frac{\partial u}{\partial y} + \frac{\partial v}{\partial x} = \frac{\partial v}{\partial z} + \frac{\partial w}{\partial y} = \frac{\partial w}{\partial x} + \frac{\partial u}{\partial z} = 0 \tag{4.4}$$

From Eq. (4.3) we find, by integration,

$$\left.\begin{aligned} u &= -\frac{1}{\rho} zx + f_1(y, z) \\ v &= \frac{\nu}{\rho} zy + f_2(x, z) \\ w &= \frac{\nu}{2\rho} z^2 + f_3(x, y) \end{aligned}\right\} \tag{4.5}$$

Here the functions f_1, f_2, and f_3 play the role of the constants of integration. Obviously, in order to obtain Eqs. (4.3) from Eqs. (4.5), we have to assume that the function f_1 is x-independent, f_2 is y-independent, and f_3 is z-independent.

After introducing Eqs. (4.5) in Eqs. (4.4), we find that the functions f_1, f_2, f_3 are related as follows:

$$\frac{\partial f_1}{\partial y} + \frac{\partial f_2}{\partial x} = \frac{\partial f_2}{\partial z} + \frac{\partial f_3}{\partial y} + \frac{\nu}{\rho} y = \frac{\partial f_3}{\partial x} + \frac{\partial f_1}{\partial z} - \frac{1}{\rho} x = 0 \tag{4.6}$$

Differentiating these equations with respect to x, y, z, we have

$$\left.\begin{aligned} \frac{\partial^2 f_2}{\partial x^2} = \frac{\partial^2 f_2}{\partial x \partial z} + \frac{\partial^2 f_3}{\partial x \partial y} = \frac{\partial^2 f_3}{\partial x^2} - \frac{1}{\rho} = 0 \\ \frac{\partial^2 f_1}{\partial y^2} = \frac{\partial^2 f_3}{\partial y^2} + \frac{v}{\rho} = \frac{\partial^2 f_3}{\partial x \partial y} + \frac{\partial^2 f_1}{\partial y \partial z} = 0 \\ \frac{\partial^2 f_1}{\partial y \partial z} + \frac{\partial^2 f_2}{\partial x \partial z} = \frac{\partial^2 f_2}{\partial z^2} = \frac{\partial^2 f_1}{\partial z^2} = 0 \end{aligned}\right\}$$

Then, we conclude that the functions f_1, f_2, f_3 must be expressed by the formulas

$$\left.\begin{aligned} f_1(y, z) &= C_1 y + C_2 z + C_3 \\ f_2(x, z) &= C_4 x + C_5 z + C_6 \\ f_3(x, y) &= \frac{1}{2\rho} x^2 - \frac{v}{2\rho} y^2 + C_7 x + C_8 y + C_9 \end{aligned}\right\} \tag{4.7}$$

and that

$$\frac{\partial^2 f_1}{\partial y \partial z} = \frac{\partial^2 f_2}{\partial x \partial z} = \frac{\partial^2 f_3}{\partial x \partial y} = 0 \tag{4.8}$$

Substituting Eqs. (4.7) into Eq. (4.6), we have

$$C_4 = -C_1 \qquad C_8 = -C_5 \qquad C_7 = -C_2$$

Conditions (4.8), as we can easily verify, are fulfilled. Thus, the displacements are expressed by the equations

$$\left.\begin{aligned} u &= -\frac{1}{\rho} xz + C_1 y + C_2 z + C_3 \\ v &= \frac{v}{\rho} zy - C_1 x + C_4 z + C_5 \\ w &= \frac{1}{2\rho}(x^2 - v y^2 + v z^2) - C_2 x - C_4 y + C_6 \end{aligned}\right\} \tag{4.9}$$

The constants $C_1, \ldots, C_6$ can be determined from the boundary conditions. If, for instance,

$$u = v = w = 0 \qquad \frac{\partial w}{\partial x} = \frac{\partial v}{\partial x} = \frac{\partial w}{\partial y} = 0$$

i.e., if the displacements and the angles of rotation at the origin are zero, then all the constants of integration are zero as well.

The bending moment in all the cross sections is the same and can be presented as

$$M = \int_A \sigma_x z \, dA = -\frac{E}{\rho} I_y \tag{4.10}$$

where

$$I_y = \int_A z^2 \, dA = b \int_{-h/2}^{h/2} z^2 \, dz = \frac{bh^3}{12} \tag{4.11}$$

is the moment of inertia of the cross section of the bar. Hence, the curvature $1/\rho$ can be evaluated by the formula

$$\frac{1}{\rho} = -\frac{M}{EI_y} = -\frac{12M}{Ebh^3} \tag{4.12}$$

For small deflections, as we know from calculus, we can use the approximation

$$\frac{1}{\rho} \cong -\frac{\partial^2 w}{\partial x^2} \tag{4.13}$$

By integrating this equation twice, we obtain the following formula for the deflection function:

$$w(x) = \frac{M}{2EI_y} x^2 \tag{4.14}$$

This is a well-known formula for the deflection of a cantilever beam.

4.2 SIMPLE TORSION OF A CIRCULAR SHAFT

Shafts of circular cross section are the most common structural elements subjected to torsion. Such elements are quite frequently encountered in many electrical and electronic structures. We assume that (see Fig. 4.2)

$$\left.\begin{aligned} &\sigma_x = \sigma_y = \sigma_z = \tau_{yz} = 0 \\ &\tau_{xy} = -\alpha G z \qquad \tau_{xz} = \alpha G y \end{aligned}\right\} \tag{4.15}$$

where α is the twist angle per unit shaft length. These formulas are known from methods of strength of materials.

Let us determine the displacements. After substituting Eqs. (4.15) in the Hooke's law equations (2.54), we have

$$\varepsilon_x = \varepsilon_y = \varepsilon_z = \gamma_{yz} = 0 \qquad \gamma_{xy} = -\alpha z \qquad \gamma_{xz} = \alpha y$$

Then, the Cauchy formulas (2.34) yield

$$\frac{\partial u}{\partial x} - \frac{\partial v}{\partial y} = \frac{\partial w}{\partial z} = 0 \tag{4.16}$$

$$\frac{\partial v}{\partial z} + \frac{\partial w}{\partial y} = \frac{\partial v}{\partial x} + \frac{\partial u}{\partial y} + \alpha z = \frac{\partial w}{\partial x} + \frac{\partial u}{\partial z} - \alpha y = 0 \tag{4.17}$$

From Eq. (4.16) we see that the function u is x-independent, the function v is y-independent, and the function w is z-independent, so that $u = u(y, z)$,

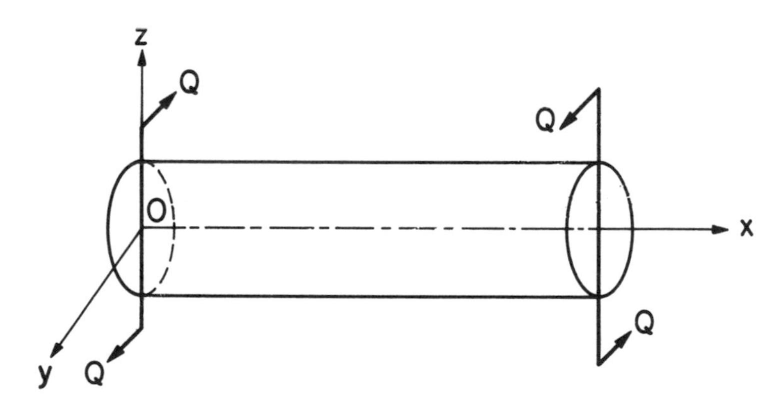

FIGURE 4.2 Pure torsion of a beam.

$v = v(x, z)$, and $w = w(x, y)$. Taking into account these relationships and differentiating Eqs. (4.17) with respect to x, y, and z, we find

$$\left.\begin{aligned}\frac{\partial^2 v}{\partial x \partial z} + \frac{\partial^2 w}{\partial x \partial y} = \frac{\partial^2 v}{\partial x^2} = \frac{\partial^2 w}{\partial x^2} = 0 \\ \frac{\partial^2 w}{\partial y^2} = \frac{\partial^2 u}{\partial y^2} = \frac{\partial^2 w}{\partial x \partial y} + \frac{\partial^2 u}{\partial y \partial z} - \alpha = 0 \\ \frac{\partial^2 v}{\partial z^2} = \frac{\partial^2 v}{\partial x \partial z} + \frac{\partial^2 u}{\partial y \partial z} + \alpha = \frac{\partial^2 u}{\partial z^2} = 0\end{aligned}\right\}$$

Then, we have

$$\frac{\partial^2 u}{\partial y^2} = \frac{\partial^2 u}{\partial z^2} = \frac{\partial^2 u}{\partial y \partial z} = 0 \qquad \frac{\partial^2 v}{\partial x^2} = \frac{\partial^2 v}{\partial z^2} = 0$$

$$\frac{\partial^2 v}{\partial x \partial z} = -\alpha \qquad \frac{\partial^2 w}{\partial x^2} = \frac{\partial^2 w}{\partial y^2} = 0 \qquad \frac{\partial^2 w}{\partial x \partial y} = \alpha$$

so that

$$\left.\begin{aligned}u &= C_1 y + C_2 z + C_3 \\ v &= -\alpha xy + C_4 x + C_5 z + C_6 \\ w &= \alpha xy + C_7 x + C_8 y + C_9\end{aligned}\right\}$$

After substituting these formulas in Eqs. (4.17), we find that the constants of integration in the preceding equations are related as $C_8 = -C_5$, $C_4 = -C_1$, $C_7 = -C_2$ and, therefore, the displacements are expressed as follows:

$$\left.\begin{aligned}u &= C_1 y + C_2 z + C_3 \\ v &= -\alpha xy - C_1 x + C_4 z + C_5 \\ w &= \alpha xy - C_2 x - C_4 y + C_6\end{aligned}\right\} \tag{4.18}$$

If, for instance,

$$u = v = w = 0 \qquad \frac{\partial v}{\partial x} = \frac{\partial w}{\partial x} = \frac{\partial w}{\partial y} = 0$$

at $x = y = z = 0$, then the constants of integration in Eqs. (4.18) are zero, and

$$u = 0 \qquad v = -\alpha xz \qquad w = \alpha xy \tag{4.19}$$

Thus, in the case of a simple torsion of a circular shaft, the plane cross sections remain plane when the twisting torque is applied ($u = 0$).

The torque can be expressed by the formula

$$Q = -\int_A \tau_{xy} z \, dA + \int_A \tau_{xz} y \, dA = \alpha G I_p \tag{4.20}$$

where

$$I_p = \int_A (y^2 + z^2) \, dA = \frac{\pi}{2} r^4 \tag{4.21}$$

is the polar moment of inertia of the shaft cross section. Hence, the twist angle during simple torsion,

$$\alpha = \frac{Q}{GI_p} \tag{4.22}$$

is directly proportional to the torque and inversely proportional to the torsional stiffness GI_p. Note that, for a hollow circular shaft of internal radius r_1 and external radius r_2, Eq. (4.21) yields

$$I_p = \frac{\pi}{2}(r_2^4 - r_1^4) \tag{4.23}$$

4.3 ALL-ROUND UNIFORM COMPRESSION

In the case of all-round ("hydrostatic") uniform compression,

$$\sigma_x = \sigma_y = \sigma_z = -\sigma \qquad \tau_{xy} = \tau_{yz} = \tau_{zx} = 0 \tag{4.24}$$

the Lamé ellipsoid (2.22) reduces to a sphere

$$\zeta^2 + \eta^2 + \xi^2 = \sigma^2$$

and the stress tensor becomes a spherical tensor. During all-round uniform compression, the form of the body remains unchanged; only its volume changes. The strain components are as follows:

$$\varepsilon_x = \varepsilon_y = \varepsilon_z = -\frac{\sigma}{E}(1 - 2\nu) \qquad \gamma_{xy} = \gamma_{yz} = \gamma_{zx} = 0$$

Then, the displacements are expressed by the formulas

$$u = -\frac{\sigma}{E}(1 - 2\nu)x \qquad v = -\frac{\sigma}{E}(1 - 2\nu)y \qquad w = -\frac{\sigma}{E}(1 - 2\nu)z \tag{4.25}$$

4.4 PRISMATIC BAR SUBJECTED TO TENSION

In the case of a prismatic bar subjected to tension, we assume

$$\sigma_x = p \qquad \sigma_y = \sigma_z = 0 \qquad \tau_{xy} = \tau_{yz} = \tau_{zx} = 0$$

The components of the strain and the displacement are expressed as follows:

$$\left.\begin{aligned} &\varepsilon_x = \frac{p}{E} \qquad \varepsilon_y = \varepsilon_z = -\frac{\nu}{E}p \qquad \gamma_{xy} = \gamma_{yz} = \gamma_{zx} = 0 \\ &u = \frac{p}{E}x \qquad v = -\frac{\nu}{E}py \qquad w = -\frac{\nu}{E}pz \end{aligned}\right\} \tag{4.26}$$

4.5 PRISMATIC BAR UNDER ITS OWN WEIGHT

The only difference between this problem and the previous one is that now the normal stress σ_x is a linear function of the coordinate x (see Fig. 4.3):

$$\sigma_x = \gamma x$$

Then, the strains are expressed by the formulas

$$\varepsilon_x = \frac{\gamma}{E}x \qquad \varepsilon_y = \varepsilon_z = -\frac{\nu\gamma}{E}x \qquad \gamma_{xy} = \gamma_{yz} = \gamma_{zx} = 0$$

The displacements can be obtained from the equations

$$\frac{\partial u}{\partial x} = \frac{\gamma}{E}x \qquad \frac{\partial v}{\partial y} = -\frac{\nu\gamma}{E}x \qquad \frac{\partial w}{\partial z} = -\frac{\nu\gamma}{E}x$$

$$\frac{\partial v}{\partial x} + \frac{\partial u}{\partial y} = \frac{\partial v}{\partial z} + \frac{\partial w}{\partial y} = \frac{\partial w}{\partial x} + \frac{\partial u}{\partial z} = 0$$

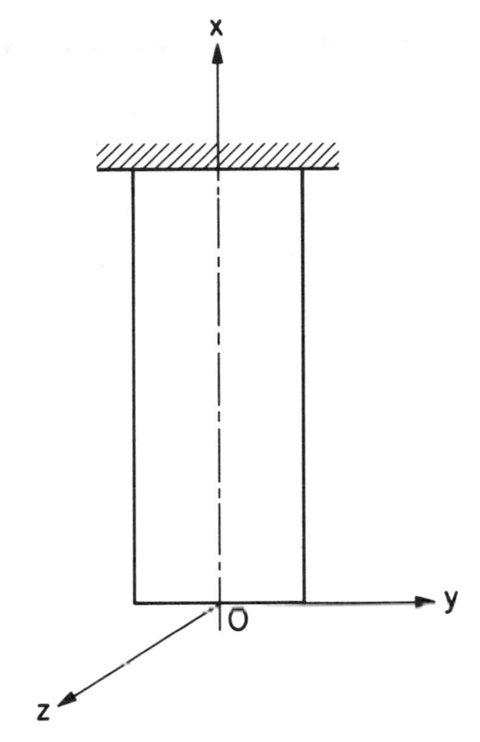

FIGURE 4.3 Prismatic bar under its own weight.

which result in the formulas

$$u = \frac{\gamma}{2E}(x^2 + \nu y^2 + \nu z^2) \qquad v = -\frac{\nu\gamma}{E}xy \qquad w = -\frac{\nu\gamma}{E}zx \tag{4.27}$$

4.6 PURE BENDING OF PLATES

Consider a rectangular plate subjected to bending moments M_1 and M_2 distributed over the edges of the plate, parallel to the y and x axes, respectively (see Fig. 4.4). The curvatures of the plate in the planes parallel to the xz and yz planes and due to the moments M_1 and M_2 are $1/\rho$ and $-(\nu/\rho)$, respectively. When bending moments act in two perpendicular directions, the curvatures of the deflection surface can be obtained by superposition. Let $1/\rho_1$ and $1/\rho_2$ be the curvatures in the planes parallel to the coordinate planes xz and yz, respectively, and let M_1 and M_2 be the bending moments per unit length applied on the edges parallel to the y and x axes,

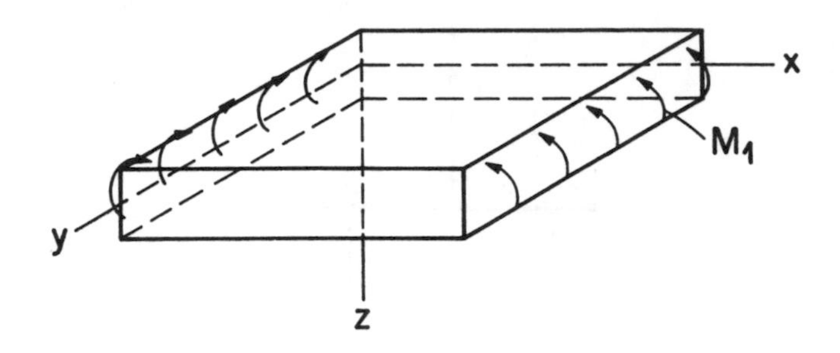

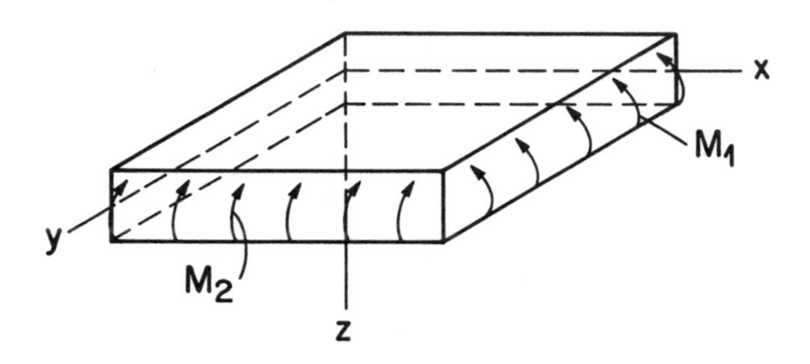

FIGURE 4.4 Pure bending of a plate.

respectively. Then, using the relationship (4.12) and the principle of superposition, we have

$$\frac{1}{\rho_1} = \frac{12}{Eh^3}(M_1 - \nu M_2) \qquad \frac{1}{\rho_2} = \frac{1}{Eh^3}(M_2 - \nu M_1)$$

The moments M_1 and M_2 are considered positive if they produce a deflection of the plate that is convex down.

Solving the preceding equations for the moments, we obtain

$$M_1 = D\left(\frac{1}{\rho_1} + \frac{\nu}{\rho_2}\right) \qquad M_2 = D\left(\frac{1}{\rho_2} + \frac{\nu}{\rho_1}\right)$$

where the constant

$$D = \frac{Eh^3}{12(1 - \nu^2)} \tag{4.28}$$

is called the *flexural rigidity* of the plate. For small deflections, approximations of the type (4.13) can be used. This results in the following formulas, which link the bending moments and the deflection function of the plate:

$$M_1 = -D\left(\frac{\partial^2 w}{\partial x^2} + \nu \frac{\partial^2 w}{\partial y^2}\right) \qquad M_2 = -D\left(\frac{\partial^2 w}{\partial y^2} + \nu \frac{\partial^2 w}{\partial x^2}\right) \tag{4.29}$$

5

Strength Theories

5.1 STRAIN ENERGY

Consider an elementary parallelepiped with the ribs dx, dy, dz located within a deformed body, and assume that all the components of the stress tensor except σ_x are zero. Then the work of the elastic forces is

$$\sigma_x \, dy \, dz \cdot \delta\varepsilon_x \, dx = \sigma_x \, \delta\varepsilon_x \, dV$$

where $\delta\varepsilon_x$ is the increase in the linear strain. If all the components of the stress tensor except τ_{xy} were zero, then the work of the elastic forces would be

$$\tau_{xy}\delta\gamma_{xy} \, dV$$

Hence, in the case of a three-dimensional state of stress, the elementary specific (per unit volume) work of deformation is

$$\delta W = \sigma_x\delta\varepsilon_x + \sigma_y\delta\varepsilon_y + \sigma_z\delta\varepsilon_z + \tau_{xy}\delta\gamma_{xy} + \tau_{yz}\delta\gamma_{yz} + \tau_{zx}\delta\gamma_{zx}$$

After substituting Eqs. (2.54) for Hooke's law into this expression, we have

$$\delta W = \frac{1}{E}\{\sigma_x\delta\sigma_x + \sigma_y\delta\sigma_y + \sigma_z\delta\sigma_z - \nu[\delta(\sigma_x\sigma_y) + \delta(\sigma_y\sigma_z) + \delta(\sigma_y\sigma_z)]\}$$

$$+ \frac{1}{G}(\tau_{xy}\delta\tau_{xy} + \tau_{yz}\delta\tau_{yz} + \tau_{zx}\delta\tau_{zx})$$

Integration of this equation yields

$$W = \frac{1}{E}\left[\tfrac{1}{2}(\sigma_x^2 + \sigma_y^2 + \sigma_z^2) - \nu(\sigma_x\sigma_y + \sigma_y\sigma_z + \sigma_z\sigma_x)\right]$$

$$+ \frac{2(1+\nu)}{E}\left[\tfrac{1}{2}(\tau_{xy}^2 + \tau_{yz}^2 + \tau_{zx}^2)\right]$$

$$= \frac{1}{2E}\left[\sigma_x^2 + \sigma_y^2 + \sigma_z^2 - 2\nu(\sigma_x\sigma_y + \sigma_y\sigma_z + \sigma_z\sigma_x)\right] + 2(1+\nu)(\tau_{xy}^2 + \tau_{yz}^2 + \tau_{zx}^2) \tag{5.1}$$

For the principal axes, this formula can be simplified:

$$W = \frac{1}{2E}\left[\sigma_1^2 + \sigma_2^2 + \sigma_3^2 - 2\nu(\sigma_1\sigma_2 + \sigma_2\sigma_3 + \sigma_3\sigma_1)\right] \tag{5.2}$$

In the case of an all-round uniform compression, when $\sigma_1 = \sigma_2 = \sigma_3 = \sigma$, we have

$$W_v = \frac{3(1-2\nu)}{2E}\sigma^2 = \frac{1-2\nu}{6E}(\sigma_1 + \sigma_2 + \sigma_3)^2 \tag{5.3}$$

Therefore, the work associated with the distorsion of the form is

$$W_f = W - W_v = \frac{1+\nu}{6E}\left[(\sigma_1 - \sigma_2)^2 + (\sigma_2 - \sigma_3)^2 + (\sigma_3 - \sigma_1)^2\right] \tag{5.4}$$

Using notation (2.28) and (2.29), we present the formulas (5.3) and (5.4) as follows:

$$W_v = \frac{3(1-2\nu)}{2E}\sigma_0^2 \qquad W_f = \frac{3(1+\nu)}{2E}\tau_0^2 \tag{5.5}$$

5.2 CLASSICAL STRENGTH THEORIES

As is known, materials science distinguishes between plastic and brittle failure, depending on whether such a failure is preceded by an appreciable plastic deformation. Although this science more or less satisfactorily explains the nature of plastic and brittle failures, it is unable to predict the quantitative characteristics of materials to an extent that would allow a mechanical engineer to perform strength calculations. Therefore, structural analysis cannot do without an experimental evaluation of strength characteristics. Since these characteristics are usually determined for simplest loading conditions, engineers have to have at their disposal a means to extend the experimental data for the elements of actual structures that experience three-dimensional stress conditions. Such a means is provided by strength theories (hypotheses).

The *first strength theory* (Galilei–Leibnitz) states that failure occurs when the maximum normal stress reaches its critical value:

$$\sigma_c = \sigma_1$$

The *second strength theory* (Mariott–Saint-Venant) assumes that failure takes place when the maximal normal strain reaches its critical value. This assumption, with Eqs. (2.54), leads to the following relationship for the critical stress value:

$$\sigma_c = \sigma_1 - \nu(\sigma_2 + \sigma_3)$$

The *third strength theory* (Coulomb) proceeds from the maximum shearing stress as strength criterion. Using the formulas (2.26), we conclude that this assumption is equivalent to the following formula for the critical shearing stress:

$$\tau_c = \frac{\sigma_1 - \sigma_3}{2}$$

The first and the second theories reflect the idea of brittle behavior on the part of the material, whereas the third theory assumes that the material behaves plastically. Since, however, the actual materials in actual structures and at actual environmental and loading conditions do not behave like ideally brittle or ideally plastic materials, the classical strength theories are currently utilized only as constituent parts of more modern and more complicated strength theories.

5.3 SOME MODERN STRENGTH THEORIES

Mohr's theory is based on a *Mohr's diagram*, which enables us to determine the normal and shearing stresses for various planes passing through the given point. If we carry out mass serial tests of specimens until failure and, based on these tests, draw *Mohr's circles*, then the envelopes for these circles can be considered as characteristics of the ultimate state of the material (Fig. 5.1). In order to evaluate the strength of a structural element, we should draw a Mohr's circle for the given material. If this circle falls within the area restricted by the envelope $\tau = f(\sigma)$, then no failure will occur. The major merits of this theory are its generality and a sound experimental basis. The major shortcomings are high volume and high cost of experimental work.

The theories most often used today are based on energy considerations. The first theory of this type was suggested by *Beltrami*, who assumed that the total strain energy of a body can be considered as a suitable criterion of strength. Since, for uniform all-round tension, $\sigma_1 = \sigma_c$ and $\sigma_2 = \sigma_3 = 0$, then, in accordance with Eq. (5.2), we have

$$W = \frac{\sigma_c^2}{2E}$$

Then, Eq. (5.2) results in the following equation for the critical stress:

$$\sigma_c = \sqrt{\sigma_1^2 + \sigma_2^2 + \sigma_3^2 - 2\nu(\sigma_1\sigma_2 + \sigma_2\sigma_3 + \sigma_3\sigma_1)} \tag{5.6}$$

Experimental verification of this equation can be carried out, for instance, in the case of an all-round uniform compression of a body by hydrostatic

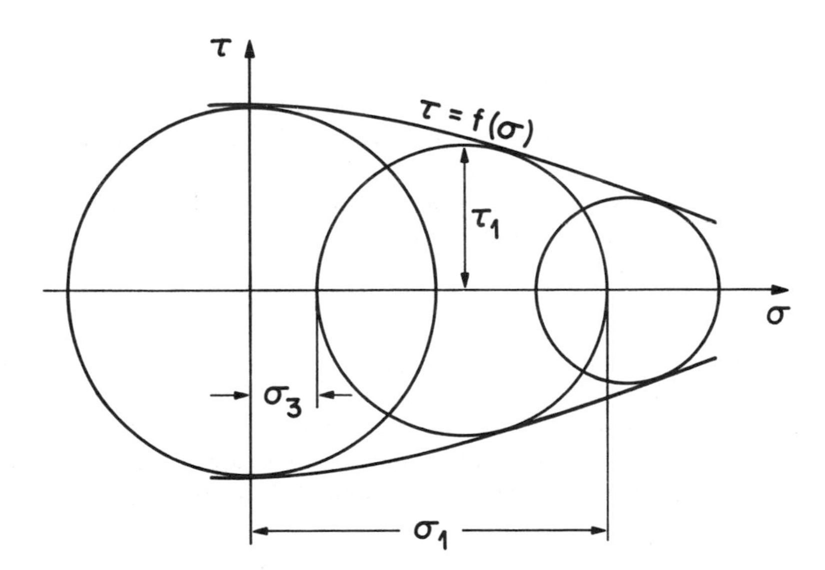

FIGURE 5.1 Envelopes for Mohr's circles.

pressure p. Putting $\sigma_1 = \sigma_2 = \sigma_3 = p$, we obtain the following formula for a critical normal stress:

$$\sigma_c = p\sqrt{3(1 - 2\nu)}$$

If the yield stress σ_Y is accepted as a critical value of the stress, then the corresponding pressure is

$$p = \frac{\sigma_Y}{\sqrt{3(1 - 2\nu)}}$$

It turns out, however, that materials are capable of withstanding much greater pressures than is predicted by this formula.

Beltrami's theory was modified by M. T. Huber, who suggested that only the energy due to the distortion of the body's form should be considered responsible for its structural strength. For uniform tension, Eq. (5.4) yields

$$W_f = \frac{1 + \nu}{3E}\sigma^2$$

so that the critical stress is expressed by the formula

$$\sigma_c = \sqrt{\tfrac{1}{2}[(\sigma_1 - \sigma_2)^2 + (\sigma_2 - \sigma_3)^2 + (\sigma_3 - \sigma_1)^2]} \tag{5.7}$$

Unlike Eq. (5.6), condition (5.7) does not contain any elastic constants of the material and, therefore, is not associated with an assumption that the material performs elastically. For this reason, this condition can be considered as a condition of plasticity (yield criterion). Such a generalization of Huber's theory was suggested by R. von Mises and H. Hencky. Putting $\sigma_c = \sigma_Y$, we obtain the *condition of plasticity* in the form

$$(\sigma_1 - \sigma_2)^2 + (\sigma_2 - \sigma_3)^2 + (\sigma_3 - \sigma_1)^2 \geq 2\sigma_Y^2 \tag{5.8}$$

The law (5.8) defines the limit of elastic behavior under any possible combination of stresses.

An experimental verification of the condition (5.8) and, hence, of the formula (5.7) can be carried out, for instance, for the case of pure shear, when $\sigma_1 = -\sigma_3 = \tau_Y$ and $\sigma_2 = 0$. Then, the formula (5.7) yields

$$\tau_Y = \frac{\sigma_Y}{\sqrt{3}} = 0.577\sigma_y \tag{5.9}$$

Experiments with steel show that the ratio between the yield stresses in shear and in tension is in good agreement with Eq. (5.9).

Some of the recent strength theories are based on the *Nádai hypothesis.* A. Nádái suggested combining the theories of Mohr and Huber–von Mises–Hencky by substituting the stresses τ and σ in the relationship $\tau = f(\sigma)$, assumed in Mohr's theory, by the octahedric stresses τ_0 and σ_0, respectively, so that the relationship $\tau_0 = f(\sigma_0)$ is applied.

Strength theories based on the considerations of the theory of elasticity and examining elastic bodies as continuous media deal with stresses that are equilibrated within volumes, which are significantly larger than the size of the crystal grains. These stresses are mean values for stresses, changing within volumes, whose size is comparable to the size of the grains. Since plastic materials are able to average the local microscopic stresses, the approaches of a continuous media mechanics are in good agreement with the actual distribution of stresses. Deterministic (nonrandom) strength theories work well in this case. However, in the case of brittle materials, it is the local strength in the most stressed areas that plays the decisive role. The stress maxima, not their mean values, are responsible for the failure of these materials. Since the locations of these local stress maxima are randomly distributed over the body's volume, then, a stochastic approach in combination with brittle mechanics methods should be used for strength evaluation in this case. Discussion of such an approach is, however, beyond the scope of this book.

6

Two-Dimensional Problem in Rectangular Coordinates

6.1 PLANE STRAIN AND PLANE STRESS

The solution to many theory-of-elasticity problems can be obtained on the basis of one of the following two special cases:

$$u = u(x, y) \qquad v = v(x, y) \qquad w = 0 \tag{6.1}$$

$$\tau_{xz} = \tau_{yz} = \sigma_z = 0 \tag{6.2}$$

called *plane strain* and *plane stress*, respectively.

In the case of a plane strain, the components u and v of the displacement are independent of the coordinate z, and the longitudinal displacement w is zero. Such a simplification is possible when the dimension of the body in the z direction is very large. The state of stress in the case of a plane stress is specified by the stresses σ_x, σ_y, τ_{xy} only. This simplification is possible when the dimension of the body in the z direction is small, i.e., in the case of a thin plate.

Both the preceding special cases result in the same mathematical problem, which is called the *two-dimensional problem* of the theory of elasticity.

The major equations of the theory of elasticity in the case of a plane strain are as follows:

Cauchy's formulas:

$$\varepsilon_x = \frac{\partial u}{\partial x} \qquad \varepsilon_y = \frac{\partial v}{\partial y} \qquad \gamma_{xy} = \frac{\partial u}{\partial y} + \frac{\partial v}{\partial x} \tag{6.3}$$

Hooke's law:

$$\left.\begin{aligned}\sigma_x &= \frac{E}{1+\nu}\left(\varepsilon_x + \frac{\nu}{1-2\nu}\theta\right)\\ \sigma_y &= \frac{E}{1+\nu}\left(\varepsilon_y + \frac{\nu}{1-2\nu}\theta\right)\\ \sigma_z &= \frac{E\nu}{(1+\nu)(1-2\nu)}\theta\\ \tau_{xy} &= G\gamma_{xy}\end{aligned}\right\} \tag{6.4}$$

where

$$\theta = \varepsilon_x + \varepsilon_y = \frac{\partial u}{\partial x} + \frac{\partial v}{\partial y} \tag{6.5}$$

or

$$\left.\begin{aligned}\varepsilon_x &= \frac{1-\nu^2}{E}\left(\sigma_x - \frac{\nu}{1-\nu}\sigma_y\right)\\ \varepsilon_y &= \frac{1-\nu^2}{E}\left(\sigma_y - \frac{\nu}{1-\nu}\sigma_x\right)\\ \gamma_{xy} &= \frac{\tau_{xy}}{G}\end{aligned}\right\} \tag{6.6}$$

Equilibrium equations for an elementary tetrahedron:

$$X_\nu = l\sigma_x + m\tau_{xy} \qquad Y_\nu = l\tau_{yx} + m\sigma_y \tag{6.7}$$

Equilibrium equations for an elementary parallelepiped:

$$\frac{\partial \sigma_x}{\partial x} + \frac{\partial \tau_{xy}}{\partial y} + X = 0 \qquad \frac{\partial \tau_{yx}}{\partial x} + \frac{\partial \sigma_y}{\partial y} + Y = 0 \tag{6.8}$$

Strain compatibility (Saint-Venant) conditions:

$$\frac{\partial^2 \varepsilon_x}{\partial y^2} + \frac{\partial^2 \varepsilon_y}{\partial x^2} = \frac{\partial^2 \gamma_{xy}}{\partial x \partial y} \tag{6.9}$$

In the case of plane stress, all the basic equations except Hooke's law turn out to be the same as for the plane strain. The Hooke's law equations in the case of plane stress are as follows:

$$\left.\begin{aligned} \sigma_x &= \frac{E}{1-\nu^2}(\varepsilon_x + \nu\varepsilon_y) \\ \sigma_y &= \frac{E}{1-\nu^2}(\varepsilon_y + \nu\varepsilon_x) \\ \tau_{xy} &= G\gamma_{xy} \end{aligned}\right\} \tag{6.10}$$

or

$$\left.\begin{aligned} \varepsilon_x &= \frac{1}{E}(\sigma_x - \nu\sigma_y) \\ \varepsilon_y &= \frac{1}{E}(\sigma_y - \nu\sigma_z) \\ \varepsilon_z &= -\frac{\nu}{E}(\sigma_x + \sigma_y) \\ \gamma_{xy} &= \frac{\tau_{xy}}{G} \end{aligned}\right\} \tag{6.11}$$

The only difference between Eqs. (6.4) and (6.10) is that different combinations of elastic constants enter these formulas and that the stress σ_z is nonzero in the case of plane strain. Formulas (6.6) and (6.11) differ also in the combinations of the elastic constants and in the fact that the strain ε_z is nonzero in the case of plane stress. From the mathematical point of view, however, these differences are not essential and, therefore, the cases of plane strain and plane stress are considered as the same two-dimensional problem of the elasticity theory.

6.2 MAURICE LÉVY THEOREM

Introducing Eqs. (6.11) in Eq. (6.9), we have

$$2(1+\nu)\frac{\partial^2\tau_{xy}}{\partial x\partial y} = \frac{\partial^2\sigma_x}{\partial y^2} + \frac{\partial^2\sigma_y}{\partial x^2} - \nu\left(\frac{\partial^2\sigma_x}{\partial x^2} + \frac{\partial^2\sigma_y}{\partial y^2}\right) \tag{6.12}$$

Assuming no body forces ($X = Y = 0$), differentiating the first equation in

Eqs. (6.8) by x and the second by y, and summing up the obtained relationships, we obtain

$$2\frac{\partial^2 \tau_{xy}}{\partial x \partial y} = -\left(\frac{\partial^2 \sigma_x}{\partial x^2} + \frac{\partial^2 \sigma_y}{\partial y^2}\right)$$

After substituting this equation into Eq. (6.12), we have

$$\frac{\partial^2 \sigma_x}{\partial x^2} + \frac{\partial^2 \sigma_y}{\partial y^2} + \frac{\partial^2 \sigma_y}{\partial x^2} + \frac{\partial^2 \sigma_x}{\partial y^2} = 0$$

or

$$\Delta(\sigma_x + \sigma_y) = \Delta\sigma_1 = 0 \tag{6.13}$$

Here,

$$\Delta = \nabla^2 = \frac{\partial^2}{\partial x^2} + \frac{\partial^2}{\partial y^2} \tag{6.14}$$

is the two-dimensional *harmonic (Laplacean) operator*. Formula (6.13), which does not contain any elastic constants, expresses the following theorem: The two-dimensional state of stress for isotropic bodies is independent of the elastic constants and, therefore, the stresses in identically loaded bodies of the same configuration of the boundary will be the same for any material.

This theorem forms, in particular, the basis of the *photoelasticity method.* This method enables us to supplement an investigation of the stress condition for nontransparent bodies (such as, for instance, metals) by the experimental investigation of the two-dimensional stress condition in transparent bodies (such as glass), which are optically sensitive to the applied stresses. In the case of constant body forces, Eq. (6.13) holds both for the cases of plane strain and plane stress.

6.3 STRESS (AIRY) FUNCTION

The number of the unknown functions in the two-dimensional problem of the theory of elasticity can be reduced if a new function $\phi(x, y)$, called the *stress*, or *Airy, function*, is introduced as follows:

$$\sigma_x = \frac{\partial^2 \phi}{\partial y^2} \qquad \sigma_y = \frac{\partial^2 \phi}{\partial x^2} \qquad \tau_{xy} = -\frac{\partial^2 \phi}{\partial x \partial y} \tag{6.15}$$

These formulas assume that there are no body forces.

In order to obtain an equation that the stress function must satisfy, we introduce Eqs. (6.15) into Eq. (6.13). This results in the following *biharmonic equation*:

$$\nabla^4 \phi = \frac{\partial^4 \phi}{\partial x^4} + 2\frac{\partial^4 \phi}{\partial x^2 \partial y^2} + \frac{\partial^4 \phi}{\partial y^4} = 0 \tag{6.16}$$

Here,

$$\Delta^2 = \nabla^4 = \frac{\partial^4}{\partial x^4} + 2\frac{\partial^4}{\partial x^2 \partial y^2} + \frac{\partial^4}{\partial y^4} \tag{6.17}$$

is the *biharmonic operator*. Functions satisfying Eq. (6.16) are called *biharmonic functions*. In addition to the biharmonic equation (6.16), they should also satisfy the boundary conditions (6.7) which, considering Eqs. (6.15), are as follows:

$$X_v = l\frac{\partial^2 \phi}{\partial y^2} - m\frac{\partial^2 \phi}{\partial x \partial y} \qquad Y_v = m\frac{\partial^2 \phi}{\partial x^2} - l\frac{\partial^2 \phi}{\partial x \partial y} \tag{6.18}$$

6.4 SOLUTION IN POLYNOMIALS

Since stresses are expressed by second-order derivatives of the stress function with respect to the conditions x and y, then this function, if sought in polynomials, must be presented by terms of second degree or higher. The second- and third-degree polynomials satisfy the biharmonic equation (6.16) identically whereas, for four- and higher-degree polynomials, the coefficients must be chosen in such a way that this equation is fulfilled.

If, for instance,

$$\phi = ax^4 + bx^3y + cx^2y^2 + dxy^3 + ey^4 + a_1x^3 + b_1x^2y + c_1xy^2 + d_1y^3 + a_2x^2 + b_2xy + c_2y^2$$

then, after substituting this formula into Eq. (6.16), we find that

$$c = -3(a + e) \tag{6.19}$$

The stresses are expressed by the formulas

$$
\left.\begin{aligned}
\sigma_x &= 2cx^2 + 6dxy + 12ey^3 + 2c_1x + 6d_1y + 2c_2 \\
\sigma_y &= 12ax^2 + 6bxy + 2cy^2 + 6a_1x + 2b_1y + 2a_2 \\
\tau_{xy} &= -3bx^2 - 4cxy - 3dy^2 - 2b_1x - 2c_1y - b_2
\end{aligned}\right\} \tag{6.20}
$$

As an example, examine the strip in Fig. 6.1, where the upper and lower edges are stress-free, the left short edge is clamped, and the right edge is subjected to shearing forces only. Then, we should put

$$
\sigma_x(L, y) = \sigma_y(x, 0) = \sigma_y(x, h) = \tau_{xy}(x, 0) = \tau_{xy}(x, h) = 0
$$

These conditions result in the following equations for the polynomial coefficients:

$$
\left.\begin{aligned}
2L^2c + 2Lc_1 + 2c_2 + 6(Ld + d_1)y + 12ey^3 &= 0 \\
2a_2 + 6a_1x + 12ax^2 &= 0 \\
2h^2c + 2hb_1 + 2a_2 + 6(hb + a_1)x + 12ax^2 &= 0 \\
b_2 + 2b_1x + 3bx^2 &= 0 \\
3h^2d + 2hc_1 + b_2 + 2(2hc + b_1)x + 3bx^2 &= 0
\end{aligned}\right\}
$$

Hence,

$$
a = a_1 = a_2 = b = b_1 = b_2 = c = e = 0
$$

$$
Lc_1 + c_2 = 0 \qquad Ld + d_1 = 0 \qquad 3hd + 2c_1 = 0
$$

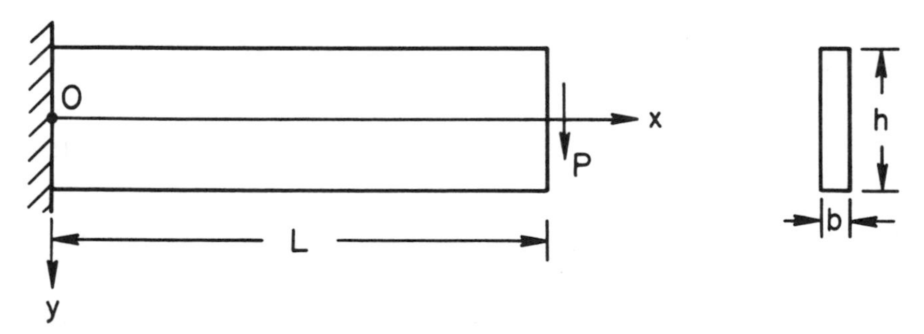

FIGURE 6.1 Cantilever beam subjected to a shearing force at the free end.

The condition (6.19) is satisfied identically, and the formulas (6.20) result in the following formulas for the stresses:

$$\sigma_x = 3d(x - L)(2y - h) \qquad \sigma_y = 0 \qquad \tau_{xy} = 3dy(h - y) \tag{6.21}$$

This solution determines the stresses in a cantilever beam of a rectangular cross section, clamped at the left end and loaded by the shearing forces along the right edge. The normal stresses σ_y are zero everywhere. The stresses σ_x are distributed linearly in any cross section x and are equal to zero on the centerline $y = h/2$. These stresses increase linearly with an increase in the distance of the given cross section from the right end. The shearing stresses $\tau_{xy} = \tau_{yx}$ change along the height h, according to the law of a parabola and are x-independent.

The obtained solution, strictly speaking, corresponds to the case in which the right end is subjected to distributed shearing forces. However, using Saint-Venant's principle, we may extend this solution for any distribution of a force P acting at the cantilever end if this force is equivalent to the actual loading, i.e., if

$$P = \tfrac{2}{3} hb\tau_{\max}$$

Here, b is the thickness of the plate, and $\tau_{\max}$ is the maximum shearing stress. The latter is equal to τ_{xy} for $y = h/2$, i.e., $\tau_{\max} = \frac{3}{4}dh^2$. Then $P = (dh^3 b)/2$. Hence, the coefficient d is $d = 2P/bh^3$. After introducing this formula in Eqs. (6.21), we have

$$\sigma_x = \frac{6P}{bh^3}(x - L)(2y - h) \qquad \tau_{xy} = \frac{6P}{bh^3}\, y(h - y)$$

For the upper and the lower extreme fibers of the cross section, i.e., for $y = h$ and $y = 0$, the shearing stress is zero, and the normal stresses σ_x are the greatest compared to the normal stresses in the inner portion of the cross section and can be presented as

$$\sigma_{\max} = \mp \frac{M(x)}{SM} \tag{6.22}$$

where $M(x) = P(L - x)$ is the bending moment in the x cross section and

$$SM = \frac{bh^2}{6} \tag{6.23}$$

is the *section modulus* of the cross section. The maximum shearing stress occurs at $y = h/2$ and is

$$\tau_{\max} = \frac{3P}{2bh} \tag{6.24}$$

The foregoing distribution of the internal forces coincides with the solution known from the strength of materials for a cantilever beam.

For complex loadings, solutions in polynomials become too cumbersome. In these cases, solutions in trigonometric series are more advisable.

6.5 SOLUTION IN TRIGONOMETRIC SERIES

We examine the state of stress in a rectangular strip, loaded as shown in Fig. 6.2, as an example of a problem to which a solution in trigonometric series can be successfully applied. We present the stress function $\phi(x, y)$ in the form (*Ribière solution*)

$$\phi(x, y) = \sum_{k=1}^{\infty} f_k(y) \cos \alpha_k x \tag{6.25}$$

where

$$\alpha_k = \frac{k\pi}{l} \qquad k = 1, 2, 3, \ldots \tag{6.26}$$

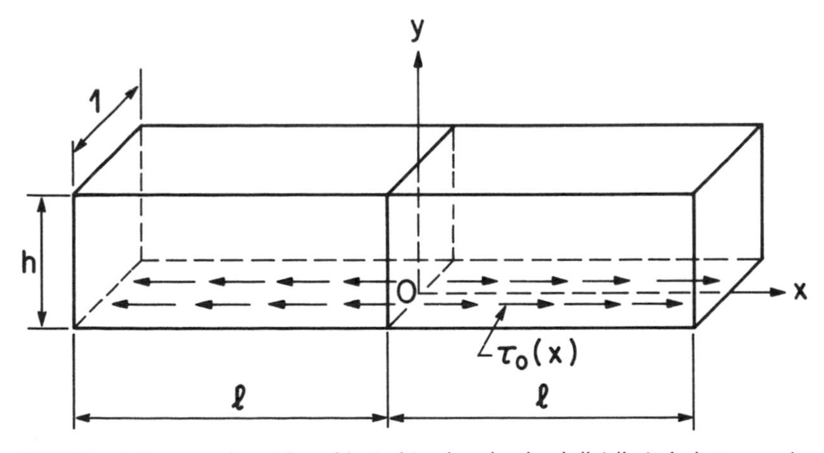

FIGURE 6.2 Rectangular strip subjected to shearing load distributed along an edge.

This solution is applicable when the shearing loads, acting along the upper and lower edges of the strip, can be presented as expansions in sines and the normal loads (if any), acting along these edges and self-equilibrated at each of them, can be presented as expansions in cosines. In addition, the shearing loads at the short edges must be zero, and the normal loads must be such that any curvature and rotation of these edges about the z axis are excluded, so that

$$u(\pm l, y) = 0 \tag{6.27}$$

This situation is usually referred to as "fixing the ends in accordance with the symmetry of the loading."

Note that a solution in the form (*Filon's solution*)

$$\phi(x, y) = \sum_{k=1}^{\infty} f_k(y) \sin \alpha_k x \tag{6.28}$$

should be applied when the shearing and the normal loadings distributed over the long edges can be presented as Fourier series in cosines and sines, respectively. In addition, the short edges $x = \pm l$ must be freely pending. The long edges may be subjected to arbitrary normal loads and to arbitrary, but self-equilibrated at each edge, shearing loads. This type of loading and the boundary conditions are known as "free pending of the strip." In this case, the tensile or compressive strains at the short edges are excluded, whereas the curvature of these edges is permitted. Generally, a combination of solutions (6.25) and (6.28) is applied.

The presentation of the stress function in the form (6.25) enables us not only to satisfy the boundary conditions at the short edges but also to replace the function $\phi(x, y)$ by a product of two functions, each depending on one variable only. After introducing Eq. (6.25) in Eq. (6.16), we obtain, instead of an equation (6.16) in partial derivatives, an ordinary differential equation for the unknown function $f_k(y)$:

$$f_k^{\mathrm{IV}}(y) - 2\alpha_k^2 f_k''(y) + \alpha_k^4 f_k(y) = 0$$

This equation has the following solution:

$$f_k(y) = (A_k + B_k y) \cosh \alpha_k y + (C_k + D_k y) \sinh \alpha_k y \tag{6.29}$$

where A_k, B_k, C_k, and D_k are constants of integration, which should be determined from the following boundary conditions on the long edges:

$$\tau_{xy}(x, 0) = \tau_0(x) \tag{6.30}$$

$$\tau_{xy}(x, h) = 0 \tag{6.31}$$

$$\left.\frac{dv}{dx}\right|_{y=0} = 0 \tag{6.32}$$

$$\left.\frac{dv}{dx}\right|_{y=h} = 0 \tag{6.33}$$

These conditions are sufficient to form four equations for the four unknown constants of integration. The conditions (6.30) and (6.31) for the shearing stresses are obvious. The boundary conditions (6.32) and (6.33) reflect an assumption that the in-plane stiffness of the strip is so great that this strip does not experience any rotation about the z axis. Obviously, we should also satisfy the condition

$$\sigma_x(\pm l, y) = 0 \tag{6.34}$$

since there are no external forces acting at the short edges in the x direction.

By introducing Eq. (6.25) in formulas (6.15), we obtain the following formulas for the stresses:

$$\left.\begin{aligned} \sigma_x &= \sum_{k=1}^{\infty} f_k''(y) \cos \alpha_k x \\ \sigma_y &= -\sum_{k=1}^{\infty} \alpha_k^2 f_k(y) \cos \alpha_k x \\ \tau_{xy} &= \sum_{k=1}^{\infty} \alpha_k f_k'(y) \sin \alpha_k x \end{aligned}\right\} \tag{6.35}$$

After introducing the formula for the stress σ_x into Eq. (6.34), we conclude that this condition will be fulfilled if the odd indices only are retained in the series (6.35). Thus, in the further analysis, we put $k = 1, 3, 5, \ldots$.

In order to be able to determine the constants of integration in the solution (6.29), we have to have the boundary conditions for the functions $f_k(y)$. Since, however, formulas (6.30) and (6.31) provide the boundary conditions for the stresses, we use the last formula in the series (6.35) to obtain the necessary conditions for the functions $f_k(y)$. After substituting this formula in condition (6.30), we have

$$\tau_0(x) = \sum_{k=1}^{\infty} \alpha_k f'_k(0) \sin \alpha_k x \tag{6.36}$$

Let us multiply both parts of this equation by $\sin \alpha_j x$ and integrate within the limits from 0 to l. Then, we have

$$\gamma_k = f'_k(0) = \frac{2}{\alpha_k l} \int_0^l \tau_0(x) \sin \alpha_k x \, dx \tag{6.37}$$

When deriving the relationship (6.37), we utilized the fact that the function $\sin \alpha_k x$, as well as the function $\cos \alpha_k x$, belongs to the class of the so-called *orthogonal functions*. In the case in question, it means, in particular, that

$$\int_0^l \sin \alpha_k x \sin \alpha_j x \, dx = \begin{cases} 0 & \text{for } k \neq j \\ \dfrac{l}{2} & \text{for } k = j \end{cases}$$

Indeed, since

$$\begin{aligned} \int_0^l \sin \alpha_k x \sin \alpha_j x \, dx &= \frac{1}{2} \int_0^l \cos (\alpha_k - \alpha_j)x \, dx - \frac{1}{2} \int_0^l \cos (\alpha_k + \alpha_j)x \, dx \\ &= \frac{1}{2(\alpha_k - \alpha_j)} [\sin (\alpha_k - \alpha_j)x]_0^l - \frac{1}{2(\alpha_k + \alpha_j)} [\sin (\alpha_k + \alpha_j)x]_0^l \\ &= \frac{l}{2(k-j)\pi} \sin (k-j)\pi - \frac{l}{2(k+j)\pi} \sin (k+j)\pi \end{aligned}$$

then this integral is always zero, if $k \neq j$. In the case $k = j$, however, only the second term in the preceding formula is zero, whereas the first term is

$$\frac{l}{2} \frac{\sin (k-j)\pi}{(k-j)\pi} = \frac{l}{2} \lim_{\beta \to 0} \frac{\sin \beta}{\beta} = \frac{l}{2}$$

The condition (6.31) for the shearing load, with consideration of the formula (6.37), results in the following obvious boundary condition for the function $f_k(x)$:

$$f'_k(h) = 0 \tag{6.38}$$

After introducing Eqs. (6.29) in Eq. (6.37) and condition (6.38), we have

$$B_k + \alpha_k C_k = \gamma_k \tag{6.39}$$

$$B_k \operatorname{cotanh} u_k + D_k = 0 \tag{6.40}$$

where

$$u_k = \alpha_k h = \frac{k\pi}{2}\frac{h}{l} \tag{6.41}$$

In order to utilize the conditions (6.32) and (6.33), we have to have a formula for the derivative dv/dx. Substituting Eqs. (6.35) into the Hooke's law equations (6.11), we obtain

$$\left.\begin{aligned} \varepsilon_x &= \frac{1}{E}\sum_k (f_k'' + \nu\alpha_k^2 f_k) \cos \alpha_k \alpha \\ \varepsilon_y &= \frac{1}{E}\sum_k (\alpha_k^2 f_k + \nu f_k'') \cos \alpha_k x \\ \varepsilon_z &= \frac{\nu}{E}\sum_k (f_k'' - \alpha_k^2 f_k) \cos \alpha_k x \\ \gamma_{xy} &= \frac{1}{G}\sum_k \alpha_k f_k' \sin \alpha_k x \end{aligned}\right\} \tag{6.42}$$

Since, as follows from Eq. (6.29),

$$f_k'(y) = [B_k + \alpha_k(C_k + D_k y)] \cosh \alpha_k y + [D_k + \alpha_k(A_x + B_k y)] \sinh \alpha_k y$$

then,

$$\begin{aligned} f_k''(y) &= [2\alpha_x D_k + \alpha_k^2(A_k + B_k y)] \cos \alpha_k y \\ &\quad + [2\alpha_k B_k + \alpha_k^2(C_k + D_k y)] \sinh \alpha_k y \\ &= \alpha_k^2 f_k(y) + 2\alpha_k(B_k \sinh \alpha_k y + D_k \cosh \alpha_k y) \end{aligned}$$

Therefore, the second formula in Eqs. (6.42) can be presented as

$$\varepsilon_y = -\frac{1}{E}\sum_k [(1 + \nu)f_k''(y) - 2\alpha_k(B_k \sinh \alpha_k y + D_k \cosh \alpha_k y)] \cos \alpha_k x$$

By integrating this formula, we find

$$v(x, y) = \int_0^y \varepsilon_y(x, y)\, dy$$

$$= -\frac{1}{E}\sum_k [(1 + \nu)f_k'(y) - 2(B_k \cosh \alpha_k y + D_k \sinh \alpha_k y)] \cos \alpha_k x \tag{6.43}$$

Note that, at the end cross sections, $(x = \pm l) \cos \alpha_k l = \cos (k\pi/2) = 0$, so that the displacement v is also zero. Introducing Eqs. (6.43) into conditions (6.32) and (6.33) and utilizing Eqs. (6.37) and (6.38), we have

$$B_k = \frac{1 + \nu}{2} \gamma_k \tag{6.44}$$

and

$$B_k \operatorname{cotanh} u_k + D_k = 0$$

so that

$$D_k = -\frac{1 + \nu}{2} \gamma_k \operatorname{cotanh} u_k \tag{6.45}$$

Equations (6.39) and (6.40) yield

$$A_k = \frac{\gamma_k}{2\alpha_k} [(1 + \nu)u_k(\operatorname{cotanh}^2 u_k - 1) - (1 - \nu) \operatorname{cotanh} u_k] \tag{6.46}$$

$$C_k = \frac{1 - \nu}{2\alpha_k} \gamma_k \tag{6.47}$$

This concludes the solution to the boundary-value problem.

We use this solution to assess the longitudinal compliance of the strip. The obtained formulas will be applied in the second volume for the analysis of bimaterial assemblies experiencing thermal loading.

Introducing the expressions for the functions $f_k(y)$ and $f_k''(y)$ into the formula for the strain ε_x in Eqs. (6.42) and integrating the obtained expression, we determine the displacements $u(x, y)$. At the edge $y = 0$, these displacements are

$$u_0(x) = u(x, 0) = \frac{1}{E} \sum_{k=1}^{\infty} \gamma_k[(1 + v)\alpha_k A_k + 2D_k] \sin \alpha_k x$$

$$= \frac{1}{4G} \sum_{k=1}^{\infty} \gamma_k[(3 - v - (1 + v)u_k \operatorname{cotanh} u_k) \operatorname{cotanh} u_k + (1 + v)u_k] \sin \alpha_k x \tag{6.48}$$

On the other hand, these displacements can be evaluated as

$$u_0(x) = -\frac{1 - v^2}{Eh} \int_0^x T(\xi)\, d\xi + \kappa\tau_0(x) \tag{6.49}$$

where the tensile force $T(x)$ acting in the x cross section is related to the loading $\tau_0(x)$ as follows:

$$T(x) = \int_{-l}^{x} \tau_0(\xi)\, d\xi \tag{6.50}$$

and κ is the coefficient of compliance.

The first term in Eq. (6.49) determines the displacement in the x cross section under an assumption that the force $T(x)$ is uniformly distributed over the cross section. The second term accounts for the actual, nonuniform distribution of this force, which results in the fact that the longitudinal displacement of the extreme fiber, where the load $\tau_0(x)$ is applied, is greater than the displacement of the main part of the cross section. Such a correction is evaluated under an assumption that only the shearing load $\tau_0(x)$ acting in this cross section may be considered. Note that, in accordance with Eq. (6.50), the force $T(x)$ is negative for tensile $\tau_0(x)$ values and, therefore, in effect, the first term in Eq. (6.49) is positive.

From Eqs. (6.36) and (6.37), we have

$$\tau_0(x) = \sum_{k=1}^{\infty} \alpha_k \gamma_k \sin \alpha_k x \tag{6.51}$$

Then, as follows from Eq. (6.50),

$$T(x) = -\sum_{k=1}^{\infty} \gamma_k \cos \alpha_k x \tag{6.52}$$

Introducing Eqs. (6.51) and (6.52) into Eq. (6.49), we obtain the following formula for the longitudinal displacement of the extreme fiber:

$$u_0(x) = \frac{1 - \nu^2}{Eh} \sum_k \frac{\gamma_k}{\alpha_k} \sin \alpha_k x + \kappa \sum_k \alpha_k \gamma_k \sin \alpha_k x \tag{6.53}$$

From Eqs. (6.48) and (6.53), we find that the compliance coefficient can be calculated as

$$\kappa = \frac{1}{E} \frac{\sum_k \gamma_k K(u_k) \sin \alpha_k x}{\sum_k \gamma_k \alpha_k \sin \alpha_k x} \tag{6.54}$$

Here the function $K(u_k)$ is expressed by the equation

$$K(u_k) = \frac{1 + \nu}{2} \left\{ [3 - \nu - (1 + \nu) u_k \operatorname{cotanh} u_k] \operatorname{cotanh} u_k + (1 + \nu) u_k - \frac{2(1 - \nu)}{u_k} \right\} \tag{6.55}$$

For very small h/l ratios (say, $h/l < 0.2$), when the u_k value is small, the following approximation can be used:

$$\operatorname{cotanh} u_k \cong \frac{1}{u_k} + \frac{u_k}{3}$$

Then, Eq. (6.55) yields

$$K(u_k) = \tfrac{2}{3}(1 + \nu) u_k = \tfrac{2}{3}(1 + \nu) \alpha_k h$$

and Eq. (6.54) reduces to the following simple formula for the coefficient of compliance:

$$\kappa = \frac{2(1 + \nu)}{3E} h = \frac{h}{3G} \tag{6.56}$$

In another extreme case, when the h/l ratio is great (say, >0.5), the compliance coefficient can be determined assuming cotanh u_k equal to unity. Then,

$$K(u_k) = \frac{(1 + \nu)(3 - \nu)}{2}$$

Retaining the first term only in the rapidly converging series in Eq. (6.54), we obtain

$$\kappa = \frac{(1 + \nu)(3 - \nu)}{\pi E} l = \frac{3 - \nu}{2\pi} \frac{l}{G} \tag{6.57}$$

Note that, for very long and narrow strips, the coefficient of compliance is length-independent whereas, for short and high two-dimensional bodies, it is height-independent. In the second volume, the formula (6.56) will be used in the analysis of the mechanical behavior of bi- and multimaterial assemblies and heteroepitaxial microelectronic structures. As to the formula (6.57), it can be used in the analysis of structures characterized by large h/l ratios such as, for instance, laser carrier-stud assemblies.

7

Two-Dimensional Problem in Polar Coordinates

7.1 BASIC RELATIONSHIPS

In discussing stresses and strains in bodies of circular shape, such as rings, disks, and solid and hollow cylinders, or the effect of circular holes on stress distribution in plates, it is advantageous to use polar coordinates.

In order to obtain equilibrium equations similar to the equilibrium equations for an elementary parallelepiped in polar coordinates, let us consider the equilibrium of an element *abcd* (Fig. 7.1) cut from the plate by two radial sections *Ob* and *Oc*, normal to the plate, and by two cylindrical surfaces *ad* and *bc*, which are also normal to the plate and whose radii are r and $r + dr$, respectively. The element in question is subjected to the radial normal stress σ_r, the tangential (circumferential) normal stress σ_θ, and the shearing stress $\tau_{r\theta}$. The sums of the projections of the forces on the direction of the radius r and in the tangential direction are as follows:

$$\left(\sigma_r + \frac{\partial \sigma_r}{\partial r}\, dr\right)(r + dr)\, d\theta - \sigma_r r\, d\theta - \left(\sigma_\theta + \frac{\partial \sigma_\theta}{\partial \theta}\, d\theta\right) dr \cdot \frac{d\theta}{2}$$
$$- \sigma_\theta\, dr \cdot \frac{d\theta}{2} + \left(\tau_{r\theta} + \frac{\partial \tau_{r\theta}}{\partial \theta}\, d\theta\right) dr \cdot \frac{d\theta}{2} - \tau_{r\theta}\, dr\, \frac{d\theta}{2} + R\, r dr\, d\theta = 0$$

$$\left(\sigma_\theta + \frac{\partial \sigma_\theta}{\partial \theta}\, d\theta\right) dr - \sigma_\theta\, dr + \left(\tau_{r\theta} + \frac{\partial \tau_{r\theta}}{\partial \theta}\, d\theta\right) dr\, \frac{d\theta}{2} + \tau_{r\theta}\, dr\, \frac{d\theta}{2}$$
$$+ \left(\tau_{\theta r} + \frac{\partial \tau_{\theta r}}{\partial r}\, dr\right)(r + dr)\, d\theta - \tau_{\theta r}\, r d\theta + \Theta\, r dr\, d\theta = 0$$

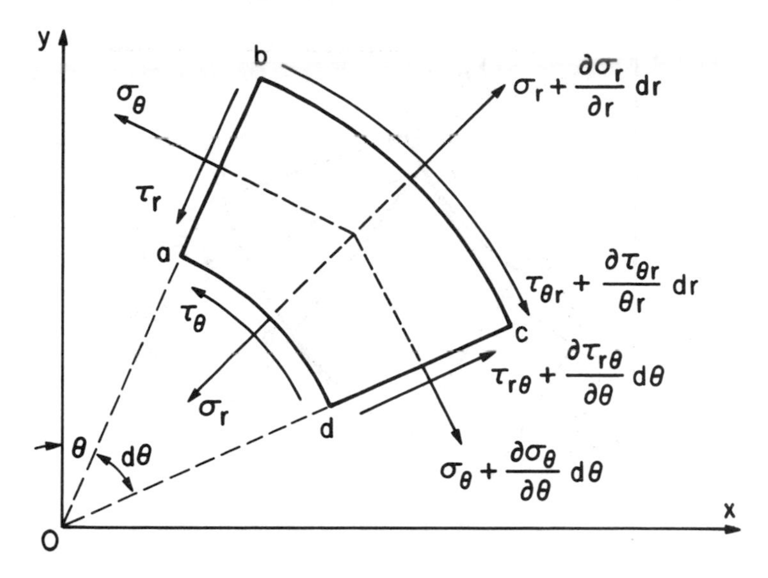

FIGURE 7.1 An element in polar coordinates with the stresses acting on it.

Here R and Θ are projections of the body forces. After neglecting, in these equations, small terms of higher orders, we obtain the following equilibrium equations:

$$\left.\begin{aligned} \frac{\partial \sigma_r}{\partial r} + \frac{1}{r}\frac{\partial \tau_{r\theta}}{\partial \theta} + \frac{\sigma_r - \sigma_\theta}{r} + R = 0 \\ \frac{1}{r}\frac{\partial \sigma_\theta}{\partial \theta} + \frac{\partial \tau_{r\theta}}{\partial r} + \frac{2\tau_{r\theta}}{r} + \Theta = 0 \end{aligned}\right\} \tag{7.1}$$

Let us obtain the formulas linking the displacements and the strains. If the radial displacement is u, then the relative elongation of the element *abcd* in the radial direction is (see Fig. 7.2)

$$\varepsilon_r = \frac{u_{bc} - u_{ad}}{dr}$$

where $u_{ad} = u$ and $u_{bc} = (u + \partial u/\partial r)\,dr$ are radial displacements of the sides *ad* and *bc*, respectively. Hence,

$$\varepsilon_r = \frac{\partial u}{\partial r} \tag{7.2}$$

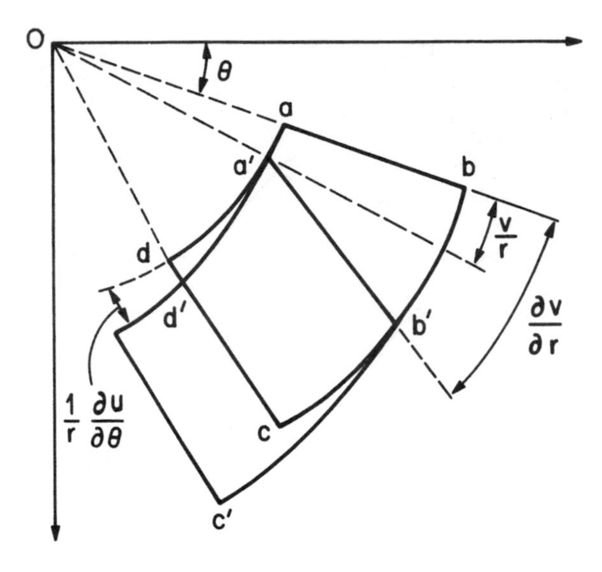

FIGURE 7.2 A plane element in the unstrained (*abcd*) and strained (*a′b′c′d′*) state (in polar coordinates).

If points a and d were subjected to radial displacements only, then the new length of the side ad would be $(r + u)\,d\theta$ and, therefore, the tangential strain would be

$$\varepsilon_\theta' = \frac{(r + u)\,d\theta - r\,d\theta}{r\,d\theta} = \frac{u}{r}$$

This value has, however, to be corrected with consideration of the tangential displacement v. Since the difference of the tangential displacements for the sides ab and cd of the element $abcd$ is $(dv/d\theta)\,d\theta$, then the corresponding tangential strain due to the displacement v is

$$\varepsilon_\theta'' = \frac{(dv/d\theta)\,d\theta}{r\,d\theta} = \frac{1}{r}\frac{dv}{d\theta}$$

The total tangential strain is

$$\varepsilon_\theta = \varepsilon_\theta' + \varepsilon_\theta'' = \frac{u}{r} + \frac{1}{r}\frac{\partial v}{\partial \theta} \tag{7.3}$$

In order to determine the shearing strain $\gamma_{r\theta}$, we examine the deformed configuration $a'b'c'd'$ of the element $abcd$ (Fig. 7.2). The angle between the directions ad and $a'd'$ can be determined as a difference of the radial displacements du of the points a and d at the length $ad = r\,d\theta$ and, therefore, is equal to $\partial u/r\partial\theta$. The angle between the directions ab and $a'b'$ can be determined as the difference of tangential displacements dv of the points a and b at the length $ab = dr$ and, therefore, is equal to dv/dr. A negative correction has to be introduced, however, into this value. Such a correction is due to the rotation of the element $abcd$ as a whole about an axis passing through the point O and normal to the plane of the figure. Thus, the total change in the rectangular angle dab is

$$\gamma_{r\theta} = \frac{1}{r}\frac{\partial u}{\partial \theta} + \frac{\partial v}{\partial r} - \frac{v}{r} \tag{7.4}$$

The Hooke's law formulas can be presented as

$$\varepsilon_r = \frac{1}{E}(\sigma_r - \nu\sigma_\theta) \qquad \varepsilon_\theta = \frac{1}{E}(\sigma_\theta - \nu\sigma_r) \qquad \gamma_{r\theta} = \frac{1}{G}\tau_{r\theta} \tag{7.5}$$

for the case of plane stress. In the case of plane strain, the elastic constants E and ν in these formulas should be substituted by $E/(1 - \nu^2)$ and $\nu/(1 - \nu)$, respectively.

The Maurice Lévy theorem can be presented as

$$\Delta(\sigma_r + \sigma_\theta) = 0 \tag{7.6}$$

where

$$\Delta = \frac{\partial^2}{\partial r^2} + \frac{1}{r}\frac{\partial}{\partial r} + \frac{1}{r^2}\frac{\partial^2}{\partial \theta^2}$$

is the harmonic operator in polar coordinates.

The stress function $\phi(r, \theta)$ is introduced as follows:

$$\left.\begin{aligned} \sigma_r &= \frac{1}{r}\frac{\partial \phi}{\partial r} + \frac{1}{r^2}\frac{\partial^2 \phi}{\partial \theta^2} \\ \sigma_\theta &= \frac{\partial^2 \phi}{\partial r^2} \\ \tau_{r\theta} &= -\frac{\partial}{\partial r}\left(\frac{1}{r}\frac{\partial \phi}{\partial \theta}\right) \end{aligned}\right\} \tag{7.7}$$

Then, we find

$$\sigma_r + \sigma_\theta = \frac{1}{r}\frac{\partial \phi}{\partial r} + \frac{1}{r^2}\frac{\partial^2 \phi}{\partial \theta^2} + \frac{\partial^2 \phi}{\partial r^2} = \Delta \phi$$

Introducing this formula in Eq. (7.6), we have the following biharmonic equation for the function ϕ:

$$\Delta^2 \phi = \nabla^4 \phi = \left(\frac{\partial^2}{\partial r^2} + \frac{1}{r}\frac{\partial}{\partial r} + \frac{1}{r^2}\frac{\partial^2}{\partial \theta^2}\right)\left(\frac{\partial^2 \phi}{\partial r^2} + \frac{1}{r}\frac{\partial \phi}{\partial r} + \frac{1}{r^2}\frac{\partial^2 \phi}{\partial \theta^2}\right) = 0 \quad (7.8)$$

In order to obtain the relationships between the stress components in the rectangular and polar coordinates, we write the formulas (2.7) and (2.8) for the case of a two-dimensional problem:

$$\left.\begin{aligned} \sigma_x &= \sigma_r l^2 + \sigma_\theta m^2 + 2\tau_{r\theta} lm \\ \sigma_y &= \sigma_r l_1^2 + \sigma_\theta m_1^2 + 2\tau_{r\theta} l_1 m_1 \\ \tau_{xy} &= \tau_{yx} = \sigma_r l l_1 + \sigma_\theta m m_1 + \tau_{r\theta}(l m_1 + l_1 m) \end{aligned}\right\} \quad (7.9)$$

where the direction cosines are expressed as follows:

$$l = \cos(r, x) = \cos\theta \qquad m = \cos(\theta, x) = \cos\left(\frac{\pi}{2} + \theta\right) = -\sin\theta$$

$$l_1 = \cos(r, y) = \cos\left(\frac{\pi}{2} - \theta\right) = \sin\theta \qquad m_1 = \cos(\theta, y) = \cos\theta$$

Introducing these formulas in Eq. (7.9), we have

$$\left.\begin{aligned} \sigma_x &= \sigma_r \cos^2\theta + \sigma_\theta \sin^2\theta - 2\tau_{r\theta}\sin\theta\cos\theta \\ \sigma_y &= \sigma_r \sin^2\theta + \sigma_\theta \cos^2\theta + 2\tau_{r\theta}\sin\theta\cos\theta \\ \tau_{xy} &= (\sigma_r - \sigma_\theta)\sin\theta\cos\theta + \tau_{r\theta}(\cos^2\theta - \sin^2\theta) \end{aligned}\right\} \quad (7.10)$$

7.2 STRESSES AND DISPLACEMENTS IN A CIRCULAR RING

In this section, we consider the states of stress and strain for a circular ring (Fig. 7.3) and present the stress function in the form

$$\phi(r, \theta) = \frac{a}{2} r\theta \sin\theta + \left(br^3 + \frac{c}{r} + dr \ln r\right)\cos\theta \quad (7.11)$$

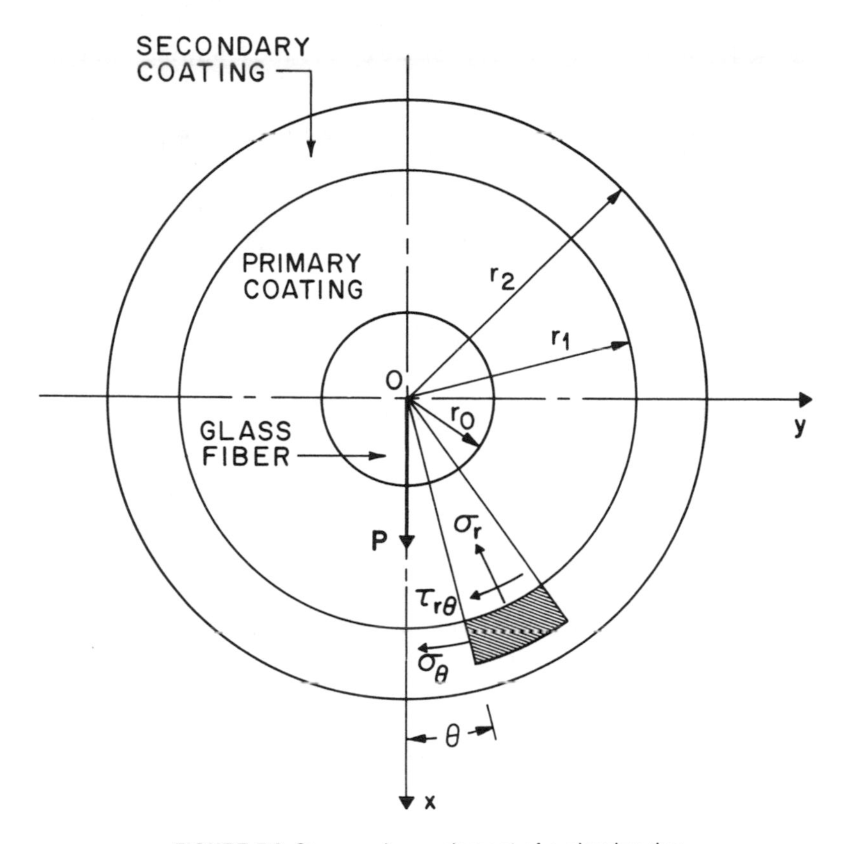

FIGURE 7.3 Stresses in an element of a circular ring.

This formula, after being introduced in Eqs. (7.7), results in the following equations for the stresses:

$$\left.\begin{aligned}\sigma_r &= \left(\frac{a}{r} + 2br - \frac{2c}{r^3} + \frac{d}{r}\right)\cos\theta \\ \sigma_\theta &= \left(6br + \frac{2c}{r^3} + \frac{d}{r}\right)\cos\theta \\ \tau_{r\theta} &= \left(2br - \frac{2c}{r^3} + \frac{d}{r}\right)\sin\theta\end{aligned}\right\} \tag{7.12}$$

Using the Hooke's law equations (7.5), we obtain the following formulas for the strains:

$$\left.\begin{aligned} \varepsilon_r &= \frac{1+\nu}{E}\left[(1-\nu)\frac{a}{r} + 2(1-4\nu)br - \frac{2c}{r^3} + (1-2\nu)\frac{d}{r}\right]\cos\theta \\ \varepsilon_\theta &= \frac{1+\nu}{E}\left[-\nu\frac{a}{r} + 2(3-4\nu)br + \frac{2c}{r^3} + (1-2\nu)\frac{d}{r}\right]\cos\theta \\ \gamma_{r\theta} &= 2\frac{1+\nu}{E}\left(2br - \frac{2c}{r^3} + \frac{d}{r}\right)\sin\theta \end{aligned}\right\} \quad (7.13)$$

Substituting the formula for the strain ε_r into Eq. (7.2) and integrating, we have

$$u = \frac{1+\nu}{E}\left[(1+\nu)a\ln r + (1-4\nu)br^2 + \frac{c}{r^2} + (1-2\nu)d\ln r\right]\cos\theta + f'(\theta) \quad (7.14)$$

where $f(\theta)$ is so far an arbitrary function of the angle θ. After substituting Eq. (7.14) and the formula for the strain ε_θ from Eqs. (7.13) into Eq. (7.3), we obtain

$$\begin{aligned} \frac{\partial v}{\partial \theta} = r\varepsilon_\theta - u = \frac{1+\nu}{E}\bigg[&-\nu a - (1-\nu)a\ln r + (5-4\nu)br^2 + \frac{c}{r^2} \\ &+ (1-2\nu)d - (1-2\nu)d\ln r\bigg]\cos\theta - f'(\theta) \end{aligned} \quad (7.15)$$

By integration, we find

$$\begin{aligned} v = \frac{1+\nu}{E}&\left[-\nu a - (1-\nu)a\ln r + (5-4\nu)br^2 + \frac{c}{r^2} + (1-2\nu)d\ln r\right]\sin\theta \\ &- f(\theta) + g(r) \end{aligned} \quad (7.16)$$

where $g(r)$ is a function of the radius r. Substituting the formulas (7.15) and (7.16) into Eq. (7.4), we obtain the following equation for the angle strain:

$$\begin{aligned} \gamma_{r\theta} = \frac{2(1+\nu)}{E}&\left[-\frac{1-2\nu}{2}\frac{a}{r} + 2br - \frac{2c}{r^3} - (1-2\nu)\frac{d}{r}\right]\sin\theta \\ &+ \frac{1}{r}[f''(\theta) + f(\theta)] - \left[g'(r) - \frac{g(r)}{r}\right] \end{aligned}$$

After comparing this equation with the third formula in Eqs. (7.13), we conclude that the coefficients a and d are related as

$$(1-2\nu)a + 4(1-\nu)d = 0 \tag{7.17}$$

and that the following equations must take place:

$$f''(\theta) + f(\theta) = 0 \qquad g'(r) - \frac{g(r)}{r} = 0$$

The latter equations have the following solutions:

$$f(\theta) = A \sin\theta + B\cos\theta \qquad g(r) = Cr \tag{7.18}$$

where A, B, and C are constants of integration. Using Eqs. (7.17) and (7.18), we present the formulas (7.12) for the stresses and the formulas (7.14) and (7.16) for the displacements as follows:

$$\left.\begin{aligned}
\sigma_r &= \left[\frac{3-2\nu}{4(1-\nu)}\frac{a}{r} + 2br - \frac{2c}{r^3}\right]\cos\theta \\
\sigma_\theta &= \left[-\frac{1-2\nu}{4(1-\nu)}\frac{a}{r} + 6br + \frac{2c}{r^3}\right]\cos\theta \\
\tau_{r\theta} &= \left[-\frac{1-2\nu}{4(1-\nu)}\frac{a}{r} + 2br - \frac{2c}{r^3}\right]\sin\theta
\end{aligned}\right\} \tag{7.19}$$

$$\left.\begin{aligned}
u &= \frac{1+\nu}{E}\left[\frac{3-4\nu}{4(1-\nu)}a\ln r + (1-4\nu)br^2 + \frac{c}{r^2}\right]\cos\theta + A\cos\theta - B\sin\theta \\
v &= \frac{1+\nu}{E}\left[-\frac{1}{4(1-\nu)}a - \frac{3-4\nu}{4(1-\nu)}a\ln r + (5-4\nu)br^2 + \frac{c}{r^2}\right]\sin\theta \\
&\quad - A\sin\theta - B\cos\theta + Cr
\end{aligned}\right\} \tag{7.20}$$

The constants a, b, c, A, B, C in these formulas can be determined from the boundary conditions, depending on a particular problem.

7.3 A CIRCULAR RING LOADED AT ITS INNER BOUNDARY

As an example, let us consider a circular ring subjected to a distributed load applied to its inner boundary. This load is due to the normal stress (Fig. 7.3)

$$\sigma_r(r_0, \theta) = \sigma_r^0(\theta) = \left[\frac{3-2\nu}{4(1-\nu)}\frac{a}{r_0} + 2br_0 - \frac{2c}{r_0^3}\right]\cos\theta$$

and the shearing stress

$$\tau_{r\theta}(r_0, \theta) = \tau_{r\theta}^0(\theta) = \left[-\frac{1-2\nu}{4(1-\nu)} \frac{a}{r_0} + 2br_0 - \frac{2c}{r_0^3} \right] \sin \theta$$

The resulting force is, therefore,

$$\begin{aligned} P &= \int_0^{2\pi} (\tau_{r\theta}^0 \sin \theta - \sigma_r^0 \cos \theta) r_0 \, d\theta \\ &= \frac{1-2\nu}{4(1-\nu)} a \int_0^{2\pi} \sin^2 \theta \, d\theta - \frac{3-2\nu}{4(1-\nu)} a \int_0^{2\pi} \cos^2 \theta \, d\theta = -\pi a \end{aligned} \tag{7.21}$$

The purpose of the analysis we are going to carry out is to determine the spring constant of the ring material. The obtained formula will be used in the second volume in the analysis of low-temperature microbending of dual-coated fibers with compliant primary coating.

Assuming that both the outer and inner boundaries are nondeformable and that the outer boundary is restricted from any displacement, we have the following conditions for the displacements:

$$u(r_1, \theta) = 0 \qquad v(r_1, \theta) = 0 \tag{7.22}$$

$$u(r_0, \theta) = \Delta \cos \theta \qquad v(r_0, \theta) = -\Delta \sin \theta \tag{7.23}$$

Here Δ is the displacement of the inner boundary in the x direction. These conditions require that $B = C = 0$ and result in the following equations for the constants a, b, c, and A:

$$\left.\begin{aligned} &\frac{1+\nu}{E}\left[\frac{3-4\nu}{4(1-\nu)} a \ln r_1 + (1-4\nu) b r_1^2 + \frac{c}{r_1^2}\right] + A = 0 \\ &\frac{1+\nu}{E}\left[-\frac{1}{4(1-\nu)} a - \frac{3-4\nu}{4(1-\nu)} a \ln r_1 + (5-4\nu) b r_1^2 + \frac{c}{r_1^2}\right] - A = 0 \\ &\frac{1+\nu}{E}\left[\frac{3-4\nu}{4(1-\nu)} a \ln r_0 + (1-4\nu) b r_0^2 + \frac{c}{r_0^2}\right] + A = \Delta \\ &\frac{1+\nu}{E}\left[-\frac{1}{4(1-\nu)} a - \frac{3-4\nu}{4(1-\nu)} a \ln r_0 + (5-4\nu) b r_0^2 + \frac{c}{r_0^2}\right] - A = -\Delta \end{aligned}\right\} \tag{7.24}$$

After excluding the unknowns A, b, and c from these equations and considering Eqs. (7.21), we obtain the following formula for the spring constant:

$$K = \frac{P}{\Delta} = \frac{4\pi E(1-\nu)(3-4\nu)}{(1+\nu)\{(3-4\nu)^2 \ln(r_1/r_0) - [(r_1/r_0)^2 - 1/(r_1/r_0)^2 + 1]\}} \tag{7.25}$$

This formula was first obtained by Vangheluwe by using Muskhelishvili's method of complex representation of displacements. Obviously, if there is a need to evaluate the stresses and displacements, we can find the constants of integration from Eqs. (7.24) and then use the general formulas (7.19) and (7.20).

7.4 AXISYMMETRIC PROBLEM

In some problems, the stresses are independent of the angle θ. In this case, Eqs. (7.7) yield

$$\sigma_r = \frac{1}{r}\frac{\partial \phi}{\partial r} \qquad \sigma_\theta = \frac{\partial^2 \phi}{\partial r^2} \qquad \tau_{r\theta} = 0 \tag{7.26}$$

Eqs. (7.1), in the absence of the body forces, result in the equation

$$\frac{d\sigma_r}{dr} + \frac{\sigma_r - \sigma_\theta}{r} = 0 \tag{7.27}$$

the displacements v are zero, and the formulas for the strains reduce to the following simple formulas:

$$\varepsilon_r = \frac{\partial u}{\partial r} \qquad \varepsilon_\theta = \frac{u}{r} \qquad \gamma_{r\theta} = 0 \tag{7.28}$$

From the Hooke's law formulas (7.5) for the plane stress, only two remain:

$$\varepsilon_r = \frac{1}{E}(\sigma_r - \nu\sigma_\theta) \qquad \varepsilon_\theta = \frac{1}{E}(\sigma_\theta - \nu\sigma_r) \tag{7.29}$$

or

$$\sigma_r = \frac{E}{1-\nu^2}(\varepsilon_r + \nu\varepsilon_\theta) \qquad \sigma_\theta = \frac{E}{1-\nu^2}(\varepsilon_\theta + \nu\varepsilon_r) \tag{7.30}$$

Introducing Eqs. (7.28) into Eqs. (7.30), we have

$$\sigma_r = \frac{E}{1-\nu^2}\left(\frac{du}{dr} + \nu\frac{u}{r}\right) \qquad \sigma_\theta = \frac{E}{1-\nu^2}\left(\frac{u}{r} + \nu\frac{du}{dr}\right) \tag{7.31}$$

Substituting these expressions in Eq. (7.27), we obtain the following equation for the unknown displacements u:

$$\frac{d^2u}{dr^2} + \frac{1}{r}\frac{du}{dr} - \frac{u}{r^2} = 0$$

This equation has the following solution:

$$u = Ar + \frac{B}{r} \tag{7.32}$$

where the constants of integration A and B can be determined from the boundary condition of the particular problem. The formulas (7.31) can be,

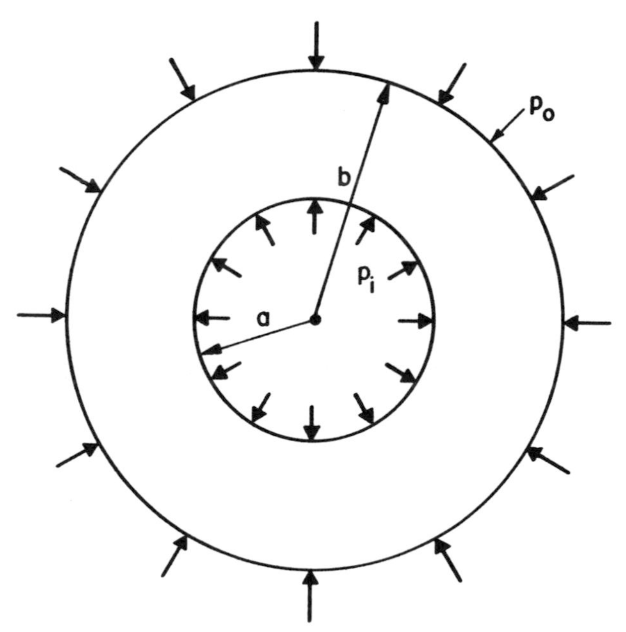

FIGURE 7.4 Flat ring under the action of external (p_e) and internal (p_i) pressure.

considering Eq. (7.32), presented as

$$\sigma_r = \frac{E}{1-\nu^2}\left[(1+\nu)A - (1-\nu)\frac{B}{r^2}\right] \qquad \sigma_\theta = \frac{E}{1-\nu^2}\left[(1+\nu)A + (1-\nu)\frac{B}{r^2}\right] \tag{7.33}$$

7.5 THICK-WALLED TUBE

We use the results of the previous section to obtain a solution to the *Lamé problem* for a thick-walled tube subjected to external pressure p_e and internal pressure p_i (Fig. 7.4). Using the boundary conditions

$$\sigma_r(r_1) = -p_e \qquad \sigma_r(r_0) = -p_i$$

we find

$$A = -\frac{1-\nu}{E}\frac{p_e r_1^2 - p_i r_0^2}{r_1^2 - r_0^2} \qquad B = -\frac{1+\nu}{E}\frac{r_0^2 r_1^2 (p_e - p_i)}{r_1^2 - r_0^2} \tag{7.34}$$

Introducing these formulas in Eqs. (7.33), we have

$$\left.\begin{aligned} \sigma_r &= \frac{1}{r_1^2 - r_0^2}\left[\frac{r_0^2 r_1^2}{r^2}(p_e - p_i) - (p_e r_1^2 - p_i r_0^2)\right] \\ \sigma_\theta &= \frac{1}{r_1^2 - r_0^2}\left[-\frac{r_0^2 r_1^2}{r^2}(p_e - p_i) - (p_e r_1^2 - p_i r_0^2)\right] \end{aligned}\right\} \tag{7.35}$$

In the absence of the opening ($r_0 = 0$), these formulas yield $\sigma_r = \sigma_\theta = -p_e$ and, therefore, the stress concentration factors due to this opening are

$$\left.\begin{aligned} \alpha_r &= -\frac{\sigma_r}{p_e} = \frac{1}{1-(r_0/r_1)^2}\left[-\left(\frac{r_0}{r}\right)^2\left(1 - \frac{p_i}{p_e}\right) + 1 - \frac{p_i}{p_e}\left(\frac{r_0}{r_1}\right)^2\right] \\ \alpha_\theta &= \frac{\sigma_\theta}{p_e} = \frac{1}{1-(r_0/r_1)^2}\left[\left(\frac{r_0}{r}\right)^2\left(1 - \frac{p_i}{p_e}\right) + 1 - \frac{p_i}{p_e}\left(\frac{r_0}{r_1}\right)^2\right] \end{aligned}\right\} \tag{7.36}$$

In another special case, when the internal pressure p_i is zero and the inner radius is significantly smaller than the outer radius, we have

$$\alpha_r = 1 - \frac{r_0^2}{r^2} \qquad \alpha_\theta = 1 + \frac{r_0^2}{r^2} \tag{7.37}$$

At the boundary of the opening ($r = r_0$), we have $\alpha_r = 0$, $\alpha_\theta = 2$. Tangential stress rapidly decreases with an increase in the distance from the opening. As one can see from Eqs. (7.37), when $r/r_0 = 5$, the α_θ value becomes $\alpha_\theta = 1.04$, so that the opening does not affect the stress field in a practical sense.

The radial displacements can be obtained from Eqs. (7.32) and (7.34) and are as follows:

$$u = -\frac{1}{E(r_1^2 - r_0^2)}\left[(1 - \nu)(p_e r_1^2 - p_i r_0^2)r + (1 + \nu)r_0^2 r_1^2(p_e - p_i)\frac{1}{r}\right] \quad (7.38)$$

In the case of a plane strain, substituting $E/(1 - \nu^2)$ for E and $\nu/(1 - \nu)$ for ν in the latter formula, we obtain the following formula for the radial displacement:

$$u = -\frac{1 + \nu}{E(r_1^2 - r_0^2)}\left[(1 - 2\nu)(p_e r_1^2 - p_i r_0^2)r + r_0^2 r_1^2(p_e - p_i)\frac{1}{r}\right] \quad (7.39)$$

7.6 THERMAL STRESS IN COAXIAL CYLINDERS

As an example of the application of the solution to the Lamé problem, we examine thermal stress in coaxial (biannular) cylinders (Fig. 7.5), assuming that Young's modulus E_1 of the outer hollow cylinder ("coating") is significantly smaller than Young's modulus E_0 of the inner solid cylinder, whereas the coefficient of thermal expansion (α_1) of the outer cylinder material is significantly greater than that (α_0) of the inner cylinder. The obtained results will be used in the second volume to analyze the mechanical performance of single-coated optical fibers at low-temperature conditions.

Let the structure shown in Fig. 7.5 be fabricated at an elevated temperature and subsequently cooled down by the temperature ΔT. Since the axial dimension of the axisymmetric body under examination is large, a plane strain rather than a plane stress approximation should be applied. Then, using Eq. (7.39), we present the radial displacements u_0 of the inner cylinder and the radial displacements u_1 of the inner boundary of the outer cylinder as follows:

$$\left.\begin{aligned} u_0 &= -\alpha_0 \Delta T r_0 - \frac{(1 + \nu_0)(1 - 2\nu_0)}{E_0} p r_0 \\ u_1 &= -\alpha_1 \Delta T r_0 + \frac{1 + \nu_1}{E_1}\frac{1 - (1 - 2\nu_1)\gamma^2}{1 - \gamma^2} p r_0 \end{aligned}\right\}$$

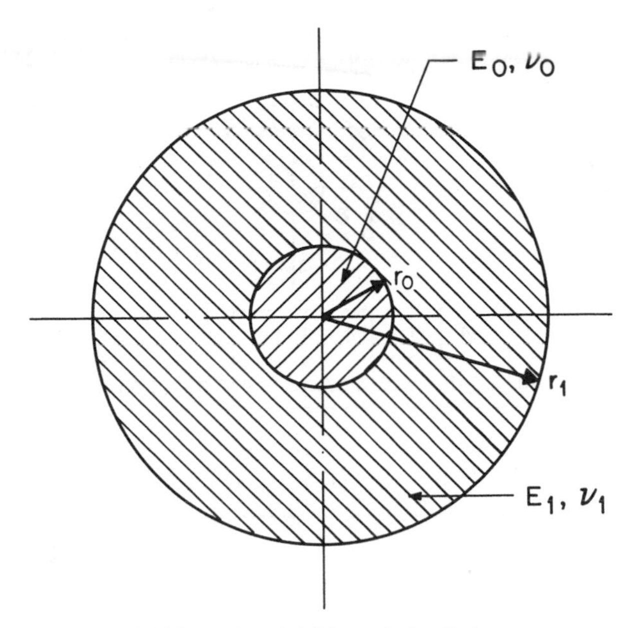

FIGURE 7.5 Coaxial (biannular) cylinders.

Here, $\gamma = r_0/r_1$ is the ratio of the cylinder radii, and p is the interfacial pressure. The first terms in these formulas are unrestricted thermal contractions, and the second terms are due to the pressure p.

Using compatibility condition $u_0 = u_1$ and considering that $E_0 \gg E_1$ and $\alpha_1 \gg \alpha_0$, we obtain the following approximate formula for the interfacial pressure:

$$p = E_1 \alpha_1 \chi_p(\gamma) \Delta T \tag{7.40}$$

where the function

$$\chi_p(\gamma) = \frac{1 - \gamma^2}{(1 + \nu_1)[1 - (1 - 2\nu_1)\gamma^2]}$$

reflects the effect of the radii ratio γ. This function is plotted in Fig. 7.6 for $\nu_1 = \frac{1}{3}$ versus the ratio of the coating thickness δ to the radius r_0 of the inner cylinder. As evident from this figure, the interfacial pressure can be significantly reduced if the coating thickness is sufficiently small (say, smaller than half the radius of the inner cylinder). A very small thickness, however, can result in high stresses in the coating material. In accordance with Eqs.

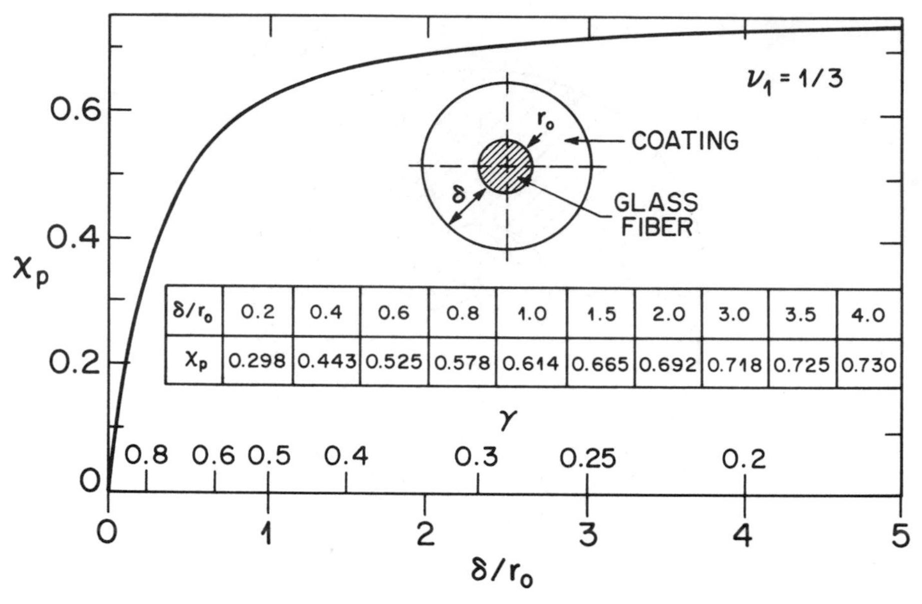

δ/r_0	0.2	0.4	0.6	0.8	1.0	1.5	2.0	3.0	3.5	4.0
χ_p	0.298	0.443	0.525	0.578	0.614	0.665	0.692	0.718	0.725	0.730

FIGURE 7.6 Effect of the coating thickness on the interfacial pressure in coaxial cylinders.

(7.35), these stresses can be evaluated by the formulas

$$\sigma_r = p\frac{r_0^2}{r_1^2 - r_0^2}\left(1 - \frac{r_1^2}{r^2}\right) \qquad \sigma_\theta = -p\frac{r_0^2}{r_1^2 - r_0^2}\left(1 + \frac{r_1^2}{r^2}\right) \tag{7.41}$$

and change from

$$\sigma_r = -p \qquad \sigma_\theta = -p\frac{r_1^2 + r_0^2}{r_1^2 - r_0^2} = -p\frac{1 + \gamma^2}{1 - \gamma^2}$$

at the interface ($r = r_0$) to

$$\sigma_r = 0 \qquad \sigma_\theta = -2p\frac{r_0^2}{r_1^2 - r_0^2} = -2p\frac{\gamma^2}{1 - \gamma^2}$$

at the outer boundary ($r = r_1$).

Using Eqs. (7.41), we present the preceding formulas for the tangential stress σ_θ in the form

$$\sigma_\theta = -E_1\alpha_1\chi_i(\gamma)\Delta T \tag{7.42}$$

at the interface and in the form

$$\sigma_\theta = -E_1\alpha_1\chi_0(\gamma)\Delta T \tag{7.43}$$

at the outer boundary. The functions

$$\chi_i(\gamma) = \frac{1+\gamma^2}{(1+\nu_1)[1-(1-2\nu_1)\gamma^2]} \qquad \chi_0(\gamma) = \frac{2\gamma^2}{(1+\nu_1)[1-(1-2\nu_1)\gamma^2]}$$

reflect the effect of the radii ratio (coating thickness) and are plotted for $\nu_1 = \frac{1}{3}$ in Fig. 7.7 versus δ/r_0 ratios. As evident from this figure, the tangential stress in the coating increases with a decrease in its thickness, especially for very thin coatings. Therefore, the coating thickness should be sufficient to withstand these stresses. Obviously, there is an incentive to use high-strength coating materials.

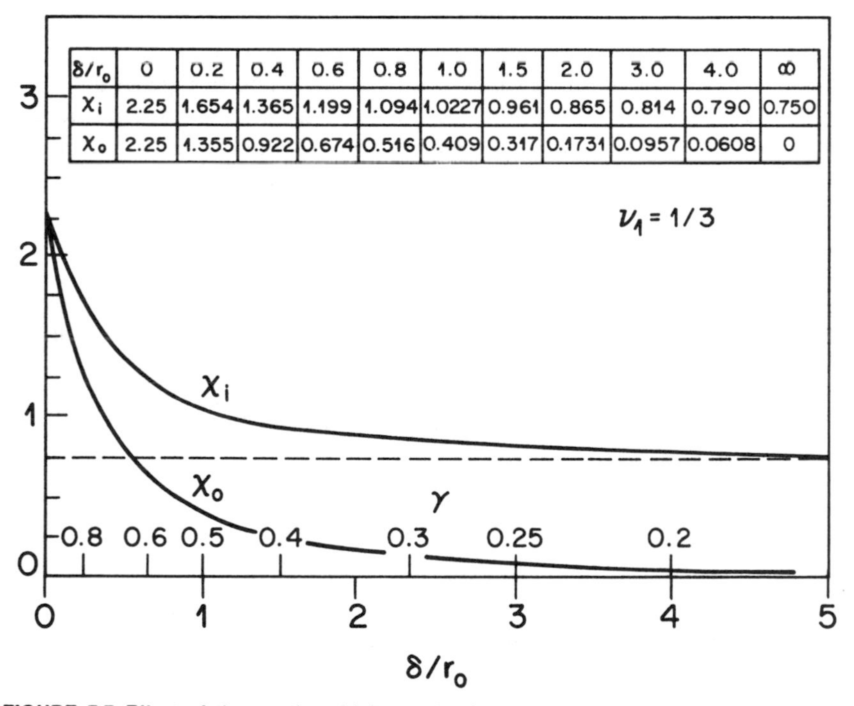

δ/r_0	0	0.2	0.4	0.6	0.8	1.0	1.5	2.0	3.0	4.0	∞
χ_i	2.25	1.654	1.365	1.199	1.094	1.0227	0.961	0.865	0.814	0.790	0.750
χ_0	2.25	1.355	0.922	0.674	0.516	0.409	0.317	0.1731	0.0957	0.0608	0

FIGURE 7.7 Effect of the coating thickness in the tangential stress at the interface (χ_i) and at the outer boundary (χ_0).

8

Torsion

8.1 SAINT-VENANT'S TORSION FUNCTION

In the problem of simple torsion of a circular shaft examined in Section 4.2, we obtained the following formulas for the displacements v and w in the lateral plane:

$$v = -\alpha xz \qquad w = \alpha xy \tag{8.1}$$

and found that the axial displacement u is always zero. In a general case of a noncircular cross section, however, the axial displacements are nonzero and, in accordance with *Saint-Venant's hypothesis*, can be presented in the form

$$u = \alpha S(y, z) \tag{8.2}$$

The function $S(y, z)$ considers the deplanation of the cross section and is called *Saint-Venant's torsion function.*

Introducing the preceding formulas for the displacements into the Lamé equations (7.4), we have $Y = Z = 0$. This results in the following equation for the function $S(y, z)$:

$$\Delta S(y, z) = \frac{\partial^2 S}{\partial y^2} + \frac{\partial^2 S}{\partial z^2} = -\frac{X}{\alpha G}$$

In the further analysis, we assume that $X = 0$, so that

$$\Delta S(y, z) = \frac{\partial^2 S}{\partial y^2} + \frac{\partial^2 S}{\partial z^2} = 0 \tag{8.3}$$

After substituting Eqs. (8.1) and (8.2) into Cauchy's formulas (6.32), we have

$$\left.\begin{aligned} &\varepsilon_x = \varepsilon_y = \varepsilon_z = \gamma_{yz} = 0 \\ &\gamma_{xy} = \alpha\left(-z + \frac{\partial S}{\partial y}\right) \qquad \gamma_{zx} = \alpha\left(y + \frac{\partial S}{\partial z}\right) \end{aligned}\right\} \tag{8.4}$$

Substitution of these relationships into the Hooke's law formulas (6.50) results in the equations

$$\left.\begin{aligned} &\sigma_x = \sigma_y = \sigma_z = \tau_{yz} = 0 \\ &\tau_{xy} = \alpha G\left(-z + \frac{\partial S}{\partial y}\right) \qquad \tau_{zx} = \alpha G\left(y + \frac{\partial S}{\partial z}\right) \end{aligned}\right\} \tag{8.5}$$

Finally, after introducing the latter equations into the equations of equilibrium of an elementary tetrahedron (2.5), we find that the surface forces X_ν, Y_ν, and Z_ν are expressed as follows:

$$\left.\begin{aligned} X_\nu &= \alpha G\left(m\frac{\partial S}{\partial y} + n\frac{\partial S}{\partial z} - mz + ny\right) \\ Y_\nu &= \alpha G\left(-z + \frac{\partial S}{\partial y}\right)l \\ Z_\nu &= \alpha G\left(y + \frac{\partial S}{\partial z}\right)l \end{aligned}\right\}$$

Since, however, there are no surface forces in the problem in question, we must put $X_\nu = Y_\nu = Z_\nu = 0$. The straight generators of the lateral surface of the shaft are parallel to the x coordinate, so that $l = 0$, and the conditions $X_\nu = Z_\nu = 0$ are fulfilled automatically. The requirement $X_\nu = 0$ results in the following relationship for the function $S(y, z)$:

$$m\frac{\partial S}{\partial y} + n\frac{\partial S}{\partial z} = mz - ny \tag{8.6}$$

The derivative of the function S along the normal ν to the contour of the shaft is

$$\frac{\partial S}{\partial \nu} = m\frac{\partial S}{\partial y} + n\frac{\partial S}{\partial z}$$

and, therefore, Eq. (8.6) can be rewritten as

$$\frac{\partial S}{\partial v} = mz - ny \tag{8.7}$$

Thus, Saint-Venant's torsion function $S(y, z)$ must satisfy Laplace's equation (8.3) and the boundary condition (8.7).

8.2 PRANDTL'S STRESS FUNCTION

Another way to approach the problem in question is to introduce, instead of the function $S(y, z)$, a stress function $\Phi(y, z)$ as follows:

$$\tau_{xy} = \frac{\partial \Phi}{\partial z} \qquad \tau_{zx} = -\frac{\partial \Phi}{\partial y} \tag{8.8}$$

This is called *Prandtl's stress function* and, as follows from Eqs. (8.5) and (8.8), is related to the Saint-Venant's torsion function S as

$$\frac{\partial \Phi}{\partial z} = \alpha G\left(-z + \frac{\partial S}{\partial y}\right) \qquad \frac{\partial \Phi}{\partial y} = -\alpha G\left(y + \frac{\partial S}{\partial z}\right) \tag{8.9}$$

By differentiating the first of these equations by z and the second equation by y, and summing up the results, we have

$$\Delta\Phi(y, z) = -2\alpha G \tag{8.10}$$

Thus, Prandtl's stress function $\Phi(y, z)$ must satisfy *Poisson's equation* (8.10), whereas the function $S(y, z)$ can be determined from a simpler Laplacean equation (8.3). We will see, however, that the boundary condition for the function $\phi(y, z)$ is much simpler.

In order to find this condition, let us solve Eqs. (8.9) for the derivatives $\partial S/\partial y$ and $\partial S/\partial z$ and substitute the obtained formulas into Eq. (8.6). Then, we have

$$m\frac{\partial \Phi}{\partial z} - n\frac{\partial \Phi}{\partial y} = 0 \tag{8.11}$$

Since the direction cosines m and n can have arbitrary values, the condition (8.11) will be fulfilled if

$$\frac{\partial \Phi}{\partial z} = \frac{\partial \Phi}{\partial y} = 0$$

i.e., if the function Φ is constant on the surface of the shaft:

$$\Phi(y, z) = \text{const}$$

From Eqs. (8.8), we can see that if a constant is added to, or subtracted from, the function Φ, the stresses remain unchanged. Therefore, the preceding condition for the function Φ can be replaced by an even simpler condition,

$$\Phi(y, z) = 0 \tag{8.12}$$

Obviously, this condition is significantly simpler than the condition (8.7) for the function $S(y, z)$.

8.3 PRANDTL'S FORMULA

The torque applied to the shaft is as follows:

$$Q = -\int_A \tau_{xy} z \, dA + \int_A \tau_{zx} y \, dA$$

where A is the cross-sectional area of the shaft. Introducing Eqs. (8.8) into this formula, we have

$$Q = -\int_A \frac{\partial \Phi}{\partial z} z \, dA - \int_A \frac{\partial \Phi}{\partial y} y \, dA \tag{8.13}$$

By integrating the first integral by parts, we can express it as

$$-\int_A \frac{\partial \Phi}{\partial z} z \, dA = -\int_y dy \int_z \frac{\partial \Phi}{\partial z} z \, dz = -\int_y dy \left([z\Phi]_z - \int_z \Phi \, dz \right)$$

$$= -\int_y [z\Phi]_z \, dy + \int_A \Phi \, dA \tag{8.14}$$

As we can see from Fig. 8.1,

$$n = \cos(v, z) = \cos(y, dL) = \frac{dy}{dL}$$

Then, the first integral in Eqs. (8.14) yields

$$-\int_y [z\Phi]_z \, dy = -\int_L z\Phi n \, dL$$

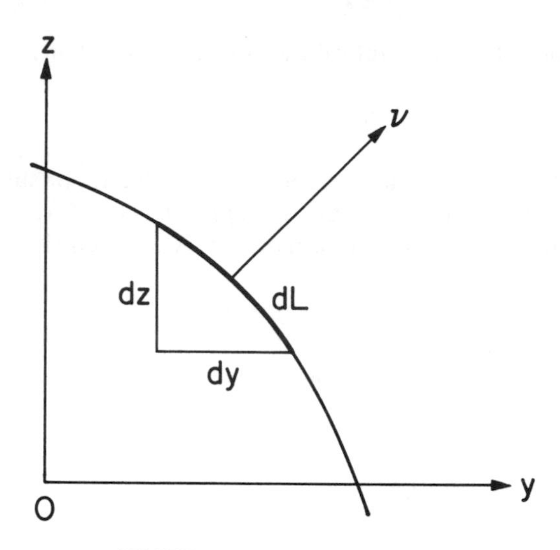

FIGURE 8.1 An element of a curve.

so that

$$-\int_A \frac{\partial \Phi}{\partial z} z\, dA = -\int_L z\Phi n\, dL + \int_A \Phi\, dA$$

A similar formula can be obtained for the second integral in Eq. (8.13),

$$-\int_A \frac{\partial \Phi}{\partial y} y\, dA = -\int_L y\Phi m\, dL + \int_A \Phi\, dA$$

Then, Eq. (8.13) results in the following equation for the torque:

$$Q = -\oint_L \Phi(my + nz)\, dL + 2\int_A \Phi\, dA \tag{8.15}$$

For a simply connected region (see Section 2.12), the line integral in this formula, in accordance with the boundary condition (8.12), is zero. Then, the torque is related to the Prandtl's stress function ϕ by a simple formula,

$$Q = 2\int_A \Phi\, dA \tag{8.16}$$

This relationship is called *Prandtl's formula.*

8.4 PRANDTL'S FORMULA FOR A MULTIPLY CONNECTED REGION

As has been indicated in Section 2.12, a region within which any closed curve can be contracted in a continuous manner into a point without intersecting the boundaries of the region is called a *simply connected region*; otherwise, such region is called a *multiply connected region.* Obviously, a simply connected region does not contain any hollows. A region with one hollow is a *doubly connected region*, a region with two hollows is a *triply connected region*, and so on. In Fig. 8.2, a *quadriconnected region* is shown. It is also obvious that a simply connected region can be obtained from a doubly connected region by means of one cut, from a triply connected region by two cuts, and so on.

Let us generalize Eq. (8.16) for the case of a multiply connected region containing l hollows. Let L_0 be the outer boundary of the region, L_k ($k = 1, 2, \ldots, l$) the kth inner boundary, A_0 the total area restricted by the contour L_0 (including the hollows, i.e., the "gross" area), A_k the area restricted by the contour L_k, Φ_0 the value of Prandtl's function at the outer boundary L_0, and Φ_k the value of Prandtl's function at the kth inner contour L_k. Paying attention to the fact that integration around the inner contours (Fig. 8.2) is carried out in directions opposite to the directions of integration around the outer contour, we present the line integral in Eq. (8.15) as follows:

$$\begin{aligned}
-\oint_L \Phi(my + nz)\, dL &= -\oint_{L_0} \Phi_0(my + nz)\, dL_0 + \sum_{k=1}^{l} \oint_{L_k} \Phi_k(my + nz)\, dL_k \\
&= -\Phi_0 \oint_{L_0} (my + nz)\, dL_0 + \sum_{k=1}^{l} \Phi_k \oint_{L_k} (my + nz)\, dL_k \\
&= -\Phi_0 \oint_{L_0} (z\, dy - y\, dz) + \sum_{k=1}^{l} \Phi_k \oint_{L_k} (z\, dy - y\, dz)
\end{aligned}$$

The line integrals in this expression can be transformed into surface integrals using *Green's formula*:

$$\oint_L [P(y, z)\, dz + Q(y, z)]\, dy = \int_A \left(\frac{\partial Q}{\partial z} - \frac{\partial P}{\partial y} \right) dA \tag{8.17}$$

Assuming $P(y, z) = -y$ and $Q(y, z) = z$, we have

$$-\oint_L \Phi(my + nz)\, dL = -2\Phi_0 A_0 + 2 \sum_{k=1}^{l} \Phi_k A_k$$

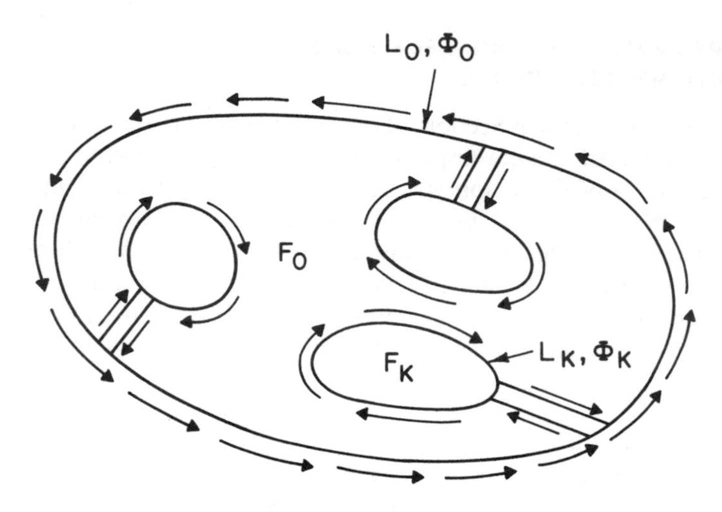

FIGURE 8.2 A hollow cross section with many boundaries (multiply connected region).

Then, Eq. (8.15) can be presented as follows:

$$Q = -2\Phi_0 A_0 + 2 \sum_{k=1}^{l} \Phi_k A_k + 2 \int_A \Phi \, dA \tag{8.18}$$

where A is the cross-sectional area of the shaft, excluding hollows ("net" area):

$$A = A_0 - \sum_{k=1}^{l} A_k$$

Formula (8.18) is the *generalized Prandtl's formula* for the case of a multiply connected region. Note that one of the $l + 1$ constants, $\Phi_0, \Phi_1, \ldots, \Phi_l$, entering this formula can be chosen in an arbitrary manner. In the majority of cases, it is assumed that $\Phi_0 = 0$.

8.5 TORSION OF AN ELLIPTICAL SHAFT

In the case of an elliptical shaft (Fig. 8.3), we assume

$$\Phi(y, z) = C\left(\frac{y^2}{a^2} + \frac{z^2}{b^2} - 1\right) \tag{8.19}$$

Introducing this expression in Eq. (8.10), we find that the constant C is expressed as

$$C = -\alpha G \frac{a^2 b^2}{a^2 + b^2}$$

Introducing Eq. (8.19) into Eqs. (8.8), we have

$$\tau_{xy} = \frac{2C}{b^2} z \qquad \tau_{zx} = -\frac{2C}{a^2} y$$

The total shearing stress can be expressed as

$$\tau_x = -\sqrt{\tau_{xy}^2 + \tau_{yz}^2} = -2C\sqrt{\frac{y^2}{a^4} + \frac{z^2}{b^4}}$$

The total shearing stress reaches its maximum value at the ends of the small axis

$$\tau_{\max} = -\frac{2A}{b} = 2\alpha G \frac{a^2 b}{a^2 + b^2} = \frac{\alpha G a^2 b A}{2I_p} \tag{8.20}$$

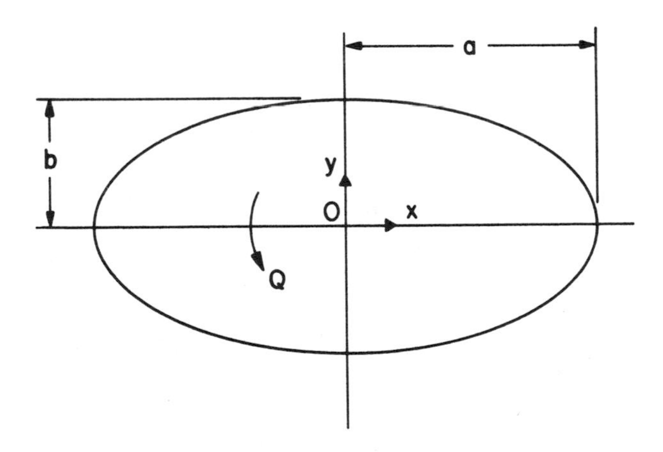

FIGURE 8.3 Elliptical cross section of a prismatic bar subjected to torsion.

Here $A = \pi ab$ is the area of the ellipse, and

$$I_p = \frac{\pi}{4} ab(a^2 + b^2)$$

is its centrifugal moment of inertia.

Formula (8.20) is often applied not only for elliptical cross sections but for other smoothly shaped cross sections as well. Of course, the appropriate A and I_p values have to be used in this case.

The torque is as follows:

$$Q = 2 \int_A \Phi \, dA = 2C \int_{-a}^{a} \int_{-b}^{b} \left(\frac{y^2}{a^2} + \frac{z^2}{b^2} - 1 \right) dy \, dz = -\pi Cab$$

The value

$$SM_t = \frac{Q}{\tau_{\max}} \tag{8.21}$$

is called the *torsion section modulus*. For an elliptical cross section, it is expressed as

$$SM_t = \frac{\pi}{2} ab^2 \tag{8.22}$$

Let us now determine the displacement $u(y, z)$ in the axial direction. From Eqs. (8.9), we have

$$\frac{\partial S}{\partial y} = \frac{1}{\alpha G} \frac{\partial \Phi}{\partial z} + z = -\frac{a^2 - b^2}{a^2 + b^2} z$$

$$\frac{\partial S}{\partial z} = -\frac{1}{\alpha G} \frac{\partial \Phi}{\partial y} - y = -\frac{a^2 - b^2}{a^2 + b^2} y$$

By integrating these formulas, we obtain the following equation for the Saint-Venant's function:

$$S = -\frac{a^2 - b^2}{a^2 + b^2} yz$$

Then, the formula (8.2) yields

$$u = -2\frac{a^2 - b^2}{a^2 + b^2}yz \tag{8.23}$$

In the case of a circular shaft, when $a = b$, the axial displacements u are zero; i.e., the cross sections of such a shaft remain undistorted in their own plane. For a noncircular shaft, the deplanation of the cross sections increases with an increase in the deviation of the cross section from a circular form, i.e., with an increase in the a/b ratio.

8.6 TORSION OF A RECTANGULAR SHAFT

Let a shaft of a rectangular cross section be subjected to a torque Q. We limit our analysis to the case when one side of the cross section is significantly greater than the other (Fig. 8.4). The expression for the Prandtl's function

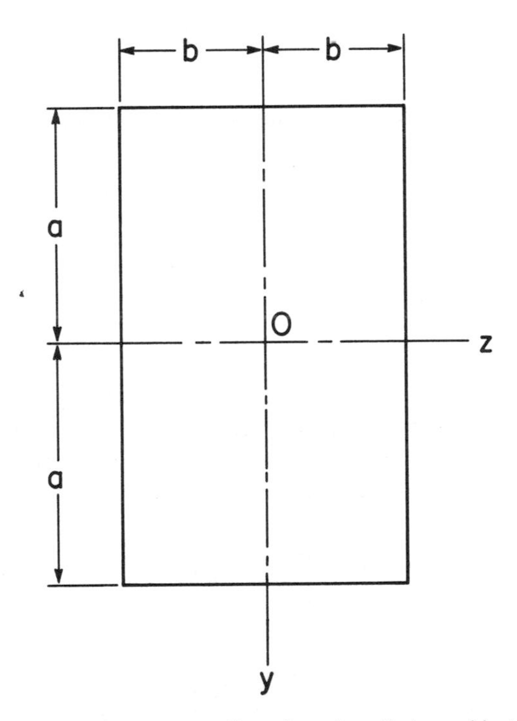

FIGURE 8.4 Rectangular cross section of a prismatic bar subjected to torsion.

we assume in the form

$$\Phi(y, z) = C(z^2 - b^2) \tag{8.24}$$

This expression satisfies the boundary conditions $\Phi(y, \pm b) = 0$ at the long sides, but violates the conditions $\Phi(\pm a, z) = 0$ at the short sides. It is believed, however, that this can be permitted because of the small lengths of the short sides and their insignificant influence on the stress field.

Introducing Eq. (8.24) into Eq. (8.10), we have

$$C = -\alpha G$$

Formulas (8.8) for the stresses yield

$$\tau_{xy} = 2Cz = -2\alpha Gz \qquad \tau_{zx} = 0$$

The maximum shearing stresses act along the long edges and are as follows:

$$\tau_{\max} = \pm 2\alpha Gb$$

Finally, the torque and the centrifugal section modulus are expressed as

$$Q = 2\int_A \phi\, dA = 2C\int_A (z^2 - b^2)\, dA = -\frac{16}{3} Cab^3 = \frac{16}{3}\alpha Gab^3$$

$$SM_t = \frac{Q}{\tau_{\max}} = \frac{8}{3}ab^2$$

The preceding formulas can be used when determining torsion characteristics of thin-walled structures of an "open" profile fabricated of thin rectangular strips ($a_k \gg b_k$). In this case, the following evident generalized formulas for the torque and the shearing stress can be applied:

$$Q = \frac{16}{3}\alpha G \sum_k a_k b_k^3 \tag{8.25}$$

$$\tau_k = 2\alpha G b_k = \frac{3Qb_k}{8\sum_k a_k b_k^3} \tag{8.26}$$

The latter formula results in the following equation for the angle of twist:

$$\alpha = \frac{3Q}{16G \sum_k a_k b_k^3} \tag{8.27}$$

8.7 BREDT'S THEOREM

In this section, we examine Bredt's theorem of the circulation of the shearing stress during torsion.

The line integral

$$\Gamma = \oint \bar{a} \cdot d\bar{L}$$

is called the *circulation* of a vector $\bar{a}$ along the contour L. When the vector $\bar{a}$ is represented by a shearing stress $\bar{\tau}_x$, then the preceding line integral is

$$\Gamma = \oint_L \bar{\tau}_x \cdot d\bar{L} = -\oint_L (\tau_{xy}\, dy + \tau_{zx}\, dz) = -\oint_L \left(\frac{\partial \Phi}{\partial z} dy - \frac{\partial \Phi}{\partial y} dz\right)$$

Let us transform this integral by using Green's formula (8.17). Putting $P(y, z) = -(\partial\Phi/\partial y)$ and $Q(y, z) = -(\partial\Phi/\partial z)$, we have

$$\Gamma = -\int_A \left(\frac{\partial^2 \Phi}{\partial y^2} + \frac{\partial^2 \Phi}{\partial z^2}\right) dy\, dz = -A\Delta\Phi = 2\alpha G A$$

The formula

$$\oint_L \bar{\tau}_x \cdot d\bar{L} = 2\alpha G A \tag{8.28}$$

expresses *Bredt's theorem of the circulation of the shearing stress*: The circulation of the shearing stress along a closed contour which is located within the cross-sectional area and does not intersect its boundaries is equal to the area, restricted by this contour, multiplied by the factor $2\alpha G$. This theorem is true for both simply and multiply connected regions. In the latter case, we can write as many equations of the type (8.28) as there are inner hollows in the cross section. These equations enable us to determine Prandtl's functions ϕ_k at the inner contours. It could be shown that, in effect, this ensures the uniqueness of the determination of the displacements $u(y, z)$.

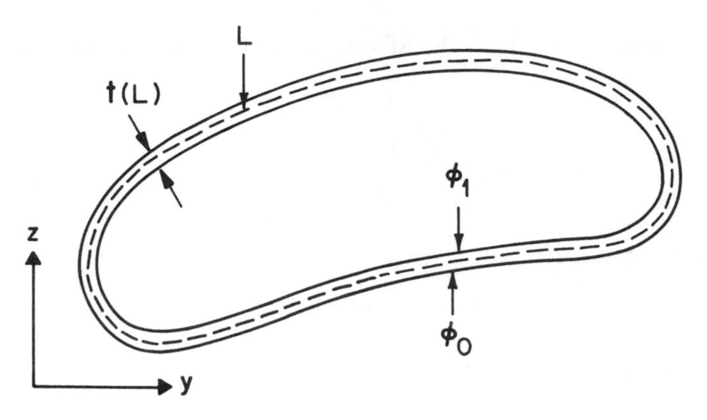

FIGURE 8.5 Closed thin-walled section (doubly connected region) of a prismatic bar subjected to torsion.

8.8 TORSION OF A CLOSED THIN-WALLED SECTION

Examine the torsion of a closed thin-walled section (Fig. 8.5). Because of the thinness of the wall, we may assume that Prandtl's function changes in the through-thickness direction linearly and that, therefore, its mean value at the normal to the contour is

$$\Phi = \frac{\Phi_0 + \Phi_1}{2}$$

Since one of the Φ_k values can be chosen in an arbitrary manner, we put $\Phi_0 = 0$. Then,

$$\Phi = \frac{\Phi_1}{2}$$

For a doubly connected section, Eq. (8.18), where $\Phi_0 = 0$, yields

$$Q = 2\Phi_1 A_1 + 2\int_A \Phi\, dA = 2\Phi_1 A_1 + \Phi_1 A = 2\Phi_1(A_0 - A) + \Phi_1 A$$
$$= \Phi_1(2A_0 - A) \cong 2\Phi_1 A_0 \tag{8.29}$$

The derivative of Prandtl's function Φ along the external normal ν to the contour can be found, considering Eqs. (8.8), as

$$\frac{\partial \Phi}{\partial v} = m\frac{\partial \Phi}{\partial y} + n\frac{\partial \Phi}{\partial z} = -m\tau_{xz} + n\tau_{xy} = -\tau_x \tag{8.30}$$

Because of the thinness of the wall, we may assume

$$\frac{\partial \Phi}{\partial v} \cong \frac{\Phi_0 - \Phi_1}{t(L)} = -\frac{\Phi_1}{t(L)} \tag{8.31}$$

From Eqs. (8.30) and (8.31), we have

$$\tau_x = \frac{\Phi_1}{t(L)} \tag{8.32}$$

After excluding the Φ_1 value from Eqs. (8.29) and (8.32), we obtain the following formula for the shearing stress:

$$\tau_x = \frac{Q}{2A_0 t(L)} \tag{8.33}$$

The maximum shearing stress occurs in the thinnest part of the section:

$$\tau_{\max} = \frac{Q}{2A_0 \tau_{\min}} \tag{8.34}$$

The Φ_1 value can be found by introducing Eq. (8.32) into Eq. (8.28):

$$\Phi_1 = \frac{\alpha C}{2A_0} \tag{8.35}$$

where

$$C = \frac{4GA_0^2}{\oint_L [dL/t(L)]} \tag{8.36}$$

In order to determine the physical meaning of the factor C, we substitute

Eq. (8.35) into Eqs. (8.29). Then, we obtain the following formula for the torque:

$$Q = \alpha C \tag{8.37}$$

Hence, the C value is the *torsional rigidity* of the cross section.

As an example, examine a thin-walled circular tube whose outer diameter is $2R$ and whose thickness t is constant. The torsional rigidity determined by Eq. (8.36) is

$$C = \frac{4GA_0^2 t}{4} = 2\pi G t R^3 \tag{8.38}$$

The twist angle is then

$$\alpha = \frac{Q}{C} = \frac{QL}{4GA_0^2 t} = \frac{Q}{2\pi G t R^3} \tag{8.39}$$

and the shearing stress is

$$\tau = \frac{Q}{2A_0 t} = \alpha G R \tag{8.40}$$

Note that a circular tube has the largest torsional rigidity of all the profiles having the same length of contour because a circle has the maximum area for the given length of the contour.

8.9 TORSION OF A MULTICONNECTED SECTION

The calculation of torsion for a multiconnected section is, in principle, the same as for a doubly connected one. For each of the contours L_k, we should determine the shearing stress by the formula

$$\tau_k = \frac{\Phi_k - \Phi_j}{t(L_k)} \tag{8.41}$$

where Φ_k is the value of Prandtl's function at the contour L_k, and Φ_j is the value of this function at the adjacent contour L_j. Formula (8.41) corresponds to the gradient $\partial\Phi/\partial v$ calculated for the inner normal to the contour L_k.

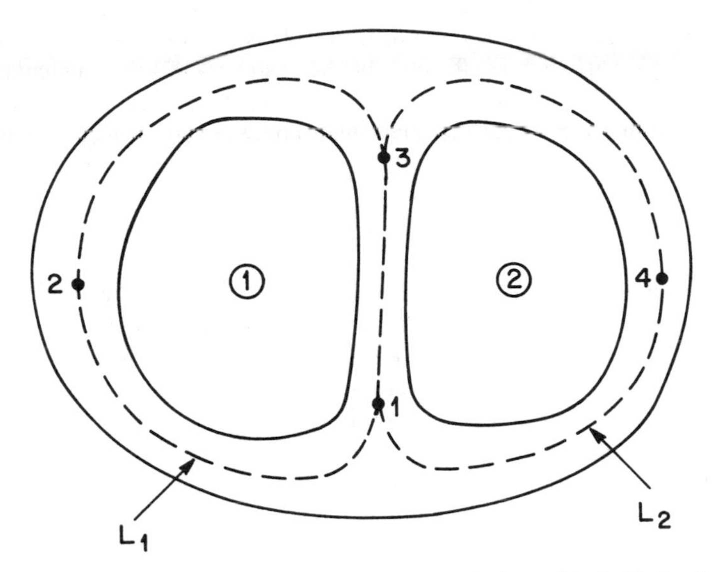

FIGURE 8.6 Triply connected section of a prismatic bar subjected to torsion.

As an example, examine the triply connected section in Fig. 8.6. For this section, we have

$$\left.\begin{aligned} \oint_{L_1} \tau_x \, dL &= \int_{1-2-3} \frac{\Phi_1 \, dL}{t(L)} + \int_{3-1} \frac{\Phi_1 - \Phi_2}{t(L)} \, dL = 2\alpha G A_1 \\ \int_{L_2} \tau_x \, dL &= \int_{3-4-1} \frac{\Phi_2 \, dL}{t(L)} + \int_{1-3} \frac{\Phi_2 - \Phi_1}{t(L)} \, dL = 2\alpha G A_2 \end{aligned}\right\}$$

or

$$\left.\begin{aligned} \Phi_1 \oint_L \frac{dL}{t(L)} - \Phi_2 \int_{3-1} \frac{dL}{t(L)} &= 2\alpha G A_1 \\ \Phi_1 \int_{1-3} \frac{dL}{t(L)} + \Phi_2 \oint_L \frac{dL}{t(L)} &= 2\alpha G A_2 \end{aligned}\right\} \tag{8.42}$$

Since

$$\int_{i-j} \frac{dL}{t(L)} = \int_{j-i} \frac{dL}{t(L)}$$

the system of Eqs. (8.42) is a canonical one. This can be used to ensure that there was no error when coefficients of the system (8.42) were calculated.

The system (8.42) determines all the functions Φ_k through the twist angle α. The equation establishing the relationship between the twist angle and the torque is given by Prandtl's formula,

$$Q = 2(\Phi_1 A_1 + \Phi_2 A_2) \tag{8.43}$$

The obtained relationships can be used to determine the torsional displacements and shearing stresses in various parts of microelectronic equipment, such as walls, bulkheads, covers, and chassis. As is known, insufficient torsional rigidity or shearing strength of these structures can result in excessive twisting of the unit, as well as in its ultimate fatigue or buckling failure. In the second volume, we will apply the theory discussed in this section to the analysis of a chassis subjected to torsion.

9

Fracture Mechanics

9.1 THE GRIFFITH THEORY

The presence of geometric discontinuities, such as holes and abrupt changes in the size of the cross section, can lead to high stress concentrations. Examples are notches in the form of scratches or machining marks and, in the extreme case, cracks. The latter can have disastrous effects on the strength and life of microelectronic and fiber-optic materials and structures.

The original work on which the majority of contemporary fracture mechanics analyses is based is due to A. A. Griffith (1924). Griffith calculated the change in stored energy caused by a through-thickness crack in a thin, biaxially loaded plate. In Griffith's solution, the total change in elastic strain energy per unit thickness is

$$U = \frac{\sigma^2 \pi l^2}{E} \tag{9.1}$$

where l is the half-length of the crack, σ is the uniform stress normal to the crack plane at infinity, and E is Young's modulus.

The structure of Eq. (9.1) can be obtained on the basis of the following simple considerations. When all the stress components but $\sigma_1 = \sigma$ are zero, the strain energy per unit volume is, in accordance with Eqs. (5.2), $W = \sigma^2/2E$. Assuming that the thickness of the plate is equal to unity, we have to multiply this value by the area of the plate to obtain the total strain energy. A crack of length $2l$, oriented perpendicularly to the direction of tension, does not change the stress in most of the plate but, in the vicinity of the crack, reduces the stresses despite the stress concentration at the ends of the crack. The total strain energy of the plate becomes smaller by a certain value U. Since the only linear dimension characterizing the crack is its length $2l$, and the U value must represent a product of the specific energy of the plate by some area, then the energy U should be proportional to the product

Wl^2, i.e., can be presented in the form

$$U = k\frac{\sigma^2}{2E}(2l)^2 = 2k\frac{\sigma^2 l^2}{E}$$

Griffith determined that, in the case of an elliptical crack, $k = \pi/2$.

Since the growth of a crack results in a smaller strain energy, the crack would keep growing without any limit if the formation of free surfaces of a crack did not lead to an increase in the surface tension energy. The latter can be evaluated as

$$U_s = 4lT$$

where T is the surface tension or surface energy per unit area. Griffith argued that instability occurs when a small increment of crack extension, dl, results in a positive value of the derivative dU/dU_s:

$$\frac{dU}{dU_s} > 0$$

Using the preceding relationships for the energies U and U_s, we have

$$\frac{dU}{dU_s} = \frac{dU/dl}{dU_s/dl} = \frac{\sigma^2 \pi l}{2ET} > 0$$

This leads to the well-known Griffith relationship for the critical stress:

$$\sigma_c = \sqrt{\frac{2ET}{\pi l_c}} \tag{9.2}$$

where l_c is the critical value of the half-length of the crack.

Thus, in accordance with Griffith's theory, the brittle strength of materials is due to the presence of microcracks. These propagate if their length is greater than $2l_c$ for the given tensile stress σ_c or when the applied stress is greater than σ_c for the given crack length $2l_c$. Note that Eq. (9.2) is appropriate to plane stress conditions. In the case of a plane strain, the $E/(1 - \nu^2)$ value should be substituted for E.

Formula (9.2) agrees satisfactorily with the experimental data for very brittle materials, such as, for instance, amorphous glass. For polycrystalline materials, however, especially when fracture in such materials is preceded by plastic flow, Griffith's theory should be modified. The appropriate extension of this theory, suggested independently by Irwin and Orowans, resulted in the following modification of Griffith's formula:

$$\sigma_c = \sqrt{\frac{2E\gamma}{\pi l_c}} \tag{9.3}$$

Here $\gamma = T + p$, where p is the work done in plastic deformation in the crack-tip region. This formula was obtained under the assumption that the near-tip plastic zone is small compared to the crack size and plate thickness. Since analysis of the plastic state of stress near the crack tip and the theoretical evaluation of the work against plastic deformation are extremely difficult, the γ value should be treated as a material constant and be determined experimentally. The modified Griffith theory agrees much better with experimental results for plastic materials, such as, for instance, ductile steel.

9.2 IRWIN'S STRESS INTENSITY APPROACH

A series of linear elastic fracture mechanics solutions was developed by Irwin, who showed that the stress field at the tip of a crack is characterized by a singularity of stress that decreases in proportion to the inverse square root of the distance from the crack tip. Irwin also demonstrated that this stress field can be regarded as the sum of three invariant stress patterns taken in

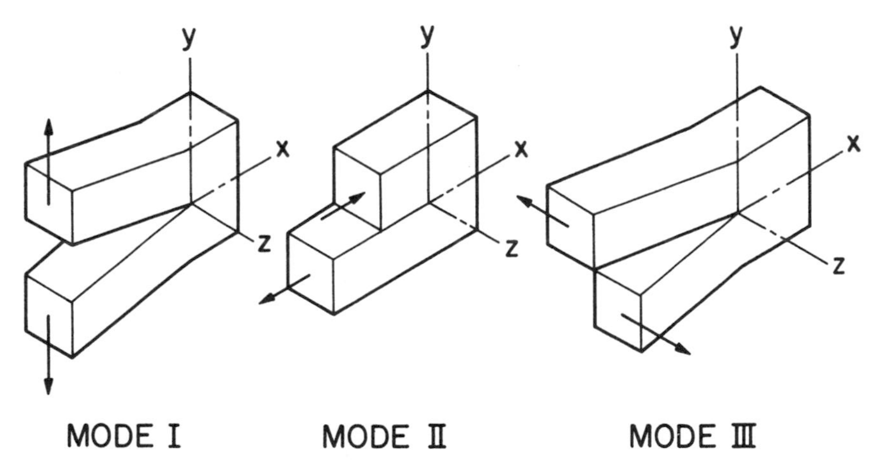

FIGURE 9.1 Modes of fracture: the opening mode (Mode I), the shearing mode (Mode II), and the tearing (antiplane) mode (Mode III).

proportions that depend on loads, dimensions, and shape factors. According to Irwin, these patterns (Fig. 9.1) are due to:

1. Crack surface displacements normal to the crack plane (the *opening mode*, or Mode I)
2. Crack surface displacements in the crack plane and normal to the crack border (the *forward shear mode*, or Mode II)
3. Crack surface displacements in the crack plane and parallel to the crack border (the *parallel* or *antiplane shear mode*, or Mode III)

The singular stress fields for the case of a rectilinear crack in the x–z plane may be characterized by the following formulas (see Fig. 9.2):

Mode I:

$$\left.\begin{aligned}
\sigma_x &\cong \frac{K_{\mathrm{I}}}{\sqrt{2\pi r}} \cos\frac{\theta}{2}\left(1 - \sin\frac{\theta}{2}\sin\frac{3\theta}{2}\right) \\
\sigma_y &= \frac{K_{\mathrm{I}}}{\sqrt{2\pi r}} \cos\frac{\theta}{2}\left(1 + \sin\frac{\theta}{2}\sin\frac{3\theta}{2}\right) \\
\tau_{xy} &= \frac{K_{\mathrm{I}}}{\sqrt{2\pi r}} \cos\frac{\theta}{2}\sin\frac{\theta}{2}\cos\frac{3\theta}{2} \\
\sigma_z &= \nu(\sigma_x + \sigma_y) \text{ for plane strain} \\
\sigma_z &= 0 \text{ for plane stress}
\end{aligned}\right\} \tag{9.4}$$

Mode II:

$$\left.\begin{aligned}
\sigma_x &= \frac{K_{\mathrm{II}}}{\sqrt{2\pi r}} \sin\frac{\theta}{2}\left(2 + \cos\frac{\theta}{2}\cos\frac{3\theta}{2}\right) \\
\sigma_y &= \frac{K_{\mathrm{II}}}{\sqrt{2\pi r}} \sin\frac{\theta}{2}\cos\frac{\theta}{2}\cos\frac{3\theta}{2} \\
\tau_{xy} &= \frac{K_{\mathrm{II}}}{\sqrt{2\pi r}} \cos\frac{\theta}{2}\left(1 - \sin\frac{\theta}{2}\sin\frac{3\theta}{2}\right) \\
\sigma_z &= \nu(\sigma_x + \sigma_y) \text{ for plane strain} \\
\sigma_z &= 0 \text{ for plane stress}
\end{aligned}\right\} \tag{9.5}$$

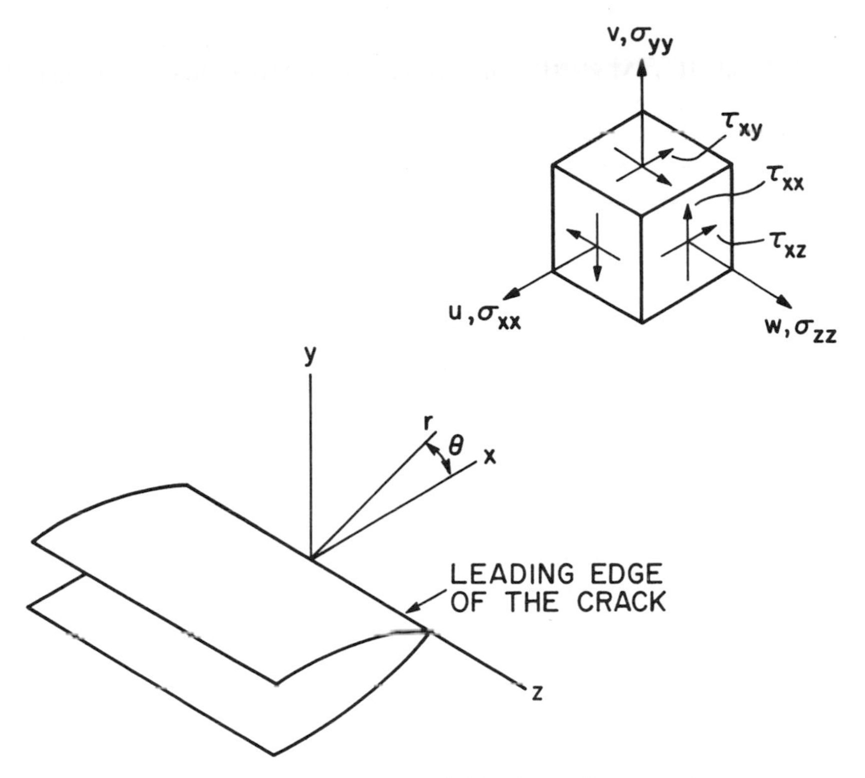

FIGURE 9.2 Singular stress fields of a rectilinear crack.

Mode III:

$$\left.\begin{aligned} &\sigma_x = \sigma_y = \sigma_z = \tau_{xy} = 0 \\ &\tau_{xz} = -\frac{K_{\mathrm{III}}}{\sqrt{2\pi r}} \sin\frac{\theta}{2} \qquad \tau_{yz} = \frac{K_{\mathrm{III}}}{\sqrt{2\pi r}} \cos\frac{\theta}{2} \end{aligned}\right\} \tag{9.6}$$

In these formulas, K_{I}, K_{II}, and K_{III} are *stress intensity factors* for the corresponding types of stress fields. These factors are independent of the coordinates r and θ, as well as of the distribution of the stresses, but depend on the magnitude of the loading forces and the crack size.

For flat tensile *fractures*, commonly regarded as brittle, it is usually assumed that crack extension occurs in a primarily opening mode (Mode I), i.e., $K_{\mathrm{II}} = K_{\mathrm{III}} = 0$. This assumption can be justified for the case of brittle crack extension in the plane of a crack in a flat, infinite plate subjected to remote tensile stresses normal to the crack plane. In practice, fractures under strictly Mode II or Mode III conditions are rare. More often, cracks extend out of their own plane, which generally indicates fracture under mixed mode conditions.

Since K provides a one-parameter measure of the severity of the crack-tip environment, it is logical to characterize resistance to fracture by a critical

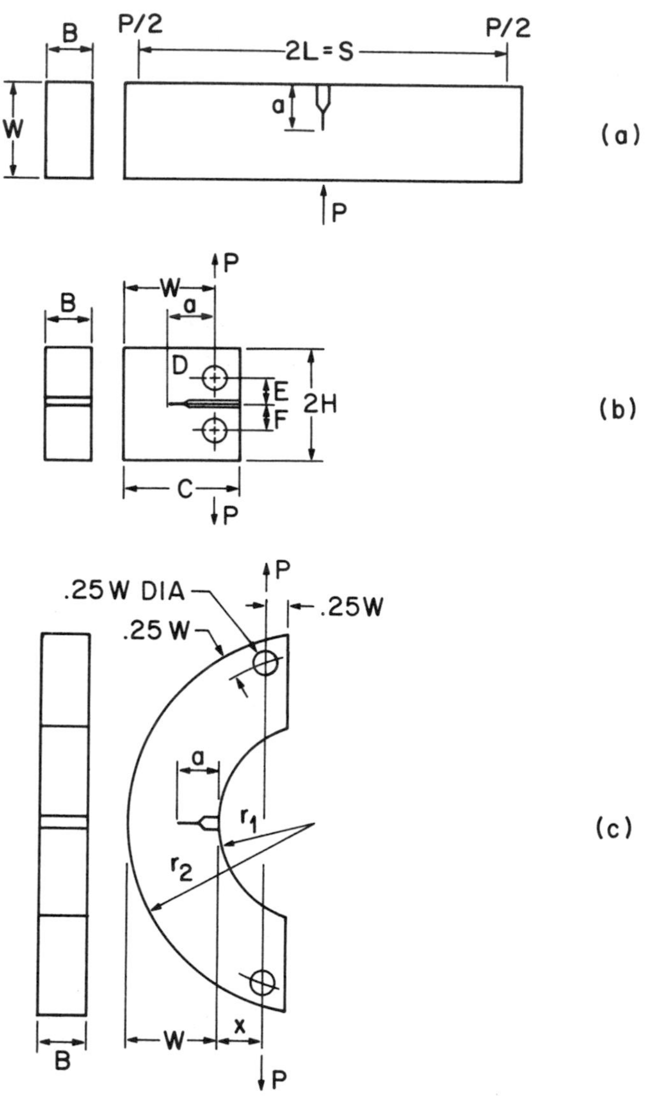

FIGURE 9.3 Specimen designs recommended by ASTM E399-78 standard test method.

value K_c, which can be determined in laboratory tests. Obviously, the laboratory specimen should be a good representation of the structural element of interest.

9.3 K_{IC} TESTS

The K_{IC} value characterizes the resistance of a material to fracture in the presence of a sharp crack under severe tensile constraint. This constraint results in the facts that the state of stress near the crack front approaches tritensile plane strain and the crack-tip plastic region is small compared to the crack size and crack plane dimensions. A K_{IC} value is believed to represent a lower limiting value of fracture toughness. The most recent version of a K_{IC} test method is the ASTM E399-78 standard test method. Various specimen designs currently recommended are shown in Fig. 9.3. All the specimens have single-edge notches that are extended by low-stress-intensity fatigue cracking. The single-edge notch bend specimen is deformed in three-point bending, whereas the compact tension specimen is subjected to two-point loading through pins on a line above the crack tip. C-shaped specimens are intended for tests on portions of hollow cylinders.

9.4 FRACTURE TOUGHNESS

Some unstable fracture behaviors are shown schematically in Fig. 9.4. This figure also demonstrates the relevance of the *linear-elastic fracture mechanics* (LEFM) and *yielding fracture mechanics* (YFM) approaches to the classic ductile-to-brittle fracture transitions that occur with decreasing temperature in common structural steels. The figure shows that LEFM and K_{IC} relate to the predominantly elastic or brittle fractures that occur in the lower shelf region of the fracture transition, whereas YFM applies to the quasibrittle fractures between the upper and lower shelves. Similarly, Fig. 9.4 shows how the ductile/brittle-transition temperature ranges and, therefore, the relevance of the LEFM and YFM approaches, are often affected by section thickness and strain rate.

Truly brittle materials, which fracture in accordance with the original Griffith relationship, are probably limited to such materials as diamonds, and also silica at very low temperatures. In crystalline materials such as metals, what is generally referred to as brittle or quasibrittle fracture often occurs on a definite crystallographic plane, which is known as the *cleavage plane.* In the common structured steels, the ductile/brittle-transition temperatures are a consequence of the change from shear to cleavage fracture. This change in fracture mode is affected by any condition that raises the maximum

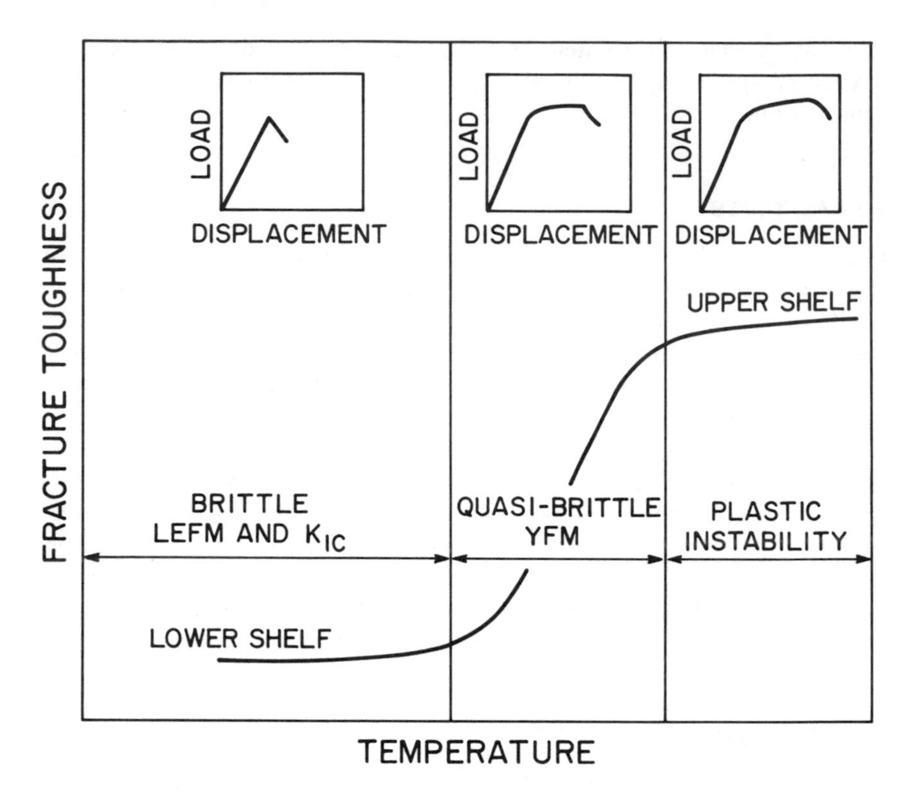

FIGURE 9.4 Unstable fracture behaviors at different temperature conditions.

shearing stress (decrease in temperature, tritensile constraint, strain hardening, high strain rates, etc.).

The stress-intensity approach has been very successful in predicting fracture under predominantly linear-elastic plane strain conditions in many different materials. There are, however, situations in which linear-elastic fracture analysis is invalidated by significant plastic deformation. In these situations, application of yield fracture mechanics is necessary.

10

Plasticity

10.1 UNIAXIAL PLASTIC DEFORMATION

A typical stress-strain curve for a carbon-steel specimen subjected to uniaxial loading within and beyond the elastic range is shown in Fig. 2.5 and is discussed in detail in Section 2.14 in connection with the application of Hooke's law. The linearly elastic range covered by this law terminates at the yield stress σ_Y. Beyond this stress, especially for very plastic materials allowing for very large plastic deformations without exceeding the ultimate strength limit, it is necessary, when evaluating strains, to consider the change in length of the specimen during stressing. Basing deformation on the actual "instantaneous" length of the specimen rather than on its original length, we should define the strain as follows:

$$\varepsilon = \int_{L_0}^{L} \frac{dL}{L} = \ln \frac{L}{L_0}$$

where L is the current length and L_0 the initial length of the specimen. Assuming that the volume of the specimen remains unchanged, we present the "instantaneous" cross-sectional area of the specimen as follows:

$$A = A_0 \frac{L_0}{L} = A_0 e^{-\varepsilon}$$

where A_0 is the original cross-sectional area. Then, the stress experienced by the specimen subjected to the external tensile force P is

$$\sigma = \frac{P}{A} = \sigma_0 e^{\varepsilon}$$

where $\sigma_0 = P/A_0$ is the *nominal (engineering) stress.* As evident from this formula, the "*true stress*" σ can be significantly larger than the nominal stress if the "*true strain*" ε is large enough and results in a substantial reduction in the cross-sectional area of the specimen during plastic deformation. Since even large strains are much smaller than unity, the preceding formula for the stress can be presented as

$$\sigma = \sigma_0(1 + \varepsilon) \cong \sigma_0(1 + \varepsilon_0)$$

where ε_0 is the *nominal (engineering) strain*, which is related to the true strain ε as follows:

$$\varepsilon = \ln(1 + \varepsilon_0)$$

The true stress-strain curve for many plastic materials in the range of *uniform plastic flow* (Fig. 10.1) can be presented in the form of a power law

$$\sigma = K\varepsilon^n \tag{10.1}$$

where the empirical constants K and n are called the *strength coefficient* and the *strain-hardening coefficient*, respectively. The uniform plastic flow

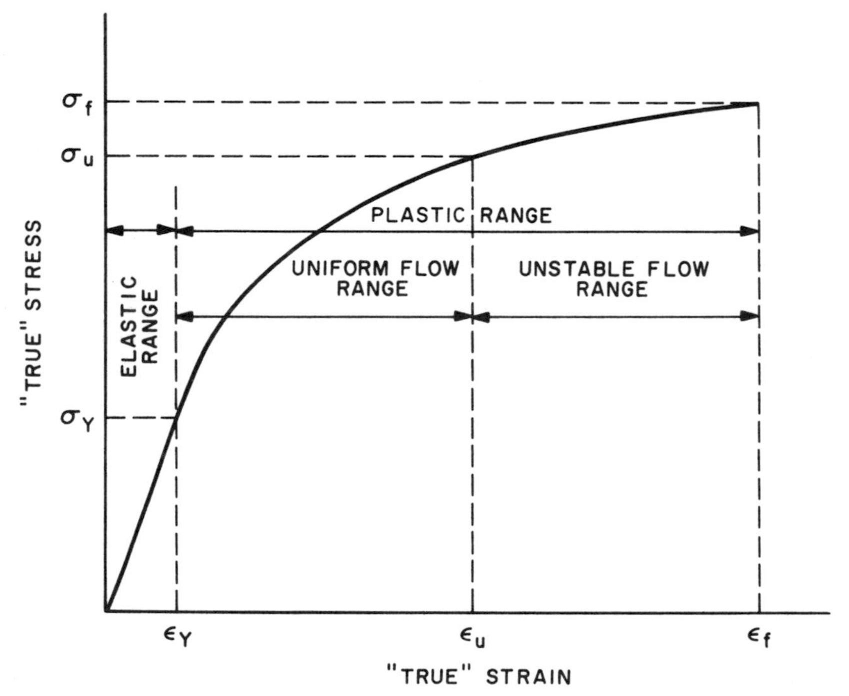

FIGURE 10.1 "True" stress-strain diagram during uniaxial elastoplastic deformation.

starts at the yield stress σ_Y. At the ultimate stress σ_u, *unstable plastic flow* initiates. At this stress, there is a balance between the effect of strain hardening, which tends to increase the strength of the material, and the decreasing cross-sectional area of the specimen, which tends to weaken the material. When unstable plastic flow occurs, the rate of increase of load-carrying capacity due to strain hardening is less than the rate of decrease in the load-carrying capacity due to the decreased cross-sectional area.

The load

$$P = \sigma A = A_0 \sigma e^{-\varepsilon}$$

is a function of both the true stress σ and true strain ε. Its differential is

$$dP = \frac{\partial P}{\partial \sigma} d\sigma + \frac{\partial P}{\partial \varepsilon} d\varepsilon = A_0 e^{-\varepsilon}(d\sigma - \sigma\, d\varepsilon)$$

At the instant of instability, $dP = 0$ and, therefore, this instability can be defined by the relation

$$\sigma = \frac{d\sigma}{d\varepsilon} \tag{10.2}$$

On the other hand, from Eq. (10.1), we have

$$\frac{d\sigma}{d\varepsilon} = Kn\varepsilon^{n-1} = K\varepsilon^n \frac{n}{\varepsilon} = \sigma \frac{n}{\varepsilon}$$

Comparing this formula with Eq. (10.2), we conclude that, at the instant of the instability of flow, the strain-hardening coefficient n is equal to the true strain ε:

$$n = \varepsilon$$

10.2 MULTIAXIAL PLASTIC DEFORMATION: SMALL ELASTOPLASTIC STRAINS

The theory of small elastoplastic multiaxial strains establishes the relationships between the stresses and strains, assuming small plastic strains. This theory is based on the following experimentally proven hypotheses:

1. Yielding is practically unaffected by a uniform all-round ("hydrostatic") tension or compression.

2. The components of the strain deviator are proportional to the corresponding components of the stress deviator.
3. The stress intensity is a function of the strain intensity that is independent of the type of the state of stress.

The first hypothesis (Bridgeman, 1944; Crossland, 1954) states that the relationship (2.57), obtained for elastic strains, is true in the case of elastoplastic strains as well. Formula (2.28) for the octahedral normal stress σ_0 and the first formula in Eqs. (2.19) for the linear stress invariant σ_I yield

$$\sigma_0 = \tfrac{1}{3}\sigma_I$$

Then, using relationship (2.57), we conclude that the octahedral normal stress σ_0 is related to the dilatation θ as

$$\sigma_0 = 3K\theta \tag{10.3}$$

where

$$K = \frac{E}{1 - 2\nu}$$

is the bulk (compressibility) modulus (see Section 2.15).

The second hypothesis states that

$$\frac{\varepsilon_x - \theta}{\sigma_x - \sigma_0} = \frac{\varepsilon_y - \theta}{\sigma_y - \sigma_0} = \frac{\varepsilon_z - \theta}{\sigma_z - \sigma_0} = \frac{\gamma_{xy}}{\tau_{xy}} = \frac{\gamma_{yz}}{\tau_{yz}} = \frac{\gamma_{zx}}{\tau_{zx}} = \Psi \tag{10.4}$$

where Ψ is constant. For elastic strains, as evident from Eqs. (2.52),

$$\Psi = \frac{1}{2G} \tag{10.5}$$

where G is the shear modulus.

In the third hypothesis, the *effective stress* σ_i is assumed equal to the critical stress σ_c in the Huber–von Mises–Hencky theory, as given by the formula (5.7):

$$\sigma_i = \sqrt{\tfrac{1}{2}[(\sigma_1 - \sigma_2)^2 + (\sigma_2 - \sigma_3)^2 + (\sigma_3 - \sigma_1)^2]} \tag{10.6}$$

Similarly, the *effective strain* ε_i is defined as

$$\varepsilon_i = \frac{1}{3}\sqrt{2[(\varepsilon_1 - \varepsilon_2)^2 + (\varepsilon_2 - \varepsilon_3)^2 + (\varepsilon_3 - \varepsilon_1)^2]} \tag{10.7}$$

The hypothesis states that the relationship

$$\sigma_i = f(\varepsilon_i) \tag{10.8}$$

is independent of the state of stress and, therefore, can be obtained from simple tension tests. For a uniaxial elastic state of stress,

$$\sigma_1 = \sigma \qquad \sigma_2 + \sigma_3 = 0 \qquad \varepsilon_1 = \varepsilon \qquad \varepsilon_2 = \varepsilon_3 = -\nu\frac{\sigma}{E} = -\nu\varepsilon$$

Then Eqs. (10.6) and (10.7) yield

$$\sigma_i = \sigma \qquad \varepsilon_i = \frac{2(1+\nu)}{3}\varepsilon = \frac{2(1+\nu)}{3E}\sigma_i = \frac{\sigma_i}{3G}$$

so that the relationship (10.8) in the elastic region is

$$\sigma_i = 3G\varepsilon_i \tag{10.9}$$

It is natural to present the relationship (10.6) in the elastoplastic region by the formula

$$\sigma_i = 3G\varepsilon_i[1 - \omega(\varepsilon_i)] \tag{10.10}$$

where the function $\omega(\varepsilon_i)$ considers the nonlinearity of the stress-strain diagram (Fig. 10.2). Obviously,

$$0 \le \omega(\varepsilon_i) \le 1 \tag{10.11}$$

If the strain is brought to point C in Fig. 10.2, the function ω can be presented as the ratio $\omega = AB/BC$. During unloading, the relationship between σ_i and ε_i becomes linear again and is determined by the straight line BD, which is parallel to OA.

In order to obtain the relationships between the stresses and strains beyond the elastic region, assuming sufficiently small plastic strain, we

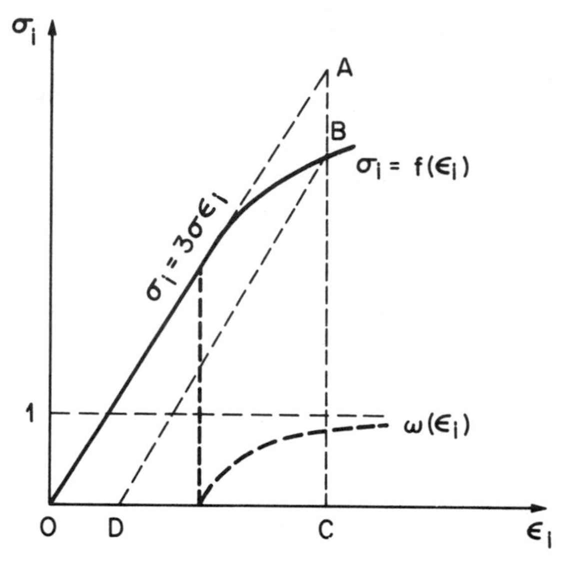

FIGURE 10.2 "Effective" stress-strain diagram during elastoplastic deformation. The function $\omega(\varepsilon_i)$ considers the nonlinearity of this diagram.

use the relationship (10.5) which, considering Eq. (10.9), can be presented as

$$\Psi = \frac{3}{2}\frac{\varepsilon_i}{\sigma_i} \tag{10.12}$$

We assume that this formula can be utilized not only within the elastic region but in the elastoplastic region as well. Then, Eqs. (10.4) yield

$$\left.\begin{aligned} \sigma_x &= \sigma_0 + \frac{2\sigma_i}{3\varepsilon_i}(\varepsilon_x - \theta) \qquad & \tau_{xy} &= \frac{\sigma_i}{6\varepsilon_i}\gamma_{xy} \\ \sigma_y &= \sigma_0 + \frac{2\sigma_i}{3\varepsilon_i}(\varepsilon_y - \theta) \qquad & \tau_{yz} &= \frac{\sigma_i}{6\varepsilon_i}\gamma_{yz} \\ \sigma_z &= \sigma_0 + \frac{2\sigma_i}{3\varepsilon_i}(\varepsilon_z - \theta) \qquad & \tau_{zx} &= \frac{\sigma_i}{6\varepsilon_i}\gamma_{zx} \end{aligned}\right\} \tag{10.13}$$

These relationships were obtained by A. A. Ilyushin. Note that they are also true when plastic deformation is accompanied by thermal strains.

Let us break down the total elastoplastic strain into elastic and plastic components. From Eqs. (10.10) and (10.12), we have

$$\Psi = \frac{1}{2G(1-\omega)} = \frac{1}{2G} + \phi \tag{10.14}$$

where the function $\phi(\varepsilon_i)$ is related to the function $\omega(\varepsilon_i)$ by the formula

$$\phi = \frac{\omega}{2G(1-\omega)} \tag{10.15}$$

Then, Eqs. (10.4) can be presented as

$$\left.\begin{array}{lll} \varepsilon_x = \varepsilon_{xe} + \varepsilon_{xp} & \varepsilon_y = \varepsilon_{ye} + \varepsilon_{yp} & \varepsilon_z = \varepsilon_{ze} + \varepsilon_{zp} \\ \gamma_{xy} = \gamma_{xye} + \gamma_{xyp} & \gamma_{yz} = \gamma_{yze} + \gamma_{yzp} & \gamma_{zx} = \gamma_{zxe} + \gamma_{zxp} \end{array}\right\} \tag{10.16}$$

where

$$\left.\begin{array}{l} \varepsilon_{xe} = \dfrac{1}{2G}(\sigma_x - \sigma_0) + \theta = \dfrac{1}{E}[\sigma_x - \nu(\sigma_y + \sigma_z)] \\ \cdots\cdots\cdots\cdots\cdots\cdots\cdots\cdots\cdots\cdots\cdots\cdots \\ \gamma_{zxe} = \dfrac{1}{2G}\tau_{zx} \end{array}\right\} \tag{10.17}$$

are elastic components, and

$$\left.\begin{array}{lll} \varepsilon_{xp} = \phi(\sigma_x - \sigma_0) & \varepsilon_{yp} = \phi(\sigma_y - \sigma_0) & \varepsilon_{zp} = \phi(\sigma_z - \sigma_0) \\ \gamma_{xyp} = \gamma\tau_{xy} & \gamma_{yzp} = \phi\tau_{yz} & \gamma_{zxp} = \phi\tau_{zx} \end{array}\right\} \tag{10.18}$$

are plastic components of the total strain.

From Eqs. (10.14) and (10.15), we have

$$\phi = \frac{3\varepsilon_i}{2\sigma_i} - \frac{1}{2G} \tag{10.19}$$

Assuming that the effective total, elastic and plastic strains have the same additive properties as the strains themselves, we present the effective strain as

$$\varepsilon_i = \varepsilon_{ie} + \varepsilon_{ip} \tag{10.20}$$

Then, Eq. (10.19) yields

$$\phi = \frac{3\varepsilon_{ip}}{2\sigma_i} \tag{10.21}$$

where the effective plastic strain ε_{ip} is expressed by the formula of the type (10.7):

$$\varepsilon_{ip} = \tfrac{1}{3}\sqrt{2[(\varepsilon_{ip} - \varepsilon_{2p})^2 + (\varepsilon_{2p} - \varepsilon_{3p})^2 + (\varepsilon_{3p} - \varepsilon_{1p})^2]} \tag{10.22}$$

The elastic strains during loading and the total strains during unloading are determined by Hooke's law. Since the decrease in the effective stress during unloading is directly proportional to the decrease in the effective strain, then the stresses and strains during unloading can be obtained on the basis of an elastic solution, assuming that the external forces acting on the body are equal to the difference between the loading forces and the forces remaining after unloading. In the case of a complete unloading, the latter are zero, so that the elastic problem should be solved for external forces loading the body. This situation is known as *Ilyushin's theorem on unloading*. The residual stresses and strains can be determined as differences between the stresses (strains) occurring during loading, and the decrease in the stresses (strains) during unloading.

10.3 ELASTOPLASTIC BENDING OF BEAMS

In the case of an elastoplastic bending of a beam, we can use the stress-strain diagram directly and the following relationship:

$$\sigma_x = E\varepsilon_x[1 - \omega(\varepsilon_x)] \tag{10.23}$$

We assume that the beam has a rectangular cross section of height h and width b and that it is subjected to the combined action of bending moments and axial tensile forces.

The *ultimate bending moment* M_Y and the *ultimate axial force* N_Y are defined by the formulas

$$M_Y = \sigma_Y \frac{bh^2}{6} \qquad N_Y = \sigma_Y bh \tag{10.24}$$

where σ_Y is the normal yield stress, $SM = bh^2/6$ the section modulus of the beam cross section, and A the cross-sectional area.

When determining the distribution of stresses in the elastoplastic region, we examine the following three major cases (Figs. 10.3a–c).

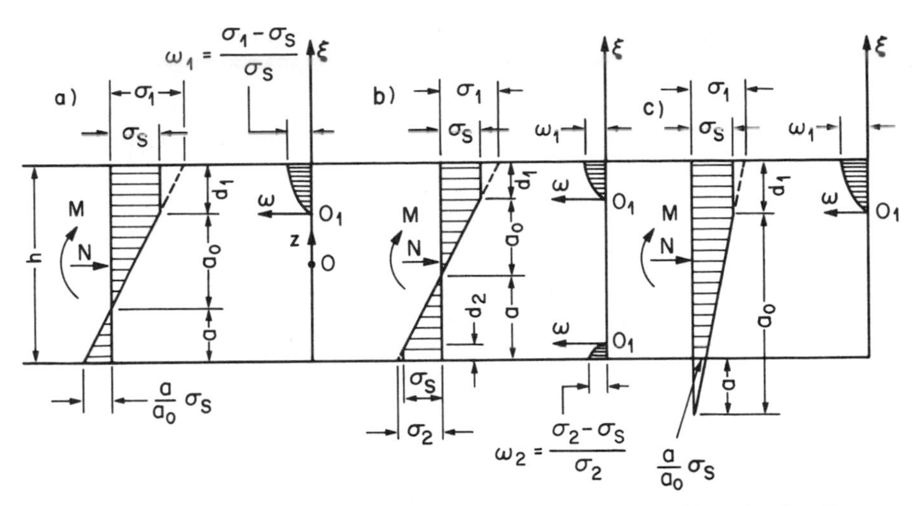

FIGURE 10.3 Distribution of elastoplastic stresses in a beam subjected to bending.

In the case shown in Fig. 10.3a, the stresses reach the yield stress σ_y only at one part of the cross section whereas, at the other part, the stresses remain elastic and have an opposite sign. In order to determine the location of the neutral axis and the size of the elastic zone, we use the following equilibrium equations:

$$\left.\begin{aligned} &N - \sigma_Y b(h - a - a_0) - \tfrac{1}{2}\sigma_Y b\left(1 + \frac{1}{a_0}\right)(a_0 - a) = 0 \\ &M + \tfrac{1}{2}Nh - \sigma_Y b(h - a - a_0)\left(h - \frac{h - a - a_0}{2}\right) - \tfrac{1}{2}\sigma_Y b a_0(a + \tfrac{2}{3}a_0) \\ &\quad + \tfrac{1}{6}\sigma_Y b\,\frac{a^3}{a_0} = 0 \end{aligned}\right\}$$

and therefore

$$\left.\begin{aligned} a_0 &= \tfrac{9}{32}h\,\frac{(2 - (M/M_Y) - 2(N/N_Y))^2}{(1 - (N/N_Y))^3} \\ a &= \tfrac{3}{4}h\,\frac{2 - (M/M_Y) - 2(N/N_Y)}{1 - (N/N_Y)}\left[1 - \frac{3}{8}\,\frac{2 - (M/M_Y) - 2(N/N_Y)}{(1 - (N/N_Y))^2}\right] \end{aligned}\right\} \quad (10.25)$$

In the case shown in Fig. 10.3b, the stresses reach the yield stress σ_y both in the upper and lower parts of the cross section and, in addition, have an elastic region $2a_0$ high. The equations of equilibrium, in this case, are

$$\left.\begin{aligned} N - \sigma_Y b(h - 2a) = 0 \\ M - \sigma_Y b(h - 2a)[\tfrac{1}{2}h - \tfrac{1}{2}(h - 2a)] - \sigma_Y ba^2 + \tfrac{1}{3}\sigma_Y ba_0^2 = 0 \end{aligned}\right\}$$

and, therefore,

$$\left.\begin{aligned} a_0 &= \frac{\sqrt{3}}{2}h\sqrt{1 - \left(\frac{N}{N_Y}\right)^2 - \frac{M}{M_Y}} \\ a &= \tfrac{1}{2}h\left(1 - \frac{N}{N_Y}\right) \end{aligned}\right\} \tag{10.26}$$

Finally, in the case shown in Fig. 10.3c, the stresses at one part of the cross section reach their ultimate value σ_Y and are elastic at the other part. The sign of the stresses is the same for both parts, and the equilibrium conditions are as follows:

$$\left.\begin{aligned} N - \sigma_Y bh + \tfrac{1}{2}\sigma_Y b\frac{(a_0 - a)^2}{a_0} = 0 \\ M - \tfrac{1}{2}Nh - \tfrac{1}{2}\sigma_Y bh^2 + \tfrac{1}{6}\sigma_Y b\frac{(a_0 - a)^3}{a_0} = 0 \end{aligned}\right\}$$

Then,

$$\left.\begin{aligned} a_0 &= \tfrac{9}{32}h\frac{(2 - (M/M_Y) - 2(N/N_Y))^2}{(1 - (N/N_Y))^3} \\ a &= \tfrac{3}{4}h\left[\frac{3}{8}\frac{2 - (M/M_Y) - 2(N/N_Y)}{(1 - (N/N_Y))^2} - 1\right] \end{aligned}\right\} \tag{10.27}$$

In Fig. 10.3, the function ω is plotted for all three typical cases. In order to obtain the relationships for the function ω, we use the stress-strain diagram in Fig. 10.4. This diagram can be obtained from the diagram in Fig. 10.2, in which the strain beyond the elastic region is *ideally plastic*. Then, we have

$$\omega(\varepsilon) = \frac{AB}{BC} = \frac{AC - BC}{BC} = \frac{\sigma - \sigma_Y}{\sigma_Y} = \frac{\varepsilon - \varepsilon_e}{\varepsilon} \tag{10.28}$$

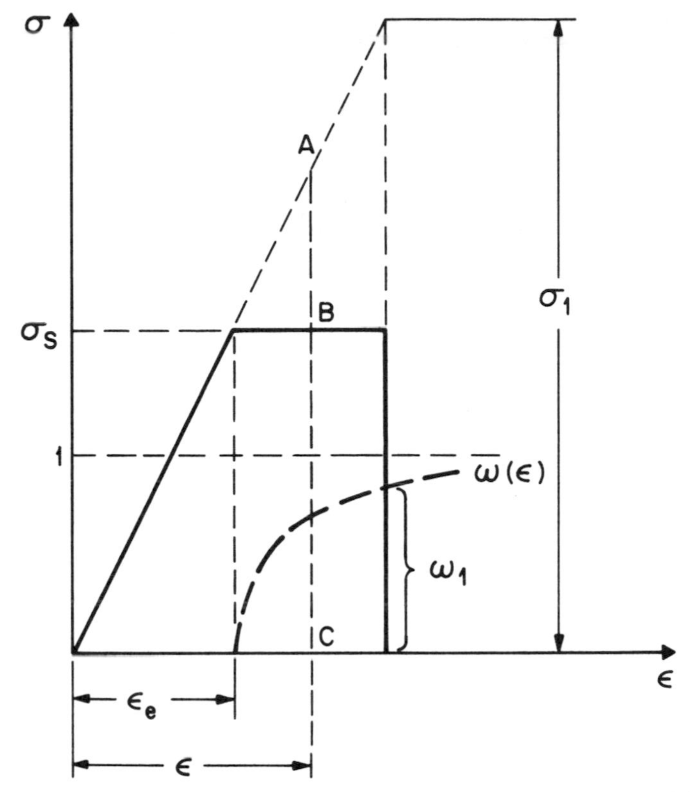

FIGURE 10.4 Evaluation of the function $\omega(\varepsilon)$ accounting for the nonlinearity of the stress-strain diagram.

for $\varepsilon \geq \varepsilon_e$. Obviously, $\omega(\varepsilon) = 0$ within the elastic region ($\varepsilon \leq \varepsilon_e$). From Eqs. (10.28), using notations in Fig. 10.3a, we have

$$\omega = \frac{\xi}{\xi + a_0} \tag{10.29}$$

The condition

$$\omega(d_1) = \omega_1 = \frac{d_1}{d_1 + a_0}$$

yields

$$a_0 = d_1\left(\frac{1}{\omega_1} - 1\right)$$

After introducing this formula in Eq. (10.29), we obtain

$$\omega(\xi) = \frac{\xi}{\xi + d_1((1/\omega_1) - 1)} \tag{10.30}$$

Similarly, one can obtain the expressions for $\omega(\varepsilon)$ in two other cases.

The longitudinal strain of the layers located at distance z from the elastic center of gravity of the cross section is

$$\varepsilon_x = \frac{\partial u}{\partial x} + z\frac{\partial^2 w}{\partial x^2}$$

and the relationship (10.23) yields

$$\sigma_x = E(1-\omega)\left(\frac{\partial u}{\partial x} + z\frac{\partial^2 w}{\partial x^2}\right) \tag{10.31}$$

Substituting this equation into the formulas

$$N = \int_A \sigma_x \, dA \qquad M = \int_A \sigma_x z \, dA$$

for the axial force and the bending moment, we have

$$\left.\begin{aligned} N &= E\int_A \left(\frac{\partial u}{\partial x} + z\frac{\partial^2 w}{\partial x^2}\right) dA - N_\omega = EA\frac{\partial u}{\partial x} - N_\omega \\ M &= E\int_A \left(\frac{\partial u}{\partial x} + z\frac{\partial^2 w}{\partial x^2}\right) z \, dA - M_\omega = EI\frac{\partial^2 w}{\partial x^2} - M_\omega \end{aligned}\right\} \tag{10.32}$$

where the components

$$\left.\begin{aligned} N_\omega &= E\int_A \left(\frac{\partial u}{\partial x} + z\frac{\partial^2 w}{\partial x^2}\right) \omega \, dA = EA_\omega\frac{\partial u}{\partial x} + ES_\omega\frac{\partial^2 w}{\partial x^2} \\ M_\omega &= E\int_A \left(\frac{\partial u}{\partial x} + z\frac{\partial^2 w}{\partial x^2}\right) z\omega \, dA = ES_\omega\frac{\partial u}{\partial x} + EI_\omega\frac{\partial^2 w}{\partial x^2} \end{aligned}\right\} \tag{10.33}$$

consider the effect of the plastic strain. In the preceding formulas, we took into account that

$$\int_A z\,dA = 0$$

and the following notations are used:

$$I = \int_A z^2\,dA \qquad F_\omega = \int_A \omega\,dA \qquad S_\omega = \int_A z\omega\,dA \qquad I_\omega = \int_A z^2\omega\,dA \tag{10.34}$$

From Eq. (10.32), we have

$$\frac{\partial u}{\partial x} = \frac{N + N_\omega}{EA} \qquad \frac{\partial^2 w}{\partial x^2} = \frac{M + M_\omega}{EI} \tag{10.35}$$

Let us transform the formulas (10.34) to a more convenient form. We have

$$F_\omega = \int_A \omega\,dA = b\int_{-h/2}^{h/2} \omega(z)\,dz = b\int_0^{d_1} \omega(\xi)\,d\xi$$

or, with consideration of Eq. (10.30),

$$F_\omega = bd_1 A_\omega \tag{10.36}$$

where

$$A_\omega = 1 + \left(\frac{1}{\omega_1} - 1\right)\ln(1 - \omega_1)$$

Similarly,

$$S_\omega = \int_A z\omega\,dA = b\int_{-h/2}^{h/2} z\omega(z)\,dz = b\int_0^{d_1}\left(\frac{h}{2} - d_1 + \xi\right)\omega(\xi)\,d\xi$$

$$I_\omega = \int_A z^2\omega\,dA = b\int_{-h/2}^{h/2} z^2\omega(z)\,dz = b\int_0^{d_1}\left(\frac{h}{2} - d_1 + \xi\right)^2\omega(\xi)\,d\xi$$

Introducing Eq. (10.30) in the latter formulas, we have

$$\left.\begin{aligned} S_\omega &= bd_1^2 B_\omega + \left(\frac{h}{2} - d_1\right) bd_1 A_\omega \\ I_\omega &= bd_1^3 C_\omega + 2\left(\frac{h}{2} - d_1\right) bd_1^2 B_\omega + \left(\frac{h}{2} - d\right)^2 bd_1 A_\omega \end{aligned}\right\} \tag{10.37}$$

where

$$B_\omega = \frac{3}{2} - \frac{1}{\omega_1} - \left(\frac{1}{\omega_1} - 1\right)^2 \ln(1 - \omega_1)$$

$$C_\omega = \frac{11}{6} - \frac{5}{2\omega_1} + \frac{1}{\omega_1^2} + \left(\frac{1}{\omega_1} - 1\right)^3 \ln(1 - \omega_1)$$

The calculated values of the factors A_ω, B_ω, and C_ω are given in the following table:

ω	0	0.05	0.10	0.20	0.30	0.40	0.50	0.60	0.70	0.80	0.90
A_ω	0	0.0259	0.0517	0.1073	0.1678	0.2219	0.3068	0.3892	0.4840	0.5976	0.7442
B_ω	0	0.0025	0.035	0.0707	0.1083	0.1494	0.1932	0.2405	0.2926	0.3506	0.4173
C_ω	0	0.00127	0.0255	0.0521	0.0801	0.1091	0.1401	0.1729	0.2079	0.2456	0.2870

Formulas (10.36) and (10.37) can also be used in the cases shown in Figs. 10.3b and 10.3c. If the plastic zones occur on both sides of the neutral axis, then the characteristics F_ω, S_ω, and I_ω should be calculated by the formulas

$$F_\omega = F_{\omega_1} + F_{\omega_2} \qquad S_\omega = S_{\omega_1} - S_{\omega_2} \qquad I_\omega = I_{\omega_1} + I_{\omega_2} \tag{10.38}$$

We can apply, generally speaking, two approaches when using the preceding relationships for the elastoplastic bending response of a beam:

1. The dimensions of the beam cross sections remain unchanged, and no plastic strains occur. However, the longitudinal forces and bending moments in the cross sections are due to "fictitious" loading,

$$N_f = N + N_\omega \qquad M_f = M + M_\omega$$

2. For the actual loading, assume a "fictitious" cross-sectional area, changing along the beam.

If the first approach is used, the force N_f and the moment M_f can be evaluated, with consideration of Eqs. (10.33), by the formulas

$$N_f = \frac{N(1 - i) + \alpha M}{1 - \mu} \qquad M_f = \frac{M(1 - f) + \beta N}{1 - \mu} \tag{10.39}$$

where

$$\alpha = \frac{S_\omega}{I} \qquad \beta = \frac{S_\omega}{A} \qquad i = \frac{I_\omega}{I} \qquad f = \frac{A_\omega}{A} \qquad \mu = i + f - if + \alpha\beta \tag{10.40}$$

Since the magnitudes of the "fictitious" loads N_f and M_f depend on the function ω, and this function, in its turn, is determined by the loads N_f and M_f, then the solution to the problem has to be obtained by iterations. This can be done using the following sequence:

1. Assuming elastic behavior of the beam, determine stresses for several cross sections, the sizes of the plastic zones in each cross section, the values ω_1 and ω_2, and find the characteristics of the cross sections using formulas (10.36) and (10.37).

   With the calculated characteristics of the cross sections, evaluate the additional axial forces and bending moments. Then, examine again an elastic beam of the same cross section as the given one, but with an unlimited elasticity. The longitudinal distribution of bending moments for such a beam can be evaluated based on the actual loading and additional bending moments applied in the middle of the initially chosen segments of the beam.

   Using such a distribution of bending moments, determine the deflections of the beam in the elastoplastic region.
2. In the second approximation, proceeding from the determined deflections, evaluate the bending moments caused by the given longitudinal and lateral loading. Then, find the new distribution of the stresses, evaluate the characteristics of the cross sections and the "fictitious" additional forces and moments, and examine an ideally elastic beam with cross sections of the same dimensions and with the same elastic characteristics as the actual beam. Then, find the deflections and the rotation angles that correspond to the elastoplastic displacements in the second approximation. The iteration process can be continued until the required accuracy is achieved.

The preceding theory can be applied, for instance, to evaluating the bearing capacity (ultimate strength) of those structural elements of electronic components that can be idealized as beams clamped at both ends, whose axial displacements are restricted.

10.4 THEORY OF PLASTIC FLOW

The theory of plastic flow treats plastic deformation as a time-dependent process, determines the relationship between the small changes in the stresses and strains, and obtains the equations for strain rates as functions of the stress magnitudes. The following major hypotheses underlie this theory:

1. Yielding is unaffected by "hydrostatic" tension or compression and, therefore, the octahedral normal stress is directly proportional to dilatation as it takes place in the theory of elasticity.
2. The components of the increments of the deviatoric plastic strains are proportional to the components of the deviatoric stresses.
3. The stress intensity is a function of an integral of the intensity of the increments of the deviatoric plastic strains, which are independent of the type of the state of stress.

The first hypothesis is the same as in the theory of small elastoplastic strains examined previously. Differentiating Eq. (10.3), we have

$$d\theta = \frac{d\sigma_0}{3K} \tag{10.41}$$

The increment of the dilatation consists of elastic $d\theta_e$ and plastic $d\theta_p$ parts:

$$d\theta = d\theta_e + d\theta_p$$

Since the elastic component follows Hooke's law, then

$$d\theta_e = \frac{d\sigma_0}{3K} \tag{10.42}$$

If we can see, from Eqs. (10.41) and (10.42), that the plastic component of the dilatation increment is zero,

$$d\theta_p = 0 \tag{10.43}$$

Hence, plastic deformation does not result in any change in the body volume and, therefore, the tensor of the plastic strain increments is a deviator (see Section 2.10).

The second hypothesis states that

$$\frac{d\varepsilon_{xp}}{\sigma_x - \sigma_0} = \frac{d\varepsilon_{yp}}{\sigma_y - \sigma_0} = \frac{d\varepsilon_{zp}}{\sigma_z - \sigma_0} = \frac{d\gamma_{xyp}}{\tau_{xy}} = \frac{d\gamma_{yzp}}{\tau_{yz}} = \frac{d\gamma_{zxp}}{\tau_{zx}} = d\lambda \quad (10.44)$$

Then, we have

$$d\varepsilon_{xp} = (\sigma_x - \sigma_0)\, d\lambda, \ldots, d\gamma_{zxp} = \tau_{zx}\, d\lambda \quad (10.45)$$

The third hypothesis can be mathematically presented as follows:

$$\sigma_i = \phi\left(\int \overline{d\varepsilon_{ip}}\right) \quad (10.46)$$

where

$$\overline{d\varepsilon_{ip}} = \tfrac{1}{3}\sqrt{2[(d\varepsilon_{1p} - d\varepsilon_{2p})^2 + (d\varepsilon_{2p} - d\varepsilon_{3p})^2 + (d\varepsilon_{3p} - d\varepsilon_{1p})^2]} \quad (10.47)$$

is the effective plastic strain increment. For a uniaxial tension, $d\varepsilon_{2p} = d\varepsilon_{3p}$ and, since $d\theta_p = 0$, we have

$$d\theta_p = d\varepsilon_{1p} + d\varepsilon_{2p} + d\varepsilon_{3p} = 0$$

so that

$$d\varepsilon_{2p} = d\varepsilon_{3p} = -\frac{d\varepsilon_{1p}}{2} = -\frac{d\varepsilon_p}{2}$$

In addition, $\sigma_1 = \sigma$, and $\sigma_2 = \sigma_3 = 0$ for a uniaxial tension. Then, from Eqs. (10.6) and (10.47), we have $\sigma_i = \sigma$, $\overline{d\varepsilon_{ip}} = d\varepsilon_p$ and, therefore,

$$\int \overline{d\varepsilon_{ip}} = \varepsilon_p$$

Hence, in the case of a uniaxial tension, formula (10.46) yields

$$\sigma = \phi(\varepsilon_p) \quad (10.48)$$

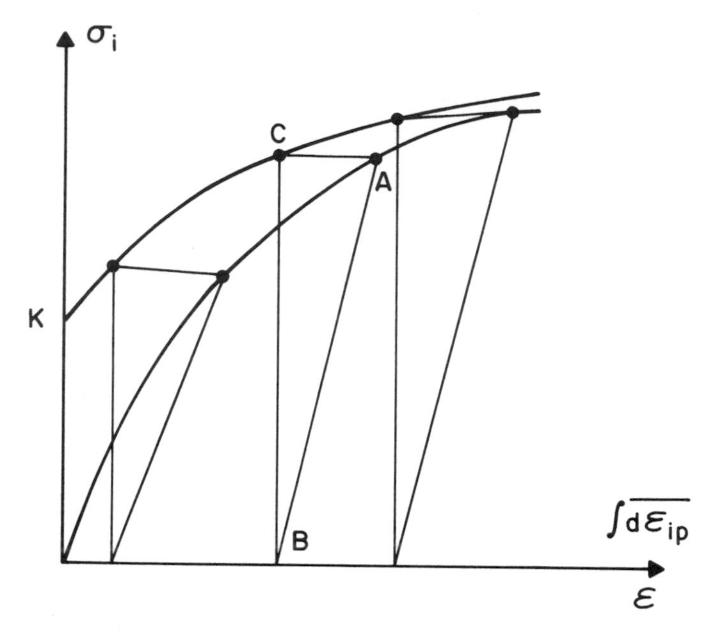

FIGURE 10.5 Evaluation of the function ϕ establishing the stress-strain relationship during plastic flow.

Thus, in order to determine the function ϕ on the basis of the stress-strain diagram, we should plot the relationship between the stress and the plastic strain. In order to do that, let us take a certain point A on the stress-strain diagram (Fig. 10.5) and pass a straight line AB parallel to the initial straight line of loading. From the point B of interaction of the line AB with the ε axis, we erect a perpendicular BC until it intersects in point C the horizontal line, passing through point A. The abscissa of point C is the plastic strain, and the ordinate represents the corresponding stress. Therefore, the curve KC is the relationship between the effective stress and the integral of the effective plastic-strain increments.

In order to evaluate the $d\lambda$ value, we introduce the relationships (10.45), written for the principal planes, into Eq. (10.47). This results in the following formula for the plastic-strain increments:

$$\overline{d\varepsilon_{ip}} = \frac{\sqrt{2}}{3} d\lambda \sqrt{(\sigma_1 - \sigma_2)^2 + (\sigma_2 - \sigma_3)^2 + (\sigma_3 - \sigma_1)^2}$$

from which, considering Eq. (10.6), we have

$$d\lambda = \frac{3}{2} \frac{\overline{d\varepsilon_{ip}}}{\sigma_i}$$

Then Eqs. (10.46) yield

$$d\varepsilon_{xp} = \frac{3}{2}\frac{\overline{d\varepsilon_{ip}}}{\sigma_i}(\sigma_x - \sigma_0), \ldots, d\gamma_{zxp} = \frac{3}{2}\frac{\overline{d\varepsilon_{ip}}}{\sigma_i}\tau_{zx} \tag{10.49}$$

The elastic-strain increments can be obtained from Hooke's law,

$$d\varepsilon_{xe} = \frac{1}{E}[d\sigma_x - \nu(d\sigma_y + d\sigma_z)], \ldots, d\gamma_{zxe} = \frac{1}{G}d\tau_{zx} \tag{10.50}$$

After summing up Eqs. (10.49) and (10.50), we obtain the following relationships:

$$\begin{aligned} d\varepsilon_x &= \frac{1}{E}[d\sigma_x - \nu(d\sigma_y + d\sigma_z)] + \frac{3}{2}\frac{\overline{d\varepsilon_{ip}}}{\sigma_i}(\sigma_x - \sigma_0) \\ &\cdots\cdots\cdots\cdots \\ d\gamma_{zx} &= \frac{1}{G}d\tau_{zx} + \frac{3}{2}\frac{\overline{d\varepsilon_{ip}}}{\sigma_i}\tau_{zx} \end{aligned} \tag{10.51}$$

known as *Prandtl–Reuss equations*. Prandtl obtained them for the case of a two-dimensional elastoplastic deformation, and A. Reuss generalized Prandtl's results for a three-dimensional case.

If the plastic-strain increments are substantially larger than the elastic-strain increments, then we may neglect the elastic terms in Eq. (10.51). Having done so, we obtain

$$\xi_x = \frac{3}{2}\frac{\xi_i}{\sigma_i}(\sigma_x - \sigma_0), \ldots, \eta_{zx} = \frac{3}{2}\frac{\xi_i}{\sigma_i}\tau_{zx} \tag{10.52}$$

where the following notation is used:

$$\xi_i = \frac{\overline{d\varepsilon_i}}{dt}$$

Formulas (10.52) are called *Saint-Venant–Lévy–von Mises equations*. They determine the relationships between the strain rates and the stress magnitudes. For the case of a plane strain, these relationships were suggested by B. Saint-Venant and then generalized by M. Lévy and R. von Mises for a three-dimensional state of strain.

In the case of a so-called *simple loading*, when deviatoric stresses are proportional to a certain parameter, the effective plastic-strain increments

$\overline{d\varepsilon_{ip}}$ is equal to the differential of the effective values $d\varepsilon_{ip}$ of the plastic strains themselves. Indeed, let

$$\sigma_x - \sigma_0 = \beta(\sigma_x^* - \sigma_0^*), \ldots, \tau_{zx} = \beta\tau_{zx}^* \tag{10.53}$$

where $\sigma_x^* - \sigma_0^*, \ldots, \tau_{zx}^*$ are certain constant values of the deviatoric stresses (say, deviatoric stresses at the end of loading), and the parameter β may be time or any other value that determines sequential values of stress. From Eqs. (10.49), with consideration of Eqs. (10.53), we find

$$d\varepsilon_{xp} = \frac{3}{2}\frac{\sigma_x^* - \sigma_0^*}{\sigma_i^*}\overline{d\varepsilon_{ip}}, \ldots, d\gamma_{zxp} = \frac{3}{2}\frac{\tau_{zx}^*}{\sigma_i^*}\overline{d\varepsilon_{ip}}$$

Then, integrating, we have

$$\varepsilon_{xp}^* = \frac{3}{2}\frac{\sigma_x^* - \sigma_0^*}{\sigma_i^*}\int\overline{d\varepsilon_{ip}}, \ldots, \gamma_{zxp}^* = \frac{3}{2}\frac{\tau_{zx}^*}{\sigma_i^*}\int\overline{d\varepsilon_{ip}} \tag{10.54}$$

Having written these formulas for the principal directions and introducing the obtained equations into Eq. (10.22), we obtain

$$\varepsilon_{ip}^* = \int\overline{d\varepsilon_{ip}} \tag{10.55}$$

so that

$$d\varepsilon_{ip}^* = \overline{d\varepsilon_{ip}}$$

The relationships (10.54), considering Eq. (10.55), can be presented as

$$\varepsilon_{xp}^* = \frac{3}{2}\frac{\varepsilon_{ip}^*}{\sigma_i^*}(\sigma_x^* - \sigma_0^*), \ldots, \gamma_{zxp}^* = \frac{3}{2}\frac{\varepsilon_{ip}^*}{\sigma_i^*}\tau_{zx}^* \tag{10.56}$$

These formulas coincide with formulas (10.18). Hence, in the case of simple loading, the theory of plastic flow and the theory of small elastoplastic strains lead to the same results. Note that these results are in good agreement with experimental data. In the case of complex loading, the theory of plastic flow agrees better with experimental results than the theory of small elastoplastic strains.

11

Viscoelasticity

11.1 VISCOELASTIC MATERIALS

The development of the theory of viscoelasticity was triggered by a wide use of materials known as *high polymers* or, more commonly, *plastics*. These materials are widely utilized in engineering, including the areas of micro-electronics and fiber optics. Some of the most widespread plastics are listed in the following table.*

Chemical Classification	Trade Name
Thermoplastic Materials	
Acetal	Derlrin, Celcon
Acrylic	Lucite, Plexiglas
Acrylonitrile-butadiene-styrene	Cycolac, Kralastic, Lustran
Cellulose acetate	Fibestos, Plastacele, Tenite I
Cellulose acetate-butyrate	Tenite II
Cellulose nitrate	Celluloid, Nitron, Pyralin
Ethyl cellulose	Gering, Ethocel
Polyamide	Nylon, Zytel
Polycarbonate	Lexan, Merlon
Polyethylene	Polythene, Alathon
Polypropylene	Avisun, Escon
Polystyrene	Cerex, Lustrex, Styron
Polytetrafluoroethylene	Teflon
Polytrifluorochlorethylene	Kel-F
Polyvinyl acetate	Gelva, Elvacet, Vinylite A
Polyvinyl alcohol	Elvanol, Resistoflex
Polyvinyl butyral	Butacite, Saflex
Polyvinyl chloride	PVC, Boltaron, Tygon, Geon
Polyvinylidene chloride	Saran
Thermosetting Materials	
Epoxy	Araldite, Oxiron
Melamine-formaldehyde	Melmac, Resimene
Phenol-formaldehyde	Bakelite, Catalin, Durez
Phenol-furfural	Durite
Polyester	Beckosol, Glyptal, Teglac
Urea-formaldehyde	Bettle, Plaskon

* This table and all the figures in this chapter are used by permission from J. H. Faupel and F. E. Fisher, *Engineering Design*, 2nd ed., John Wiley & Sons, New York, 1981.

As we can see from the table, there are two main classes of plastic materials, *thermoplastic* and *thermosetting materials* or *thermoplasts* and *thermosets*. Thermoplastic materials are able to flow under application of heat and pressure, whereas thermosetting materials undergo a structural change when subjected to such conditions and are thus rendered incapable of subsequent deformation by temperature and pressure. Thermoplasts may be formed into a variety of shapes by the simple application of heat and pressure. Thermosets can be formed only by cutting and machining methods or other specialized procedures. Plastics exhibit mechanical properties that, generally, cannot be described either by a purely elastic model or by a purely viscous fluid model. This caused the necessity of development of a more general theory, the theory of *viscoelasticity*.

In addition to the basic types, numerous new formulations of plastic materials are being developed. Many of these are used in microelectronics and fiber optics as protective coatings, substrates, and encapsulants because of their excellent thermal and electrical properties, low cost, light weight, and availability in a wide range of colors and transparencies. Many plastics show remarkable resistance to wear and corrosion. At the same time, most plastics have a relatively low Young's modulus, which is a favorable factor from the standpoint of the occurring stresses. On the other hand, most of the mechanical properties of plastics are time-dependent and are strongly affected by changes in temperature and the rate of load application. Finally, it should be pointed out that, because of the complexity of the mechanical behavior of plastics, there is often a serious shortage of reliable design data, especially in the areas of creep, stress rupture, and complex stressing. Some mechanical properties of plastic materials are given in Fig. 11.1.

11.2 TIME EFFECTS

Viscoelastic deformation of plastics is a transition type of behavior that is characterized by both elastic strain and time-dependent flow. Time-dependent properties are extremely important for plastics. The specific time-dependent behavior of plastic materials is due to the fact that they are composed of long, entangled, threadlike molecules. Under the action of applied stress, these "threads" can uncoil, straighten out, or otherwise accommodate the imposed force. Within the threadlike structure of molecules, there are bonds holding the individual molecules together.

As far as static loading is concerned, the time effects are usually reduced to creep, stress relaxation, and stress rupture. *Creep* is the deformation that occurs over a period of time in a material subjected to a constant load, and *stress relaxation* is the reduction in stress that occurs in a material when it is deformed to some specified deformation that is held constant. Thus, creep

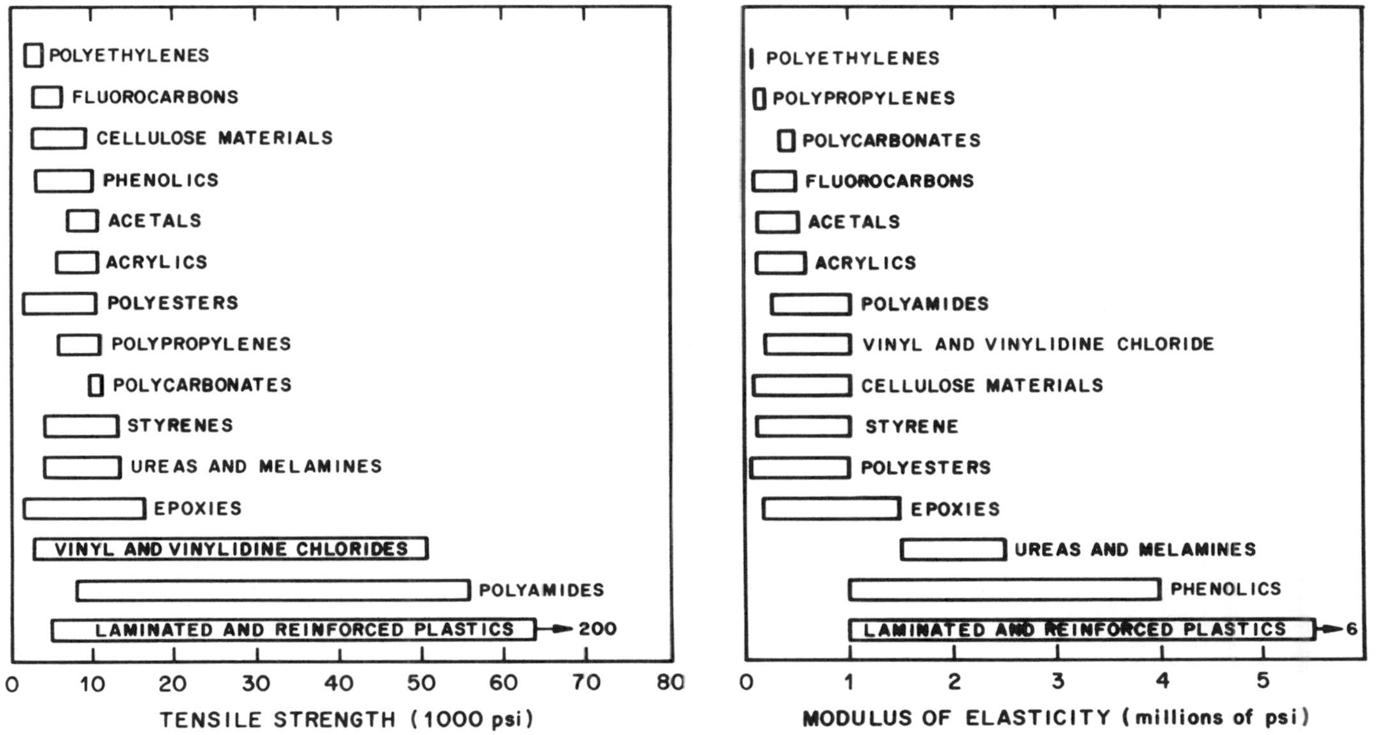

(a) TENSILE STRENGTH OF SOME PLASTICS AT ROOM TEMPERATURE

(b) MODULUS OF ELASTICITY OF SOME PLASTICS AT ROOM TEMPERATURE

FIGURE 11.1 Tensile strength, moduli of elasticity, coefficients of thermal expansion, and maximum operation temperatures of some plastics.

is deformation at constant load, and stress relaxation is decay of stress at constant strain. For most plastics, even small loads induce a continuous type of creep behavior and, depending on the state of stress and environment, fracture by stress rupture can occur at loads considered safe by conventional standards. In designing with viscoelastic materials, a compromise often has to be made between proportions that will not reach an excessive deformation during a specified time period and proportions that, under a given load, will not generate stresses that would result in failure by stress rupture in this time period. The main difficulty is that these effects can take place at ambient temperatures and are intensified at elevated temperatures. Another complication is that creep and stress rupture of plastics may vary considerably for the same material or different batches obtained from different manufacturers of the material.

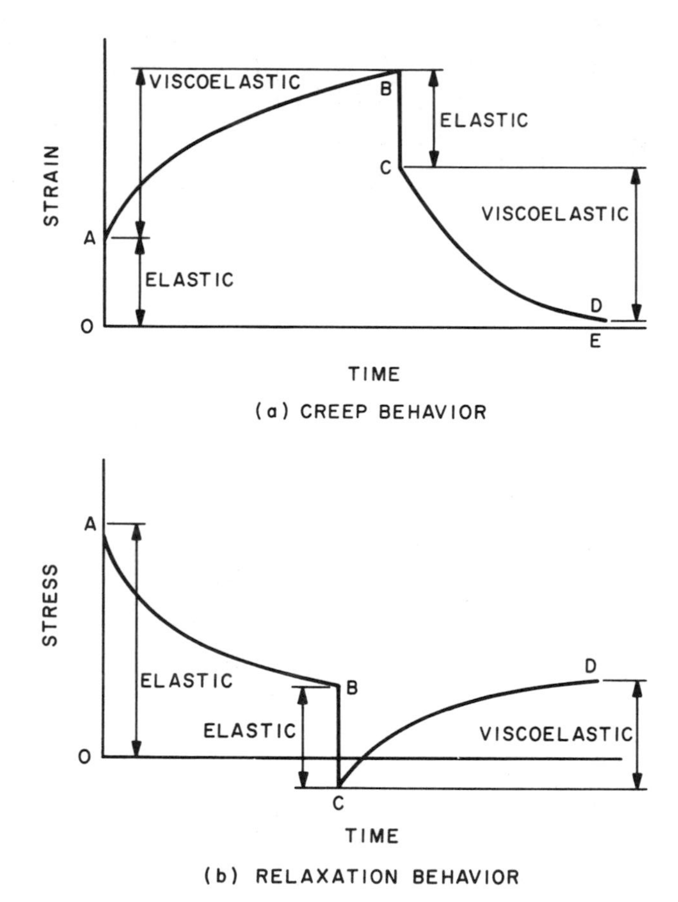

FIGURE 11.2 Typical creep and relaxation behaviors of plastics.

The effects of creep and stress relaxation are illustrated in Figs. 11.2a and b, where typical creep and stress-relaxation characteristics of plastic materials are presented. In the creep tests (Fig. 11.2a), a sample is initially loaded to some stress, which is then maintained at a constant level. It is assumed that the tests are carried out at stresses below the short-time yield strength of the material. The initial stress results in a certain strain, which is an elastic response. As the stress is maintained, the sample deforms (creeps) viscoelastic-

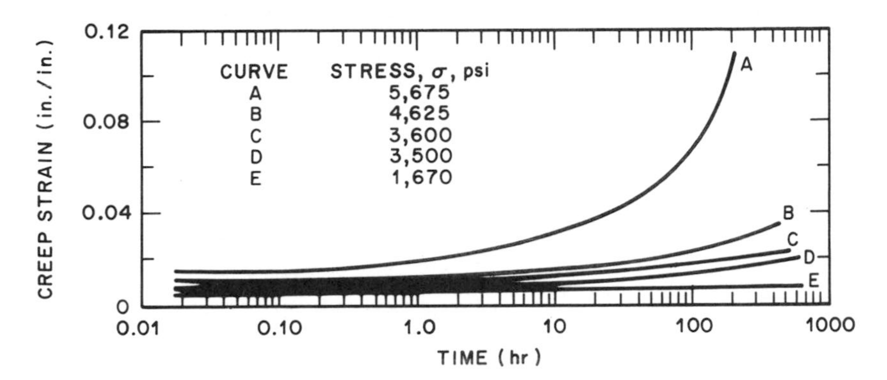

(a) CREEP BEHAVIOR OF RIGID POLYVINYL CHLORIDE AT 72°F

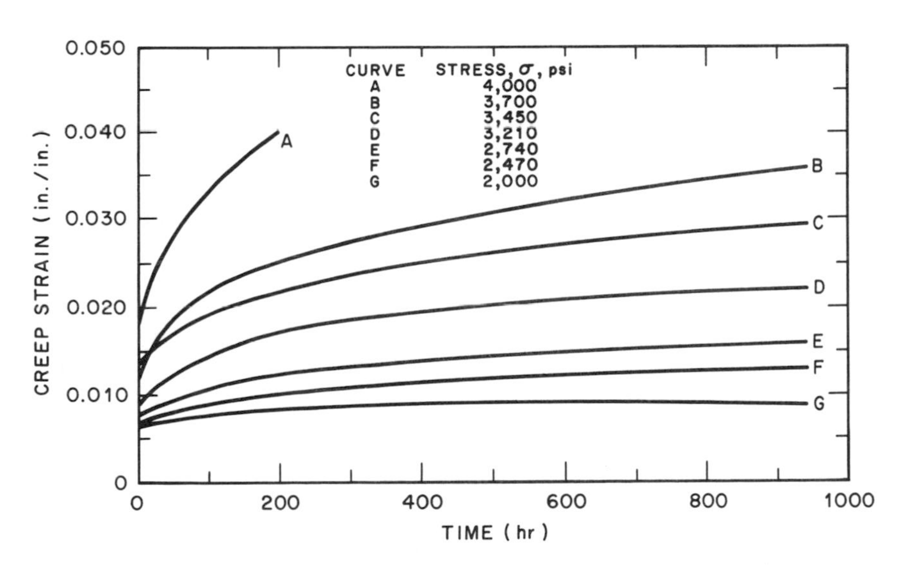

(b) CREEP OF METHACRYLATE PLASTIC AT 77°F AS A FUNCTION OF STRESS

FIGURE 11.3 Creep behavior of some plastics.

ally. If the load is removed from the sample at point *B*, there is an immediate elastic recovery to point *C*, followed by a final gradual viscoelastic recovery to point *D*, which leaves a permanent residual deformation *DE*. In the stress-relaxation test (Fig. 11.2b), a sample is instantaneously loaded by a stress to some deformation, which is then held constant. Immediately, the material starts to "relax," that is, the originally imposed stress diminishes to *B* over a period of time. If, at this time, the load is removed, there is an

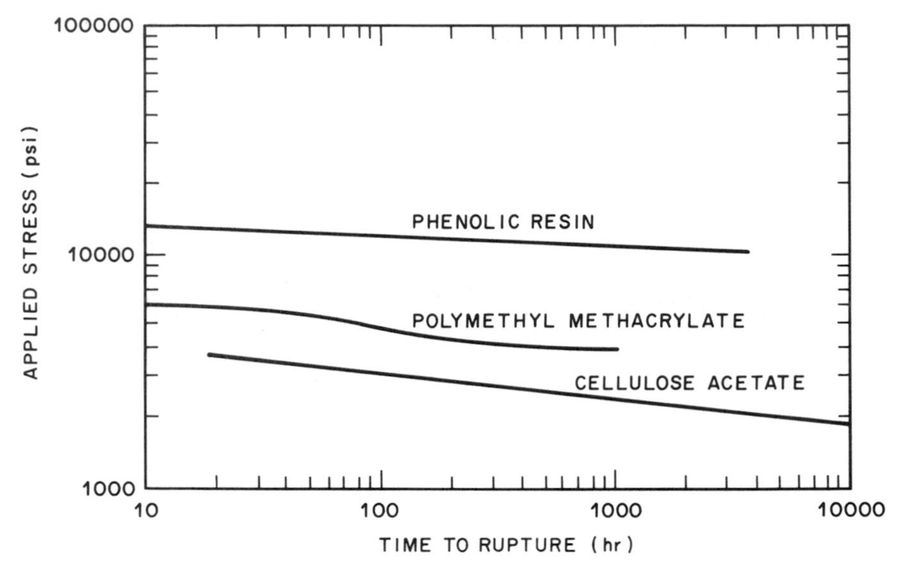

(a) TENSILE STRESS-RUPTURE OF SOME PLASTICS AT 73°C

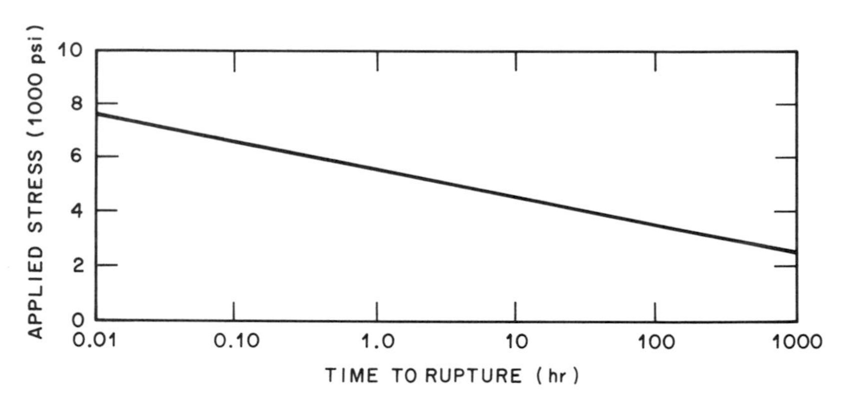

(b) TESILE STRESS-RUPTURE OF RIGID POLYVINYL CHLORIDE AT 73°C

FIGURE 11.4 Tensile stress rupture of some plastics.

immediate elastic recovery to *C*. Such a recovery may even be below zero, depending on the initial deformation. The residual stress then induces further viscoelastic deformation to *D*.

Typical tensile creep data for plastics are shown in Fig. 11.3, and some tensile stress-rupture data are given in Fig. 11.4.

11.3 MATHEMATICAL MODELS OF PRERUPTURE DEFORMATION OF PLASTICS

As has been indicated earlier, failure in plastic materials can be due to an actual rupture or to an excessive creep deformation. Obviously, designs with plastics are intended to operate for long periods of time in the prerupture zone and, therefore, in the development of mathematical models, attention is given primarily to the problem of characterizing creep or viscoelastic deformation of plastic components.

Three general types of static deformation are thought to be able to occur in plastics: linear elastic deformation ε_l, nonlinear elastic deformation ε_n, and viscous deformation ε_v. Linear elastic deformation can be evaluated on the basis of Hooke's law and is believed to be associated primarily with the molecular bonding forces. During nonlinear elastic deformation, stress is no longer proportional to strain; however, complete recovery occurs when the imposed load is removed. This type of deformation is believed to be associated with the uncoiling of strings of molecules. Viscous deformation is associated with the shipping of strings of molecules over each other. Although no time effect is involved in elastic deformation, in viscous deformation, time is a major factor, resulting in permanent (residual) strains. Most plastics exhibit combinations of behavior and, therefore, various models have been suggested to interpret and quantify the observed behavior (Fig. 11.5).

The simplest model is shown in Fig. 11.5a. It is known as a *Maxwell model* and represents behavior consisting of instantaneous elastic extension (spring) and time-dependent flow (dashpot). In this system, only the elastic strain is recoverable. In the *Voight–Kelvin model* shown in Fig. 11.5b, the spring and the dashpot are in parallel and, therefore, this model can be used to represent materials exhibiting viscoelastic recovery and no permanent set. It cannot be used, however, to describe instantaneous elastic deformation. Since most plastics exhibit elastic deformation, a more complex spring-dashpot model (Fig. 11.5c), which is a combination of the Maxwell and Voight–Kelvin models, is often utilized to describe the behavior of plastic materials. It gives the best approximation of the observed behavior of plastics exhibiting instantaneous elastic extension, time-dependent creep, strain recovery, stress relaxation, and permanent set.

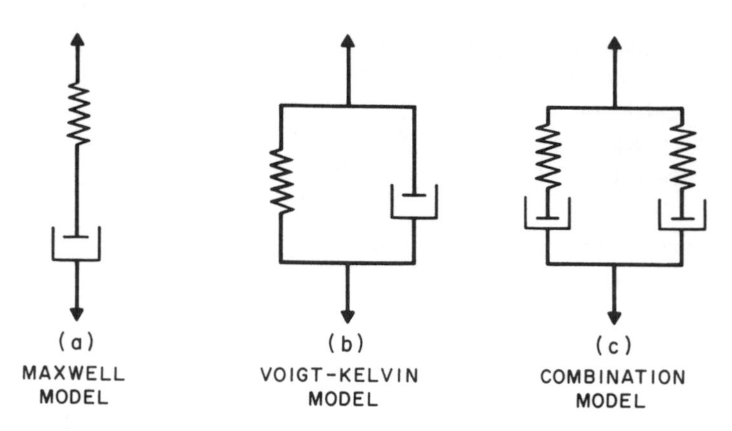

FIGURE 11.5 Models to interpret prerupture behavior of plastics.

11.4 THEORY OF LINEAR VISCOELASTICITY

The theory of linear viscoelasticity has received considerable attention during the last four decades and is now highly developed as a result of the work of T. Albrey, D. R. Bland, B. Gross, E. H. Lee, R. Slips, H. A. Stuart, and other investigators. The material can be treated as linearly viscoelastic if the stress in it is sufficiently low. It is thought that the results of this theory can be used if the stress does not exceed half the short-time yield strength.

If the Maxwell model is employed, then the basic differential equation is

$$\frac{d\varepsilon}{dt} = \frac{1}{E}\frac{d\sigma}{dt} + \frac{\sigma}{\eta} \tag{11.1}$$

where ε is the strain at any time t, σ the stress, E Young's modulus at time t, and $\eta = \sigma/(d\varepsilon/dt)$ the viscosity. In the case of constant strain, this equation has the following solution:

$$\sigma = \sigma_0 e^{-t/\tau} \tag{11.2}$$

where σ_0 is the initial tensile stress, and $\tau = \eta/E$ is the *relaxation time* or *time constant.*

In the case of the Voight–Kelvin model, the basic differential equation is

$$\sigma = \eta \frac{d\varepsilon}{dt} + E\varepsilon \tag{11.3}$$

and its integral in the case of constant stress is as follows:

$$\varepsilon = \frac{\sigma}{E}(1 - e^{-t/\tau}) \tag{11.4}$$

When the model in Fig. 11.5c is applied, the additional deformation contributed by the spring and dashpot can be found using the fact that this model consists of one nondegenerate Voight–Kelvin unit and two degenerate elements, the spring and the dashpot. When the degenerate element is the spring, we have

$$\varepsilon = \frac{\sigma}{E} = \frac{\sigma}{E_0} \tag{11.5}$$

where E_0 is the elastic (instantaneous) Young's modulus. When the dashpot is the degenerate element, the strain is

$$\varepsilon = \frac{\sigma t}{\eta} \tag{11.6}$$

Therefore, the total deformation can be presented as follows:

$$\varepsilon = \frac{\sigma}{E_0} + \frac{\sigma}{E}(1 - e^{-t/\tau}) + \frac{\sigma t}{\eta} \tag{11.7}$$

In the absence of the permanent set, the last term is zero, and the strain is due only to the creep involved in monolithic loading,

$$\varepsilon = \frac{\sigma}{E_0} + \frac{\sigma}{E}(1 - e^{-t/\tau}) \tag{11.8}$$

Note that the time τ has different significance in Eqs. (11.2) and (11.8): in the first case, this time is based on a concept of relaxation whereas, in the second case, it is based on creep. Accordingly, these two cases can be referred to as *relaxation time* and *retardation time.* One of the principal accomplishments of the phenomenological viscoelastic theory is the linking of the *relaxation function* and the *retardation function* analytically, so that creep deformation can be predicted from relaxation data, and vice versa.

From a practical point of view, the designer is concerned with the deformation behavior of the material, and deformation can be predicted by the use of the standard elastic-stress analysis formulas in which the elastic constants E and ν can be substituted for the viscoelastic equivalents:

$$E = E_0 \frac{\sigma}{\sigma_0} \qquad \nu = \frac{K - E}{2K} \tag{11.9}$$

Here E and ν are the values used to replace Young's modulus and Poisson's ratio in the elastic solution to a problem; K is the bulk modulus, expressed by Eqs. (2.57), a value that remains essentially constant throughout deformation; E_0 is the instantaneous (initial, elastic) Young's modulus; and the decay of stress σ/σ_0 can be obtained from a relaxation test. An example of such a test is shown in Fig. 11.6. For many materials, data are available on the effect of time, temperature, and strain rate on modulus. If such data are not available, creep tests should be conducted for responsible applications to obtain the essential information.

As an example of the application of the theory, examine tension (compression) of a linearly 10-in.-long viscoelastic prismatic bar subjected to a load F. This load is assumed to give an initial stress below the short-time yield strength of the material, so that linear viscoelastic models can be used. Under this condition, the initial modulus E_0 is expressed, in accordance with Eq. (11.5), as

$$E_0 = \frac{\sigma}{\varepsilon_0} \tag{11.10}$$

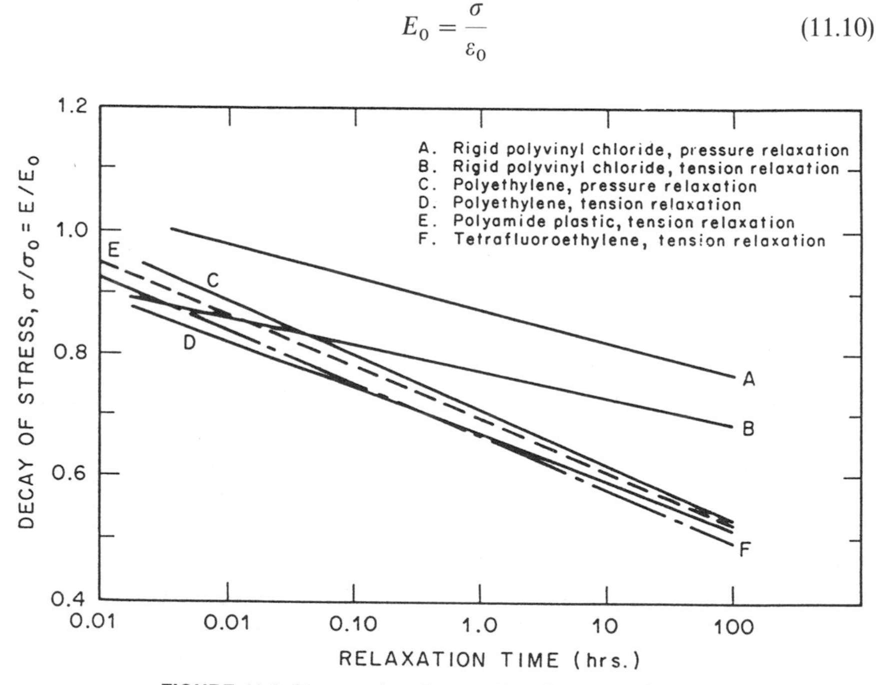

FIGURE 11.6 Stress relaxation vs. time for some plastics.

where σ is the applied stress and ε_0 is the initial strain. If the load F is sustained for the time t, the strain at the end of this time will be the sum of the initial strain ε_0 and that due to viscoelastic flow. The new or *apparent modulus* at the time t can be obtained as

$$E_t = \frac{\sigma}{\varepsilon_0 + \varepsilon_t} \tag{11.11}$$

where ε_t is the creep strain. Suppose that a 10-in. long plastic bar is subjected to a stress of 1000 psi that induces a total elongation of 0.10 in. Then, from Eq. (11.10), we find

$$E_0 = \frac{\sigma}{\varepsilon_0} = \frac{1000}{0.10/10} = 100{,}000 \text{ psi}$$

If the load is sustained for 1000 hours and the total creep experienced is 0.15 in., then, in accordance with Eq. (11.11), the modulus at 1000 hours is

$$E_t = \frac{\sigma}{\varepsilon_0 + \varepsilon_t} = \frac{1000}{0.01 + (0.15/10)} = 40{,}000 \text{ psi}$$

Thus, for linearly viscoelastic behavior, by measuring creep strains and using Eq. (11.11), time-modified modulus curves can be drawn. When such curves have been established for a particular material, the data can be used to predict behavior for other conditions. Typical time-modified modulus curves for linearly viscoelastic materials are shown in Fig. 11.7. Suppose that the bar in question is made of a polyethylene material conforming to the behavior shown in Fig. 11.7. Equation (11.11) indicates that

$$E_t = \frac{\sigma}{\varepsilon_t} \tag{11.12}$$

and Fig. 11.7 shows that the limiting modulus values are 19,000 psi and 7500 psi. Thus, by using these values in Eq. (11.12), it is seen that

$$\sigma/\varepsilon_t = 19{,}000$$

or

$$\sigma/\varepsilon_t = 7500$$

Suppose that the bar is stressed at 100 psi. The initial elastic modulus of this

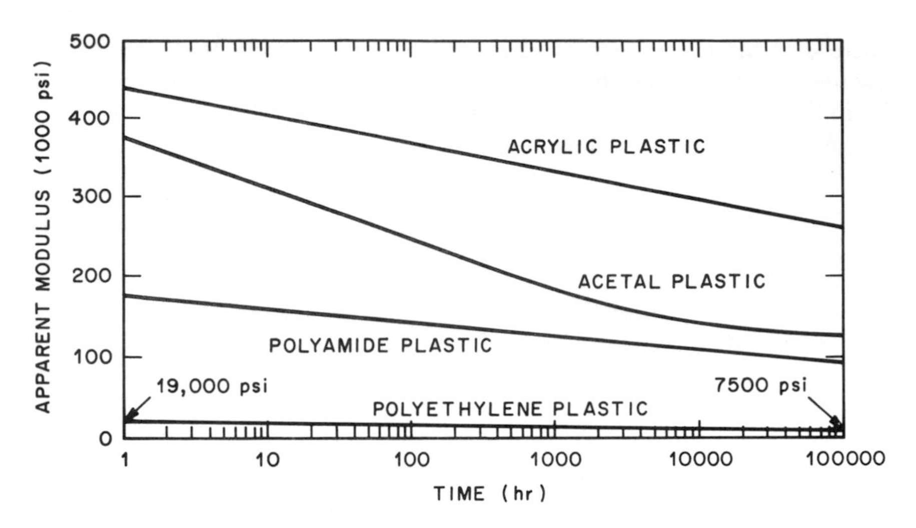

FIGURE 11.7 Time-modified modulus curves for four plastics at 73°F (limited to use at maximum of 0.5 percent strain).

material is 20,000 psi at 73°F; therefore, the initial strain ε_0 is

$$\varepsilon_0 = \frac{\sigma}{E_0} = \frac{100}{20{,}000} = 0.005$$

Then, the creep strain is

$$\varepsilon = \varepsilon_0 \frac{\sigma_0}{\sigma} = \varepsilon_0 \frac{E_0}{E}$$

Thus, for the condition at hand, the following predictions can be made:

Time, hr	Apparent Modulus, psi	Calculated Creep Strain
1	19,000	0.0053
10	16,500	0.0061
100	14,000	0.0071
1,000	12,000	0.0083
10,000	10,000	0.0100
100,000	7,500	0.0133

Note that, for relatively large loads, say, loads in excess of 100 psi, creep predictions would have to be made on the basis of the *rate theory*, which is not examined here.

11.5 THERMODYNAMICS OF A VISCOELASTIC DEFORMATION

Deformation of a viscoelastic body is accompanied by a more or less significant change in temperature and, therefore, the external work is expended not only for the deformation itself but, in accordance with the first law of thermodynamics, for the change in the thermal energy as well. Examine, for the sake of simplicity, a cubic element whose two opposite faces are subjected to tensile stresses σ. If the lengths of the element's edges are equal to unity, then the face areas are also equal to unity, and the force applied to the cube is equal to σ. Since the length of the cube's edge is unity, the absolute elongation of the cube is equal to its strain ε. If the force is increased by $d\sigma$, the elongation is increased by $d\varepsilon$, and the external work of this force is $\sigma\, d\varepsilon$. Considering also the change dQ in the thermal energy, we present the change in the total internal energy dU of the body as

$$dU = \sigma\, d\varepsilon + dQ$$

The second law of thermodynamics states that reversible processes can be characterized by a function S, called *entropy*, which is introduced as follows:

$$dQ = T\, dS$$

where T is temperature. Thus, the change in the total internal energy is

$$dU = \sigma\, d\varepsilon + T\, dS \tag{11.13}$$

The function

$$U(\varepsilon, S) = \sigma\varepsilon + TS \tag{11.14}$$

is a *thermodynamic potential*. Indeed, the total differential of the function (11.14) is

$$dU = \frac{\partial U}{\partial \varepsilon}\, d\varepsilon + \frac{\partial U}{\partial S}\, dS = \sigma\, d\varepsilon + T\, dS$$

which coincides with Eq. (11.13). Note that

$$\sigma = \frac{\partial U}{\partial \varepsilon} \qquad T = \frac{\partial U}{\partial S}$$

Similarly, other thermodynamic potentials can be introduced:

Free energy:

$$U_*(\varepsilon, T) = U - TS \qquad \sigma = \frac{\partial U_*}{\partial \varepsilon} \qquad S = -\frac{\partial U_*}{\partial T} \tag{11.15}$$

Enthalpy:

$$\Phi(\sigma, S) = \sigma\varepsilon + U \qquad \varepsilon = \frac{\partial \Phi}{\partial \sigma} \qquad T = \frac{\partial \Phi}{\partial S} \tag{11.16}$$

Free enthalpy (Gibbs' potential):

$$G(\sigma, T) = \Phi - TS \qquad \varepsilon = \frac{\partial G}{\partial \sigma} \qquad S = -\frac{\partial G}{\partial T} \tag{11.17}$$

For an *adiabatic process*, when there is no exchange in the heat with the environment, $dQ = 0$ and, therefore, $S = \text{const}$. For an *isothermic process*, $T = \text{const}$. Hence, the stress potential in an adiabatic process is the internal energy, and the strain potential in this process is the enthalpy. For an isothermic process, these potentials are the free energy and the free enthalpy, respectively. The relationship between the stresses and strains in an elastic body is unique only in the cases of an isothermic or adiabatic process. If there is an exchange in the thermal energy with the environment, such a relationship is not single-valued anymore.

Let the experimental relationship between the stress, strain, and temperature be as follows:

$$\sigma = E\varepsilon - E\alpha(T - T_0) \tag{11.18}$$

where α is the coefficient of linear thermal expansion, and T_0 is the initial temperature.

The free energy U_*, which is expressed, in accordance with Eqs. (11.15) and (11.14), as

$$U_* = \int \sigma(\varepsilon)\, d\varepsilon$$

can then be evaluated as

$$U_* = \frac{E\varepsilon^2}{2} + E\alpha\varepsilon(T - T_0) + \Psi(T)$$

where $\Psi(T)$ is a thus far unknown function of temperature. The entropy defined by the last formula in Eqs. (11.15) is then expressed as

$$S = E\alpha\varepsilon - \Psi'(T) \tag{11.19}$$

Introducing this formula in Eq. (11.14), we find

$$U = \frac{E\varepsilon^2}{2} + E\alpha\varepsilon T_0 + \Psi - T\Psi' \tag{11.20}$$

If the strain ε is zero, the internal energy is due to the heat only and is equal to $c_e T$, where c_e is the specific heat. In this case, the function Ψ can be found from the equation

$$\Psi - T\Psi' = c_e T$$

Assuming that the c_e value is constant, i.e., is independent of both the strain and the temperature, we obtain the integral of this equation in the form

$$\Psi = (C - c_e \ln T)T \tag{11.21}$$

where C is the constant of integration. Since the entropy is defined up to an arbitrary constant, we will choose it in such a way that the following boundary conditions are fulfilled:

$$S = 0 \quad \text{for } \varepsilon = 0 \qquad T = T_0$$

Then, Eq. (11.19) yields

$$\Psi'(T_0) = 0$$

and Eq. (11.21) results in the following formula for the constant C:

$$C = c_e(1 + \ln T_0)$$

Then the function Ψ is

$$\Psi = c_e T\left(1 - \ln\frac{T}{T_0}\right)$$

and its derivative entering the formula for entropy (11.19) is

$$\Psi'(T) = -c_e \ln\frac{T}{T_0}$$

Thus, the entropy given by Eq. (11.19) and the total internal energy given by Eq. (11.20) are expressed as

$$S = E\alpha\varepsilon + c_e \ln\frac{T}{T_0} \tag{11.22}$$

$$U = \frac{E\varepsilon^2}{2} + E\alpha\varepsilon T_0 + c_e T \tag{11.23}$$

Excluding temperature T from these equations, we have

$$U = \frac{E\varepsilon^2}{2} + E\alpha\varepsilon T_0 + c_e T_0 \exp\left(\frac{S - E\alpha\varepsilon}{c_e}\right) \tag{11.24}$$

In order to establish the relationship between the stress and the strain for adiabatic tension, when there is no heat exchange, we differentiate this equation with respect to ε:

$$\sigma = \frac{\partial U}{\partial\varepsilon} = E\varepsilon + E\alpha T_0\left[1 - \exp\left(\frac{S - E\alpha\varepsilon}{c_e}\right)\right] \tag{11.25}$$

As evident from this formula, the linear relationship between stress and strain taking place in isothermic conditions ceases to exist in adiabatic conditions.

In order to assess the effect of nonlinearity on the stress-strain relationship, let us assume that tension started at temperature T_0 and that entropy S is zero throughout the deformation process. With these assumptions, expanding the exponent into series and retaining the two first terms in this series, we find

$$\sigma \cong E\left(1 + \frac{E\alpha^2 T_0}{c_e}\right)\varepsilon \tag{11.26}$$

The quantity

$$E_\alpha = E\left(1 + \frac{E\alpha^2 T_0}{c_e}\right) \tag{11.27}$$

is called the *adiabatic modulus of elasticity*; whereas the difference between the adiabatic and isothermic moduli is insignificant for metals, for polymeric materials, it can be on the order of 10 percent or even greater.

Solving Eq. (11.22) for temperature T, we have

$$T = T_0 \exp\left(\frac{S - E\alpha\varepsilon}{c_e}\right)$$

Expanding the exponent into series and assuming $S = 0$, we have

$$T = T_0\left(1 - \frac{E\alpha}{c_e}\varepsilon\right) \tag{11.28}$$

This formula indicates that tension is accompanied by the drop in temperature, and compression leads to increase in temperature, which is similar to the behavior of gases. Note that many polymers exhibit just the opposite behavior, i.e., increase their temperature when subjected to tension and are cooled down under compressive stresses. This can be accounted for by changing the sign in the corresponding relationships. In addition, the material characteristics E, α, and c_e may deviate substantially from constant values, which may have quite a significant effect on the results. Finally, in the presence of plastic strains, the laws of thermodynamics of nonreversible processes should be applied.

QUESTIONS AND PROBLEMS

1. What are the major assumptions underlying the science of engineering elasticity? What is the physical essence of these assumptions?
2. What is the major goal of the engineering theory of elasticity? What is the main difference between this theory and the science of the strength of materials?
3. What are the components of a stress tensor? How can they be transformed to new coordinate axes?
4. What is the role of the equilibrium conditions for an elementary tetrahedron? What problems can be solved by using these conditions?
5. What is the role of the equilibrium conditions for an elementary parallelepiped? What problems can be solved by using these conditions?
6. Why is the problem of determining the stresses in an elastic body three times statically indeterminate?
7. What are the principal planes and principal normal stresses? How are they defined?
8. How are the linear, quadratic, and cubic invariants of the stress tensor expressed? What does the word "invariant" mean?
9. What is the physical meaning of the axes of the stress ellipsoid?
10. In what cases are all the planes in the given point the principal ones?
11. How are the principal shearing stresses related to the principal normal stresses? What is the orientation of the corresponding planes with respect to the principal planes?
12. Define the octahedral stresses and octahedral planes.
13. Define the stress deviator and spherical tensor. What do these tensors characterize?
14. The state of stress is determined by the following stress tensors:

$$\sigma_x = 200(z - 4) \qquad \sigma_y = 400_x \qquad \sigma_z = 400(y - 5)$$

$$\tau_{xy} = 40(z + 1) \qquad \tau_{xz} = 8800(y - z) \qquad \tau_{yz} = 200(x + 3)$$

Calculate the stress tensors for points $A(-3, -2, -1)$ and $B(0, 5, 4)$. Determine the total stresses, the total shearing stresses, and their directions for the three coordinate planes at point A.

15. With the input data from the previous problem, calculate the stress invariants for points A and B.
16. With the input data from problem 11, evaluate the stress deviator and spherical tensor for points A and B. Determine the invariants of the stress deviator for these points.
17. Explain the physical meaning of the following quantities:

$$\frac{\partial u}{\partial y}, \frac{\partial u}{\partial z}, \frac{\partial v}{\partial x}, \frac{\partial v}{\partial z}, \frac{\partial w}{\partial x}, \frac{\partial w}{\partial y}$$

18. What are the linear (normal) strain and the angular (shearing) strain? What is the rotation at the given point? How are these values related to the displacements?
19. What is the physical meaning of the Saint-Venant compatibility equations? What is their role in the theory of elasticity?
20. What is a simply connected region?
21. What are the components of a strain tensor?
22. How are the invariants of the strain tensor expressed?
23. What is dilation, and what does it characterize?
24. The displacements in an elastic body are given by the formulas:

$$u = 0.01(x^2 + y^2) \qquad v = 0.01(xy + z^2) \qquad w = 0.01(z^2 + y^2)$$

Evaluate the total displacement for point A (3, −2, 1), and determine its direction.
25. With the input data from the previous problem, determine the change in the distance between points A (3, −2, 1) and B (−3, 2, −1).
26. One of the components of the strain tensor is expressed as $\varepsilon_y = a(x^3y + 0.3y^2 - z^2)$, where a is a coefficient. Determine the change in the length of a straight element AB, whose endpoints A and B have the following initial coordinates: A (0, 0, 4) and B (0, 10, 4).
27. Formulate the generalized Hooke's law. What kind of problems can be solved using Hooke's law?
28. What are the physical meanings of Young's modulus and Poisson's ratio?
29. Describe the stress-strain curve for metals, and determine the notions of proportional limit, elastic limit, yield stress, and ultimate strength.
30. Determine the shear modulus and the bulk (compressibility) modulus. What do these values characterize?
31. What does the Duhamel–Neumann hypothesis state? What terms should be added to the expressions for the strains in Hooke's law to account for the thermoelastic strains?
32. The stresses in an elastic body (Young's modulus $E = 2 \times 10^{11}$ Pa, $\nu = 0.25$) are given by the equations:

$$\sigma_x = 200(z - 4) \qquad \sigma_y = 400_x \qquad \sigma_z = 400(y - 5)$$

$$\tau_{xy} = 40(z + 1) \qquad \tau_{xz} = 8800(y - z) \qquad \tau_{yz} = 200(x + 3)$$

Determine the displacements if, at the origin: $u = v = w = 0$, $\partial u/\partial y = 0$, and $\partial u/\partial z = \partial v/\partial z = 0.125$.
33. Describe the general scheme for getting the solution to the direct problem of the theory of elasticity.
34. Describe the general scheme for obtaining the solution to the inverse problem of the theory of elasticity. Consider the following three cases: the displacements are known, the stresses are known, the strains are known.
35. Explain the essence of the semi-inverse method of Saint-Venant. What is its role in the theory of elasticity?

36. Formulate the principle of Saint-Venant. Why does the application of this principle simplify obtaining the solutions to the theory-of-elasticity problems?

37. Explain the principle of superposition and how it is used when obtaining solutions to the theory-of-elasticity problems.

38. How are Beltrami–Michell's equations obtained? What is the scheme to obtain a solution to the direct problem of the theory of elasticity when these equations are used?

39. How are Lame′ equations obtained? What is the scheme to obtain a solution to the direct problem of the theory of elasticity when Lamé equations are used?

40. What are Duhamel–Neumann's equations, and when are they used?

41. Two identical rectangular plates are loaded along their short edges by normal stresses, as shown in Fig. QP.1. Using Saint-Venant's principle, determine under what conditions and in what region practically the same state of stress occurs in these plates.

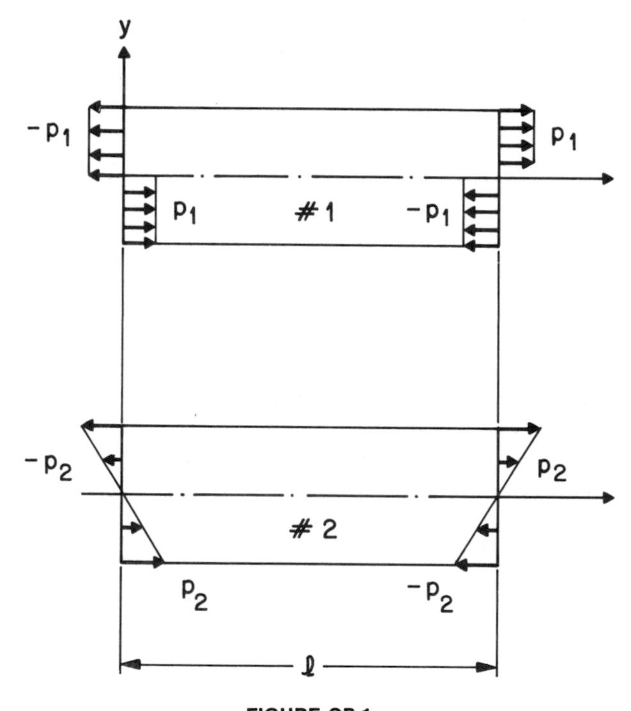

FIGURE QP.1

42. For Fig. QP.1, write the exact and integral ways the boundary conditions at the short edges are formulated.

43. The displacements in a three-dimensional elastic body are: $u = ax$, $v = ay$, $w = az$. Determine the strains, stresses, and external loading causing these displacements.

44. List the elementary problems in the theory of elasticity. Why are these problems called "elementary"?

45. The displacements in a circular shaft are as follows:

$$u = 0 \qquad v = -\frac{Q}{C} zx \qquad w = \frac{Q}{C} yx$$

Determine the stresses and strains, as well as the loading, causing this state of strain. Determine the physical meaning of the constants Q and C.

46. A circular shaft experiences shearing loads applied to its butt ends and resulting in a torque Q. Show that the cross sections of the shaft will remain plane and perpendicular to its axis after deformation. Show that the axial displacements are zero and that the twist angle is proportional to the axial coordinate.

47. What is strain energy? How is it defined?

48. What is the expression for the work associated with the distortion of the body's form? What tensors define the change in the body's volume and the distortion of its form? What is the reason for breaking down the total strain energy into two categories?

49. The external loading on an elastic body is increased by a factor of n. How will this affect the total strain energy of the body and the strain energy per unit volume?

50. Formulate the major strength hypotheses. For what are these hypotheses (theories) needed?

51. Describe the strength theory of Huber–von Mises–Hencky. Why can the critical stress in this theory be used as a condition of plasticity (yield criterion)?

52. The displacements in an elastic body are expressed as

$$u = -\frac{\nu\gamma}{E} zx \qquad v = -\frac{\nu\gamma}{E} yz \qquad w = \frac{\gamma}{2E}[z^2 + \nu(x^2 + y^2)]$$

Derive formulas for the strain energies per unit volume: total, due to uniform compression, and due to distortion of the form.

53. For a circular shaft shown in Fig. 4.2, determine the total strain energy if the displacements are given by the formulas:

$$u = \frac{Q}{GI_p} yz \qquad v = -\frac{Q}{GI_p} zx \qquad w = 0$$

Here Q is a constant value, and $I_p = (\pi/2)r^4$ is the polar moment of inertia. Determine the physical meaning of the parameter Q.

54. The state of stress in an elastic body restricted by the surfaces $x = 0$, $x = 8$, $y^2 + z^2 = 400$ is given by the equations

$$\sigma_x = 50(2 + y) \qquad \sigma_y = -50x \qquad \sigma_z = 50(2 + y)$$

$$\tau_{xy} = 0 \qquad \tau_{yz} = 0 \qquad \tau_{zx} = 50(2 + x + y)$$

Using the energy strength theory (Huber–von Mises–Hencky), determine the point at which the critical stress is the greatest.

55. Solve the previous problem using the first and the third strength theories.

56. With the input data and the solution to problem 50, determine by what factor the external loading should be increased in order that the most stressed point would experience a yield stress of $\sigma_Y = 4 \times 10^8$ Pa.

57. Under what conditions do plane strain and plane stress occur? In what cases are these approximations used?

58. Formulate the M. Lévy theorem, and explain its role in the photoelasticity method.

59. What is the stress function, and how is it related to the stresses? Why is it introduced?

60. How should the stress function be presented in polynomials?

61. How should the stress function be presented when the solution to a two-dimensional theory-of-elasticity problem is sought in a trigonometric series? In what situation is the Ribière solution used? When should Filon's solution be applied?

62. What stresses, strains, and displacements are taken as the principal unknowns when polar coordinates are used in a two-dimensional problem? Explain the basic equations of the two-dimensional theory-of-elasticity problem in polar coordinates.

63. What is stress concentration, and when does it occur? Illustrate this using an example of a thick-walled tube subjected to an internal uniform pressure.

64. The stress components in rectangular coordinates are

$$\sigma = p + \sigma_0 \frac{y}{a} \qquad \sigma = 0 \qquad \tau_{xy} = \tau_0$$

where β, τ_0, σ_0 are constant values. Determine the stresses σ_r, σ_θ, $\sigma_{r\theta}$ in polar coordinates, assuming that the origins of both systems coincide.

65. In what circumstances does simple torsion of a prismatic bar occur? Why cannot the formulas obtained from the strength of materials for simple torsion of circular shafts be applied to prismatic bars of an arbitrary cross-sectional shape?

66. What is the reason for introducing a stress function in the theory of torsion of noncircular shafts?

67. How are the boundary conditions in the problem of torsion formulated? How are they expressed through the stress function?

68. What are the merits and shortcomings of the Saint-Venant and Prandtl stress functions?

69. What is the relationship between the torque and the stress function?

70. What are the peculiarities of the solutions to the problem of the simple torsion of multiply connected regions?

71. Formulate and explain the meaning of the Bredt theorem of the circulation of shearing stress. For what purpose is this theorem used?

72. What is the torsional rigidity of a thin-walled section?
73. Why do multiply connected (tubelike) bars withstand simple torsion better than simply connected bars of the same cross-sectional area?
74. What is the most feasible shape of a cross section of a bar subjected to simple torsion? Why?
75. Derive the formula for torsional rigidity for an elliptic bar, and examine a special case of a circular shaft.
76. How are the maximum shearing stresses in a circular and square shaft of the same cross-sectional area related to the torsional rigidities of the shafts?
77. How is the total change in elastic strain energy related to the length of a crack according to Griffith's formula?
78. What is the Griffith relationship for critical stress?
79. How should Griffith's formula for critical stress be modified for the case in which fracture is preceded by plastic flow (Irwin–Orowan's formula)?
80. Describe the stress patterns in Irwin's stress-intensity approach. Characterize the stress-intensity factors for different stress fields.
81. What parameter is used to characterize fracture resistance of brittle materials? What do the specimens used in fracture toughness tests look like?
82. Characterize the area of application of yield fracture mechanics.
83. What are the major assumptions of the theory of small elastoplastic strains?
84. What are the major assumptions of the theory of plastic flow?
85. How are the stress-strain curves of actual materials approximated in plasticity theories?
86. What is the essence of the method of elastic solutions in the theory of elastoplasticity?
87. How is the ultimate bending moment determined for a prismatic beam?
88. The ultimate bending moment for a cantilever shown in Fig. 6.1 is $M_Y = 1/6\sigma_Y bh^2$, where σ_Y is the yield stress, b the width of the cantilever cross section, and h its height. Assuming that the length l of the cantilever is significantly greater than its height h, that the cantilever's material is ideally elastoplastic, and that the external moment $M = Pl$ exceeds M_Y, find the maximum deflection w_0 and the residual deflection w_r, after the force P is removed. Assume that $\gamma = M/M_Y = 11/8$.
89. Using the input data and the solution obtained in problem 88, find the height of the elastic zone in the beam cross sections and the residual stress. Plot a diagram for the distribution of the residual stress along the beam.

BIBLIOGRAPHY

Atkin, R. J., and Fox, N., *An Introduction to the Theory of Elasticity*, Longman, London, 1980.

Boresi, A. P., and Lynn, P. B., *Elasticity in Engineering Mechanics*, Prentice-Hall, Englewood Cliffs, N.J., 1974.

Broek, D., *Elementary Engineering Fracture Mechanics*, Sijthoff and Noordhoff, The Netherlands, 1978.

Chou, P. C., and Pagano, N. J., *Elasticity*, Van Nostrand, Princeton, N.J., 1967.

Den Hartog, J. P., *Advanced Strength of Materials*, McGraw-Hill, New York, 1952.

Dugdale, D. S., and Ruiz, C., *Elasticity for Engineers*, McGraw-Hill, London, 1971.

Fenner, R. T., *Engineering Elasticity*, Ellis Horwood, New York, 1986.

Ford, H., *Advanced Mechanics of Materials*, 2nd ed., Ellis Horwood, New York, 1977.

Fraeijs de Veubeke, B. M., *A Course of Elasticity*, Springer-Verlag, New York, 1979.

Fung, Y. C., *Foundations of Solid Mechanics*, Prentice-Hall, Englewood Cliffs, N.J., 1965.

Green, A. E., and Zerna, W., *Theoretical Elasticity*, University Press, Oxford, 1954.

Hahn, H. G., *Theory of Elasticity*, G. G. Teubner, Stuttgart, 1985 (in German).

Love, A. E. H., *A Treatise on the Mathematical Theory of Elasticity*, 4th ed., Cambridge Univ. Press, Cambridge, England, 1959.

Lurie, A. I., *Three-dimensional Problems of the Theory of Elasticity*, Wiley-Interscience, New York, 1964.

McClintock, F. A., and Argon, A. S., *Mechanical Behavior of Materials*, Addison-Wesley, Reading, Mass., 1966.

Owen, D. R. J., and Fawkes, A. J., *Engineering Fracture Mechanics*, Pineridge Press, London, 1983.

Pearson, C. E., *Theoretical Elasticity*, Harvard Univ. Press, Cambridge, Mass., 1959.

Prescott, J., *Applied Elasticity*, Dover, New York, 1961.

Saada, A. S., *Elasticity Theory and Applications*, Pergamon, New York, 1974.

Sechler, E. E., *Elasticity in Engineering*, Wiley, New York, 1952.

Seeley, F. B., and Smith, J. O., *Advanced Mechanics of Materials*, 2nd ed., Wiley, New York, 1957.

Sokolnikoff, I. S., *Mathematical Theory of Elasticity*, McGraw-Hill, New York, 1946.

Southwell, R. V., *An Introduction to the Theory of Elasticity for Engineers and Physicists*, Oxford Univ. Press, 2nd ed., London, 1941.

Timoshenko, S. P., and Goodier, J. N., *Theory of Elasticity*, 3rd ed., McGraw-Hill, New York, 1970.

Wang, C. C., and Truesdell, C., *Introduction to Rational Elasticity*, Noordhoff, Leyden, The Netherlands, 1973.

Wang, C. T., *Applied Elasticity*, McGraw-Hill, New York, 1953.

Williams, J. G., *Stress Analysis of Polymers*, 2nd ed., Ellis Horwood, New York, 1980.

Part 2

Fundamentals of Structural Analysis

12

Bending of Beams

A structure may be defined as a collection of bodies (structural elements) arranged and supported in such a way that it can withstand and transmit external and/or thermally induced loads. As indicated in the Introduction, the body of knowledge associated with the description and prediction of the behavior of structures is structural analysis, or structural mechanics. Unlike the theories of elasticity, plasticity, or viscoelasticity, which are concerned with the study of bodies of various shapes and even continuum media, structural mechanics, which is, in essence, an advanced strength of materials, takes advantage of geometric characteristics of certain classes of structures. Examples are: bars that are bodies whose cross-sectional dimensions are small compared with their lengths, plates that are flat bodies whose thickness is small compared with the other two dimensions, and shells that can be identified as curved plates.

Structural analysis disciplines, unlike strength of materials, deal, as a rule, not with "abstract" bodies of relatively simple geometry but only with those structural elements and joints that are used in the given branch of engineering. Examples of typical structural elements and joints employed in microelectronic and fiber-optic systems are: beams, frames, plates, various bi- and multimaterial assemblies, thin-film structures, and solder-joint interconnections. For instance, optical fibers, electrical contactors, bending test specimens, or wires in flexible test probes can be idealized as beams or bars experiencing bending or buckling. Various chassis can be treated as frames. Many caps and lids in electronic packages can be evaluated on the basis of a flat-plate theory. Accordingly, in this part, the emphasis is on the mechanical behavior of beams, frames, and plates. Note that although shell elements may also be encountered in microelectronic structures (barrels of plated-through holes, for instance, can be treated as shells of revolution), these elements are thought to be rather atypical for the areas of engineering in question and are not examined in this book.

With few exceptions, our primary concern will be with linear static problems. Unless noted otherwise, we assume that all members of the structural system are constructed of linearly elastic, homogeneous, and isotropic materials and that the displacements are small compared to the dimensions of the given element. The analyses of linearly behaving structures, as the name implies, can be based on the linear differential equations. In addition, these structures usually obey the principle of superposition, which states that the combined effect of a number of loads acting on a structure is equal to the sum of the effects of each load applied separately. Utilization of this principle simplifies significantly the analysis of structures subjected to different types of loadings and, particularly, to external (mechanical) and internal (thermally or lattice-mismatch–induced) loads.

12.1 BASIC DEFINITIONS, HYPOTHESES, AND RELATIONSHIPS

A *beam*, as well as a bar, is a body whose cross-sectional dimensions are small compared to its length. A beam carries loads that are perpendicular to its longitudinal axis and cause stresses due to internal (elastic) bending moments and shearing forces. In the analyses that follow, we will deal only with *straight prismatic* beams, i.e., with beams whose axes are straight lines and whose *cross sections* have the same shape and areas along the beam.

Loads and bending moments distributed over small lengths and having relatively high intensity are usually substituted in the theory of bending of beams for statically equivalent *concentrated forces* and *concentrated moments*. This could be done because local contact stresses and strains are not examined in this theory. These are studied, when necessary, using theory-of-elasticity methods. Obviously, the concentrated load should be equal to the area under the loading curve and should be applied to the beam under the centroid of this area.

The forces acting on a beam can be subdivided into *active forces*, which are usually known beforehand, and *reactive forces (reactions)* occurring between the beam and its *supports*. The reactive forces should be determined in the process of solving the problem. In this section, we examine supports having a small area of contact with the beam. We assume that all the supports are pointlike and that each of them is associated with just one cross section of the beam, which is called a *support cross section*. Long supports, such as elastic foundations, will be examined in Section 12.6.

If a support is connected to the support cross section of a beam by a *hinge*, the displacement of this support is equal to the displacement of the beam cross section, and a reaction force occurs between the beam and the support. This is a *simple (free, hinged) support* (Fig. 12.1a). The displacement

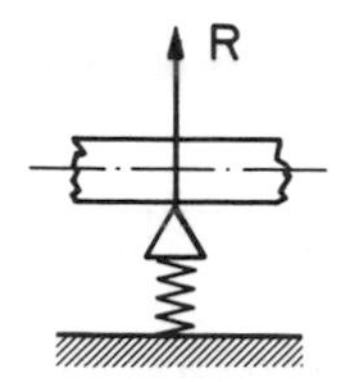

(a) HINGED (FREE) ELASTIC SUPPORT ($W \neq 0$, $R \neq 0$)

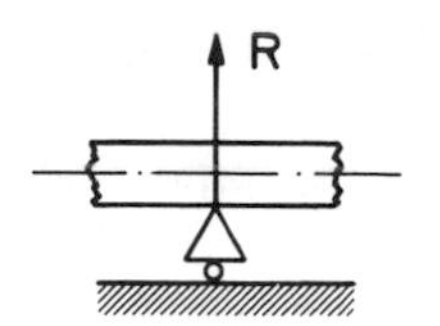

(b) HINGED RIGID SUPPORT ($W = 0$, $R \neq 0$)

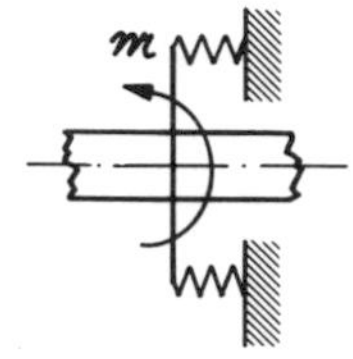

(c) ELASTICALLY CLAMPED (BUILT-IN) CROSS-SECTION ($\alpha \neq 0$, $\mathfrak{m} \neq 0$)

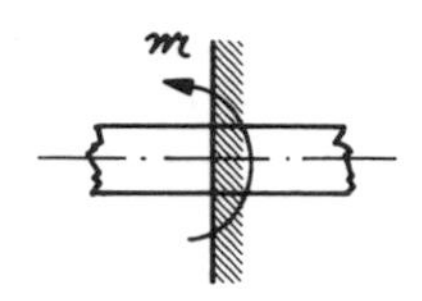

(d) RIGIDLY CLAMPED CROSS-SECTION ($\alpha = 0$, $\mathfrak{m} \neq 0$)

FIGURE 12.1 Different types of beam supports.

w_i of a linearly deformed simple support is proportional to its reaction R_i:

$$w_i = A_i R_i \qquad R_i = K_i w_i \tag{12.1}$$

Here, A_i is the *compliance of an elastic support*, and $K_i = 1/A_i$ is its *spring constant*. Supports characterized by infinitely large spring constants ($K_i \to \infty$, $A_i = 0$, $w_i = 0$) are called *rigid supports* (Fig. 12.1b). If a support is connected to the beam so that the angle of rotation of the support is equal to the angle of rotation of the beam in the support cross section whereas the linear displacement of the beam is not restricted at all (Fig. 12.1c), then a reactive moment occurs only in this cross section. This is a *clamped (built-in)* cross section. The angle α_i of rotation of a linearly deformed support in a clamped cross section is proportional to the reactive moment M_i:

$$\alpha_i = \bar{A}_i M_i \qquad M_i = \bar{K}_i \alpha_i \tag{12.2}$$

Here $\bar{A}_i$ is the compliance of the support with respect to the rotation angle, and $\bar{K}_i = 1/\bar{A}_i$ is its spring constant. In the extreme case $\bar{K}_i \to \infty$, $\bar{A}_i = 0$, $\alpha_i = 0$, and the cross section is called *rigidly clamped* (Fig. 12.1d). A rigid support and a rigidly clamped cross section are idealizations that are applied in those cases in which the linear w_i and angular α_i displacements at the support are negligibly small in comparison with the displacements of other cross sections. Clearly, if there are no supports in the ith cross section, then neither a reactive force nor a reactive moment occur in this cross section ($K_i = \bar{K}_i = 0$, $A_i = \bar{A}_i \to \infty$), whereas the displacement w_i and the rotation angle α_i have nonzero values. These are *free* cross sections.

As we know from the course in theoretical mechanics, a rigid body is in equilibrium if the resultant force and resultant couple are each equal to zero. Although, in the general case of a three-dimensional body, there are six equilibrium equations, for a coplanar system of forces acting on a beam, the following three equations of equilibrium must take place:

$$\sum F_x = 0 \qquad \sum F_z = 0 \qquad \sum M_0 = 0 \tag{12.3}$$

The first two equations state that the sum of the projections of all the forces on the coordinate axes must be zero, and the third equation states that the sum of the moments about an axis parallel to the y axis through any point 0 must be zero. If, in addition, all the loads acting on a beam are parallel to the y axis, the first condition in Eqs. (12.3) is fulfilled identically, and the number of the available static equations is reduced to two. Therefore, if the beam supports create only two reactions, the latter can be determined from the equations of statics, and such a beam is called *statically determinate.* Obviously, only the following three cases result in a statically determinate problem for the reactions in a single-span beam (Fig. 12.2):

1. Simply supported beam (hinged at two cross sections)
2. Cantilever beam, when only one cross section is elastically or rigidly clamped and is elastically or rigidly supported
3. A beam elastically or rigidly supported at one cross section and elastically or rigidly clamped at the other cross section

In all the other cases, the number of the unknown reactions is greater than two, and the beam becomes *statically indeterminate.* In this case, all the reactions but two are called *redundant unknowns* since they are not necessary from the standpoint of the equilibrium of the beam subjected to an arbitrary loading. As indicated in Section 2.5, solutions to the statically indeterminate problems are obtained by supplementing the equations of statics with additional conditions of compatibility of displacements.

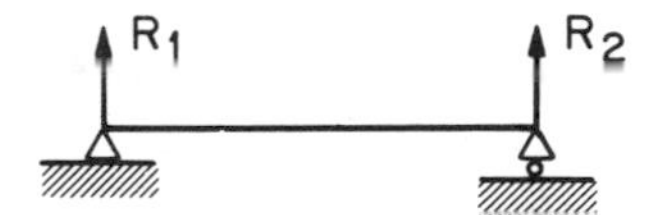

(a) FREELY SUPPORTED BEAM

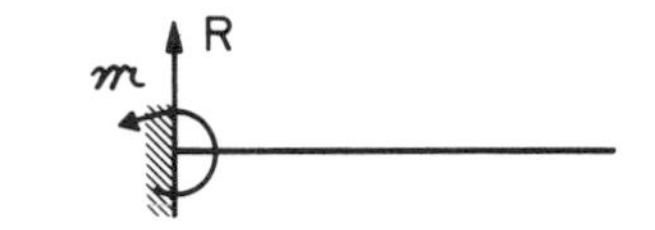

(b) CANTILEVER

(c) HINGED–CLAMPED BEAM

FIGURE 12.2 Single-span beams with different types of supports.

Once the reactions have been calculated, we can proceed to calculate the internal lateral (shearing) forces and bending moments by taking a slice of the beam and considering the loads that act on this portion. A *lateral (shearing) force* is a resultant of all the internal elastic shearing forces acting in the given cross section and equilibrating the external forces (including the reactive ones). A *bending moment* in the given cross section is the moment of all the internal elastic normal forces acting in the direction perpendicular to the cross-sectional plane. This moment is calculated about an axis passing through the centroid of the cross section. If the beam is located in the x,z plane, then the lateral force $N(x)$ is parallel to the z axis, and the bending moment $M(x)$ is evaluated about the y axis. Examples of the *diagrams* for the lateral forces and bending moments acting over the cross sections of a beam are shown in Fig. 12.3a and b.

The elementary beam theory is based on the following two basic assumptions:

1. The stresses σ_y, σ_z, and τ_{yz} acting in the planes parallel to the x axis are negligibly small compared to the stresses σ_x, τ_{xy}, and τ_{xz} acting over the cross-sectional areas and can be put equal to zero.

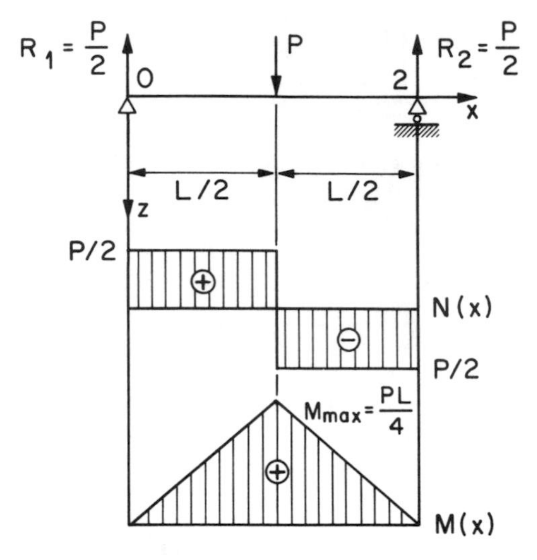

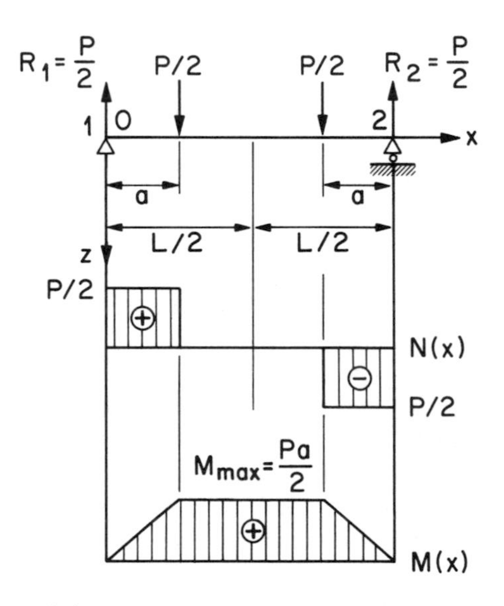

FIGURE 12.3 Beams subjected to three- and four-point bending.

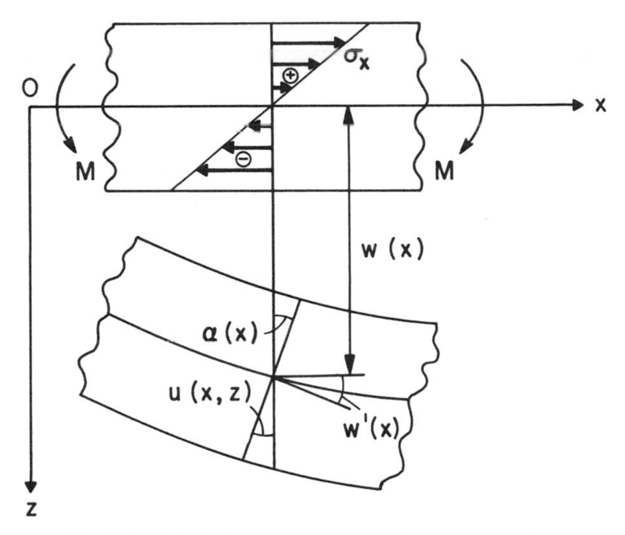

FIGURE 12.4 Deflected shape of a beam in bending.

2. The normal stresses σ_x can be evaluated under an assumption that the cross sections of the beam remain plane during bending and perpendicular to the elastic curve of the beam (*hypothesis of plane cross sections*). This assumption states that the angles $\alpha(x)$ of rotation of the cross sections are equal to the slope angles $w'(x)$ of the deflection curve $w(x)$. In accordance with the hypothesis of plane cross sections, the longitudinal (along the x axis) displacements of the points of an arbitrary cross section can be presented as (Fig. 12.4)

$$u(x, z) = -z\alpha(x) = -zw'(x) \tag{12.4}$$

This results in the normal strain

$$\varepsilon_x = \frac{\partial u}{\partial x} = -zw''(x) \tag{12.5}$$

Then, the normal stress σ_x calculated on the basis of Hooke's law (2.55) is

$$\sigma_x = E\varepsilon_x = -Ezw''(x) \tag{12.6}$$

The bending moment at the given cross section x is

$$M(x) = -\int_A \sigma_x z \, dA = EI_y(x)w''(x) \tag{12.7}$$

where

$$I_y(x) = \int_A z^2 \, dA \tag{12.8}$$

is the *moment of inertia* of the cross section. From Eqs. (12.6) and (12.7), we find that the bending normal stress σ_x is related to the bending moment $M(x)$ by the equation

$$\sigma_x(x, z) = -\frac{M(x)}{I_y(x)} z \tag{12.9}$$

Formulas (12.5) and (12.9) indicate that the normal stresses and strains in the direction of the beam axis are zero at all the points of the line $z = 0$ passing through the centroid of the cross-sectional areas. Therefore, this line is called the *neutral axis* of the cross section. The neutral axis is perpendicular to the plane xz of bending. The neutral axes of all the cross sections form a *neutral surface* or a *neutral layer* of a bent beam.

As follows from the theory-of-elasticity solution presented in Section 4.1, the conditions

$$\sigma_y = \sigma_z = \tau_{yz} = 0$$

require that the beam (bar) is prismatic and experiences pure bending, i.e., that

$$I_y(x) = \text{const} \qquad M(x) = \text{const} \qquad N(x) = 0$$

Therefore, Eq. (12.7), in which the bending moment and the moment of inertia vary along the beam, cannot be justified by the theory-of-elasticity relationships, and some of these relationships have to be sacrificed when the shearing stresses τ_{xy} and τ_{xz} are determined. It can be shown, however, that if these stresses are obtained equilibrium equations while the strain compatibility conditions are violated, then the obtained results agree satisfactorily with the more rigorous (and more complicated) theory-of-elasticity solutions.

The second and the third equilibrium conditions in Eqs. (12.3) result in the following formulas (Fig. 12.5):

$$\sum F_z = N(x) - \left[N(x) + \frac{dN(x)}{dx} dx \right] + q(x) \, dx = 0$$

$$\sum M_0 = M(x) - \left[M(x) + \frac{dM(x)}{dx} dx \right] + N(x) \, dx = 0$$

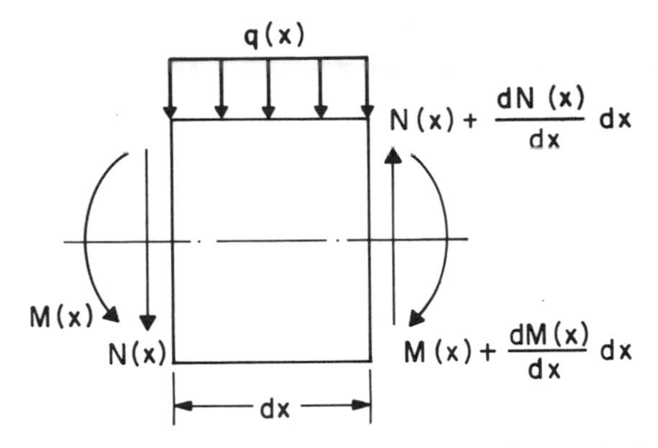

FIGURE 12.5 Forces and moments acting on an element of a beam.

where $q(x)$ is the distributed lateral load per unit beam length. After simplifying these equations, we obtain the following differential relationships:

$$\frac{dN(x)}{dx} = q(x) \qquad \frac{dM(x)}{dx} = N(x) \tag{12.10}$$

In addition to the equilibrium conditions $\Sigma F_z = 0$ and $\Sigma M_0 = 0$, we use a condition that any portion of the beam cut off by an arbitrary longitudinal cross section 1–1 or 2–2 (Fig. 12.6a) must be in equilibrium. For the portion cut off by the cross section 1–1, the equilibrium condition is

$$\int_{A_1} \left(\sigma_x + \frac{\partial \sigma_x}{\partial x} dx \right) dy\, dz - \int_{A_1} \sigma_x\, dy\, dz + \tau_{xy} \delta\, dx = 0$$

so that

$$\tau_{xy} = -\frac{1}{\delta} \int_{A_1} \frac{\partial \sigma_x}{\partial x} dy\, dz$$

From Eq. (12.9), assuming that the bending moment $M(x)$ usually changes much more rapidly along the beam than the moment of inertia $I_y(x)$ of the cross section, we have

$$\frac{\partial \sigma_x}{\partial x} \cong -\frac{z}{I_y(x)} \frac{dM(x)}{dx} = -\frac{N(x)}{I_y(x)} z$$

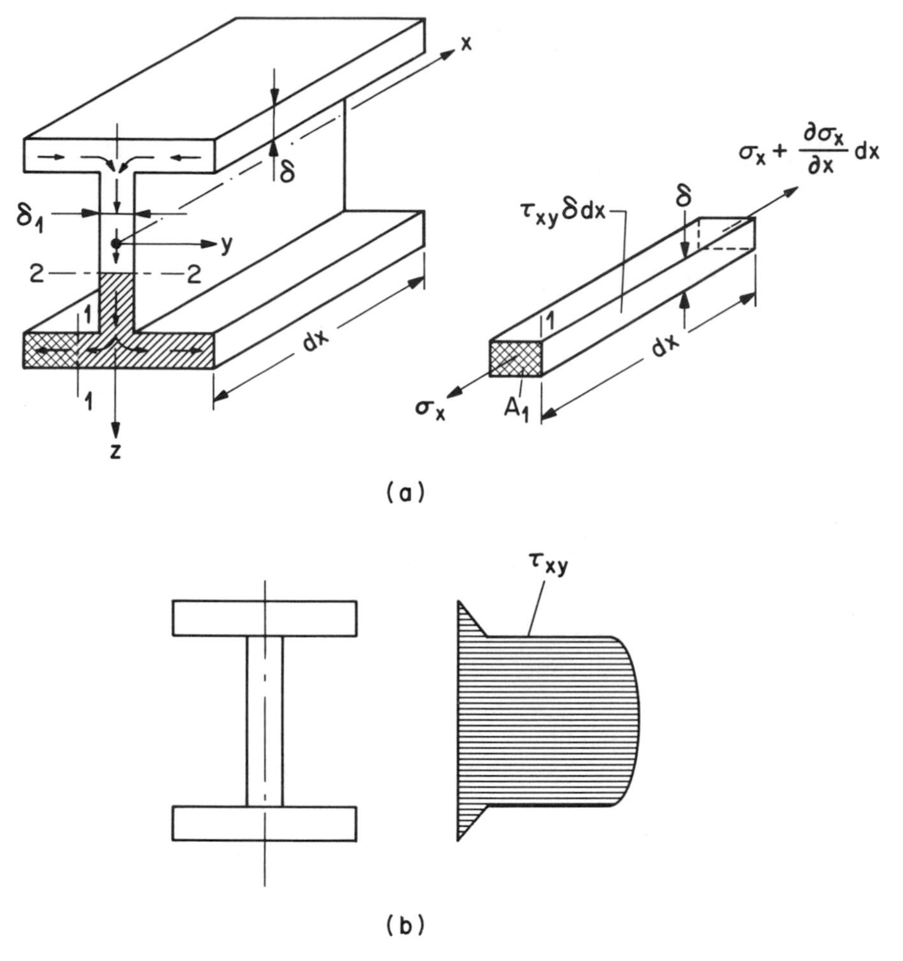

FIGURE 12.6 Shearing stress in a beam subjected to bending.

Introducing this formula into the preceding relationship for the shearing stress τ_{xy}, we have

$$\tau_{xy} = \frac{N(x)S_y}{I_y(x)\delta} \tag{12.11}$$

where

$$S_y = \int_{A_1} z \, dA_1$$

is the static moment of the area A_1 with respect to the neutral axis. Similarly, examining the equilibrium of the portion of the beam cut off by the cross section 2–2, we have

$$\tau_{xz} = \frac{N(x)S_y}{I_y(x)\delta_1} \tag{12.12}$$

Formulas (1.11) and (1.12) enable us to evaluate the shearing stress for any point of the cross section of a beam subjected to bending. The distribution of the shearing stress over the cross section of the beam is shown schematically in Fig. 12.6b.

Summarizing the obtained results, we conclude that the elementary beam theory is aimed at the evaluation of the elastic curve (deflection function), angles of rotation of the cross sections, support reactions, bending moments, shearing forces, and the resulting normal and shearing stresses.

The relationships (12.7) and (12.10) result in the following basic equation for the bending of a nonprismatic beam, i.e., a beam whose cross section is x-dependent:

$$[EI_y(x)w''(x)]'' = q(x) \tag{12.13}$$

For a prismatic beam having a constant moment of inertia $I_y(x) = I$, this equation yields

$$EIw^{\mathrm{IV}}(x) = q(x) \tag{12.14}$$

The solutions to this equation must satisfy the following obvious boundary conditions at the supports:

At an elastic support: $w_i = R_i/K_i$, $w_i'' = 0$ (the support reaction is directly proportional to the deflection, and the curvature, which is proportional to the bending moment, is zero).

At a rigid support: $w_i = 0$, $w_i'' = 0$ (both the deflection and the curvature are zero).

At an elastically clamped end: $w_i' = M_i/\bar{K}_i$, $w_i'' = 0$ (the bending moment is directly proportional to the angle of rotation, and the curvature is zero).

At a rigidly clamped end: $w_i = 0$, $w_i' = 0$ (both the deflection and the rotation angle are zero).

At a free end: $w_i'' = 0$, $w_i''' = 0$ (both the curvature, which is proportional to the bending moment, and its derivative, which is proportional to the shearing force, are zero).

12.2 SOLUTIONS BASED ON DIRECT INTEGRATION OF THE EQUATION OF BENDING

We illustrate this method using examples of beams subjected to three- and four-point bending.

The bending moment in the arbitrary cross section x of a beam shown in Fig. 12.3a is

$$M(x) = \frac{P}{2}x \qquad 0 \le x \le \frac{L}{2}$$

Then, Eq. (12.7) yields

$$w''(x) = \frac{P}{2EI}x$$

Integrating this equation twice, we have

$$w'(x) = \frac{P}{4EI}x^2 + C_1 \qquad w(x) = \frac{P}{12EI}x^3 + C_1 x + C_2 \tag{12.15}$$

The constants C_1 and C_2 of integration can be determined on the basis of the boundary conditions

$$w(0) = 0 \qquad w'\left(\frac{L}{2}\right) = 0$$

The first condition reflects the fact that the deflection of the beam at $x = 0$ is zero. The second condition is due to the symmetry of the elastic curve with respect to the mid–cross section of the beam. After substituting the expressions for the elastic curve and its derivative into the above-mentioned boundary conditions, we find

$$C_1 = -\frac{PL^2}{16EI} \qquad C_2 = 0$$

so that the elastic curve is expressed by the equation

$$w(x) = w_0 \frac{x}{L}\left(3 - 4\frac{x^2}{L^2}\right) \tag{12.16}$$

where

$$w_0 = -\frac{PL^3}{48EI} \tag{12.17}$$

is the deflection at the mid–cross section. Equation (12.7), written for the midportion of the beam in Fig. 12.3b, is

$$w''(x_1) = \frac{Pa}{2EI} \qquad 0 \le x_1 \le L - 2a$$

Integrating this equation twice, we obtain

$$w_1'(x_1) = \frac{Pa}{2EI}x_1 + D_1 \qquad w_1(x) = \frac{Pa}{4EI}x_1^2 + D_1x_1 + D_2 \tag{12.18}$$

The condition of symmetry $w_1'[(L/2) - a] = 0$ yields

$$D_1 = -\frac{Pa}{4EI}(L - 2a)$$

so that

$$w_1(0) = D_2 \qquad w_1'(0) = D_1 = -\frac{Pa}{4EI}(L - 2a)$$

The elastic curve at the portion $0 \le x \le a$ of the beam can be described by Eqs. (12.15), where the constant C_2 should be put equal to zero to satisfy the condition $w(0) = 0$. At the cross section $x = a$, we have

$$w(a) = \frac{Pa^3}{12EI} + C_1a \qquad w'(a) = \frac{Pa^2}{4EI} + C_1$$

The conditions of compatibility

$$w_1(0) = w(a) \qquad w_1'(0) = w'(a)$$

result in the following values of the constants C_1 and D_2:

$$C_1 = -\frac{Pa}{4EI}(L - a) \qquad D_2 = -\frac{Pa^2}{12EI}(3L - 4a)$$

Then the deflection function in the midportion of the beam, given by the second equation in Eqs. (12.18) is,

$$w_1(x) = w_0 \frac{4a(3L - 4a) + 12(L - 2a)x_1 - 12x_1^2}{3L^2 - 4a^2}$$

where

$$w_0 = -\frac{Pa}{48EI}(3L^2 - 4a^2) \tag{12.19}$$

is the deflection at the mid–cross section.

Let us now determine the stresses occurring in a beam of a rectangular cross section subjected to three-point bending. The maximum bending moment is in the mid–cross section and is $M_{max} = PL/4$. Then the maximum normal stress given by Eq. (12.9) is

$$\sigma_{max} = \frac{PL}{bh^3/12} \frac{h}{2} = 6 \frac{PL}{bh^2}$$

The static moment of area A_1 with respect to the neutral axis (Fig. 12.7) is

$$S_y = \int_{A_1} z \, dA_1 = A_1 z_c = \frac{b}{2}\left[\left(\frac{h}{2}\right)^2 - z^2\right]$$

The maximum shearing force is $N = P/2$ and, in accordance with Eq. (12.11), the shearing stress is distributed over the height of the beam as follows:

$$\tau_{xy}(z) = \frac{3P}{bh^3}\left(\frac{h^2}{4} - z^2\right) = \tau_{max}\left[1 - \left(\frac{2z}{h}\right)^2\right]$$

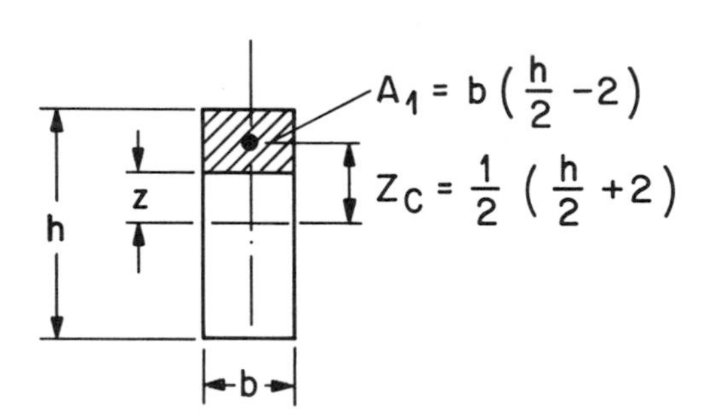

FIGURE 12.7 Definition of a static moment of an area (A_1) with respect to an axis (z).

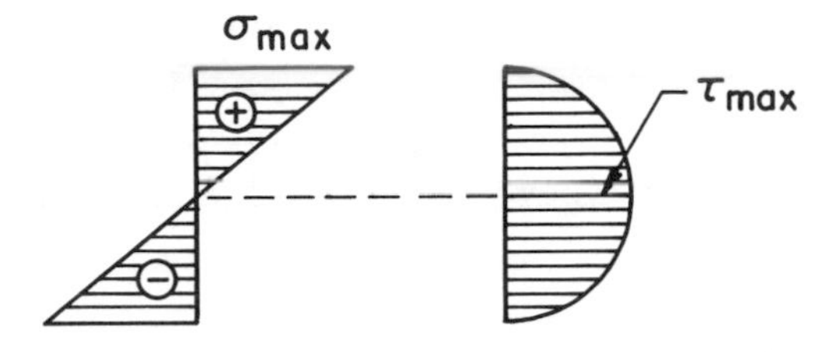

FIGURE 12.8 Distribution of the normal (σ) and shearing (τ) stresses over the height of the beam cross section.

where

$$\tau_{\max} = \frac{3P}{4bh} = \frac{h}{8L}\sigma_{\max}$$

is the maximum shearing stress. This takes place at the neutral axis $z = 0$. As evident from the latter formula, for sufficiently long beams, the shearing stress is substantially smaller than the maximum normal stress and may not be taken into account. The distribution of the normal and the shearing stress over the beam's height is shown schematically in Fig. 12.8.

12.3 METHOD OF INITIAL PARAMETERS

Equation (12.7) or (12.14) can be integrated quite easily if there are no concentrated forces and moments in the beam's span and when the distributed load per unit beam length is given by a single analytical expression. Otherwise, as shown in the preceding section, the integration has to be performed for each portion of the beam separately, and the constants of integration are determined using the compatibility conditions for the adjacent portions of the beam. This is a rather tedious and complicated procedure. It can be significantly simplified, however, by using the *method of initial parameters*.

Let a beam be loaded as shown in Fig. 12.9. Note that all the further considerations are based on an assumption that, once applied, distributed load does not change its analytical expression until the very end of the beam. If, for instance, a uniformly distributed load q is applied at the portion $a \leq x \leq b$, we will assume that this load acts at the portion $a \leq x \leq L$, whereas an additional load $-q$ acts at the portion $b \leq x \leq L$, where L is the total length of the beam.

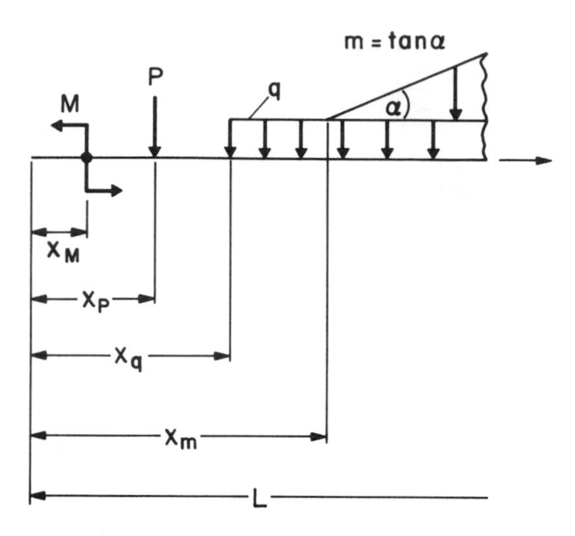

FIGURE 12.9 Different types of loads on a beam in bending.

In accordance with the method of initial parameters, the integral of the equation of bending is presented in the form

$$w(x) = w(0) + w'(0)x + \frac{M_0 x^2}{2!EI} + \frac{N_0 x^3}{3!EI}$$
$$+ \Big\|_{x_M} \frac{M(x - x_M)^2}{2!EI} + \Big\|_{x_P} \frac{P(x - x_P)^3}{3!EI} + \Big\|_{x_q} \frac{q(x - x_q)^4}{4!EI} + \Big\|_{x_\alpha} \frac{m(x - x_m)^5}{5!EI} \tag{12.20}$$

The sign $\|_{x_*}$ indicates that the corresponding term should be considered only for $x > x_*$.

In the preceding equation, the first four terms are the solution to the equation of bending for the initial portion (starting from the left end) of the beam. In this solution, $w(0)$, $w'(0)$, M_0, and N_0 are the displacement, the rotation angle, the moment, and the shearing force, respectively, at the left end $x = 0$ of the beam (*initial parameters*). The rest of the terms can be viewed as corrections, which take into account the contributions of the loads applied in the beam's span.

We illustrate the use of the method of initial parameters for the case of bending of the beam in Fig. 12.3b. Equation (12.20) yields, in this case,

$$w(x) = w'(0)x + \frac{Px^3}{12EI} - \Big\|_a \frac{P(x - a)^3}{12EI} - \Big\|_{L-a} \frac{P(x - L + a)^3}{12EI}$$

Then, we have

$$w'(x) = w'(0) + \frac{Px^2}{4EI} - \Bigg\|_a \frac{P(x-a)^2}{4EI} - \Bigg\|_{L-a} \frac{P(x-L+a)^2}{4EI}$$

The condition $w'(L/2) = 0$ results in the following formula for the initial parameter (angle of rotation) $w'(0)$:

$$w'(0) = -\frac{Pa}{4EI}(L-a)$$

The maximum deflection at $x = L/2$ is

$$w_0 = -\frac{Pa}{48EI}(3L^2 - 4a^2)$$

This formula coincides with Eq. (12.19) but is obtained in a much simpler way.

The method of initial parameters can also be used to solve statically indeterminate problems. As an example, examine a beam clamped at both ends, whose left end is displaced by the distance Δ (Fig. 12.10). Such a situation might occur, for instance, when the portion of an optical fiber located within a termination fixture experiences bending due to the offset of the fiber ends. Such an offset can be caused by the misalignment of the opening in the fixture frame and the aperture in the electronic device.

The equation of the initial parameters for the beam in question is

$$w(x) = \Delta + \frac{M_0 x^2}{2EI} + \frac{N_0 x^3}{6EI} \tag{12.21}$$

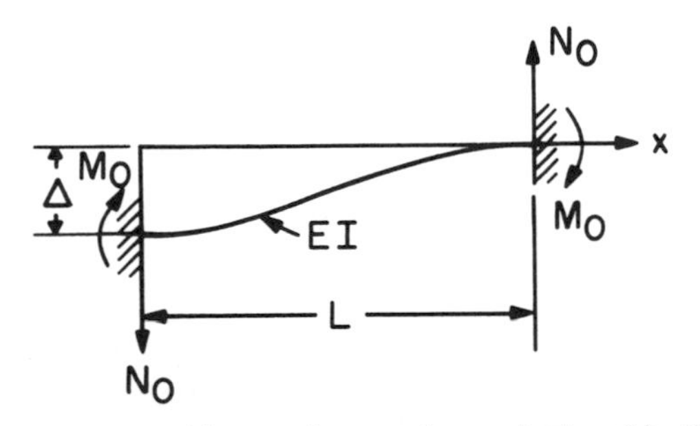

FIGURE 12.10 Clamped-clamped beam whose ends are displaced by the given distance.

Then we have

$$w'(x) = \frac{M_0 x}{EI} + \frac{N_0 x^2}{2EI}$$

Conditions $w(L) = 0$ and $w'(L) = 0$ result in the following formulas for the initial parameters M_0 and N_0:

$$M_0 = -\frac{6EI}{L^2}\Delta \qquad N_0 = \frac{12EI}{L^3}\Delta \tag{12.22}$$

Then the elastic curve of the beam is

$$w(x) = \Delta\left(1 - 3\frac{x^2}{L^2} + 2\frac{x^3}{L^3}\right)$$

The maximum curvatures responsible for the fiber strength and the possible added transmission losses occur at the end cross sections:

$$w''_{\max} = \pm 6\frac{\Delta}{L^2}$$

This formula indicates that increasing the length L of the portion of the fiber within the termination fixture is a more effective way to bring down the maximum curvature than reducing the misalignment Δ.

12.4 REFERENCE TABLES FOR BEAM DEFLECTIONS

In engineering practice, we can often obtain all the necessary parameters of bending of single-span beams simply by using reference tables for beam deflections. Some of these tables are presented in Appendix A. We can also use these tables, which have been developed for relatively simple loadings, to obtain solutions to more complicated combinations of loads by simply superposing the table data for the appropriate special cases. It is noteworthy that such an approach is based on the application of the superposition principle and, therefore, can be used only for ideally elastic beams with linearly deformed supports.

The tables for beam deflections can also be helpful when we deal with statically indeterminate problems for which no reference tables are available. In order to find the redundant reactions, we have to determine first the *degree*

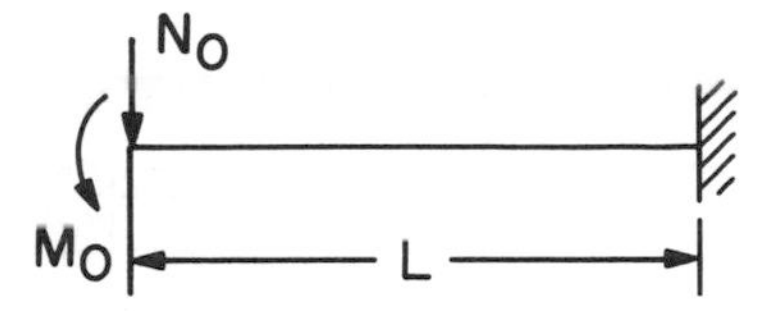

FIGURE 12.11 Cantilever beam subjected to a concentrated force and a concentrated moment at the free end.

of the statical indeterminacy of the beam and the *redundant constraints.* These are mentally removed from, and replaced by, the corresponding reactions: a moment M_i instead of a rigidly or elastically clamped cross section and a reactive force R_i instead of a rigid or an elastic support. For the obtained statically determinate beam, we should evaluate, using reference tables, the deflections of the cross sections associated with the removed constraints: the angles of rotation instead of clamped cross sections, and the deflections instead of the support cross sections. Clearly, the redundant reactions will enter the expressions for the displacements as algebraic symbols since their numerical values are not yet known. The equations for the redundant reactions can be obtained by equating these expressions with the actual displacements.

As an example, examine the problem solved in the previous section (Fig. 12.10). This beam is twice statically indeterminate. By removing the left support and replacing it with a force N_0 and a moment M_0, we obtain the beam shown in Fig. 12.11. Using the table data for a cantilever beam loaded at its end by a force and a moment, and equating the corresponding linear and angular displacements to Δ and zero, respectively, we obtain the following equations:

$$\frac{N_0 L^3}{3EI} + \frac{M_0 L^2}{2EI} = \Delta \qquad \frac{N_0 L^2}{2EI} + \frac{M_0 L}{EI} = 0$$

which lead to formulas (12.22).

12.5 SHEAR DEFORMATIONS

In the previous developments, only the deflections due to bending moments were considered. However, in short and deep beam members experiencing high lateral forces, consideration should be given to the fact that shear is accompanied by a detrusion of a beam element, which produces additional deflections. These have to be considered to obtain a more accurate description of the elastic curve. In this section, we assess the effect of shear

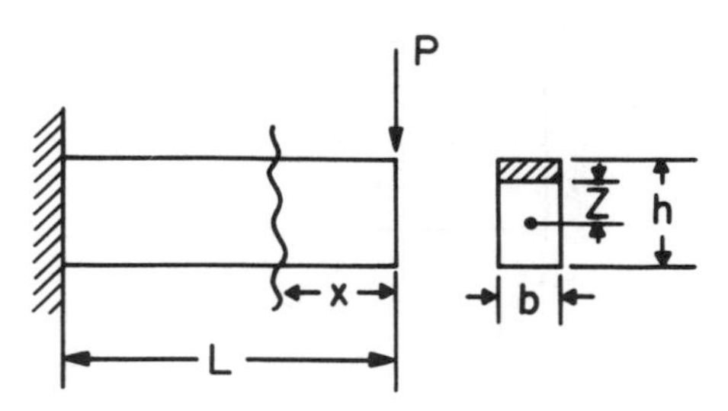

FIGURE 12.12 Cantilever beam of rectangular cross section subjected to a concentrated force at the free end.

deformations for a cantilever beam of a rectangular cross section (Fig. 12.12), using energy considerations.

The external work is

$$W = \int_0^{w_{max}} P(w)\, dw = \frac{P}{w_{max}} \int_0^{w_{max}} w\, dw = \frac{1}{2} P w_{max} \qquad (12.23)$$

where w_{max} is the maximum displacement of the beam. This formula simply states that if the force is proportional to the produced displacement and, therefore, changes, in the process of bending, from zero to P, then it can be replaced in the final expression for the external work by its mean value $P/2$. The strain energy due to bending can be evaluated using similar considerations. The end cross sections of a beam element dx subjected to bending moments $M(x)$ rotate with respect to each other by the angle $d\phi$. Then, the strain energy of this element is

$$dV_b(x) = \tfrac{1}{2} M(x)\, d\phi(x)$$

where $M(x)$ is the moment of the internal elastic forces equal to the external bending moment.

Since

$$d\phi(x) = d[w'(x)] = \frac{d}{dx}[w'(x)]\, dx = w''(x)\, dx$$

then,

$$dV_b(x) = \tfrac{1}{2} M(x) w''(x)\, dx = \frac{1}{2} \frac{M^2(x)}{EI_y(x)}\, dx$$

For the entire beam, the strain energy due to bending is

$$V_b = \frac{1}{2}\int_0^L \frac{M^2(x)\,dx}{EI_y(x)} \tag{12.24}$$

For a prismatic beam, $EI_y(x) = EI = \text{const}$. For the cantilever beam in question, $M(x) = -Px$, so that its strain energy due to bending is

$$V_b = \frac{1}{2EI}\int_0^L (-Px)^2\,dx = \frac{P^2L^3}{6EI} \tag{12.25}$$

The expression for the strain energy due to shearing stress can be obtained from Eq. (2.94), assuming that all the stresses but τ_{xz} are zero. Then the energy W per unit volume is

$$W = \frac{1+\nu}{E}\tau_{xz}^2 = \frac{1}{2G}\tau_{xz}^2$$

The strain energy per unit beam length can be expressed as

$$V_s' = \int_A \frac{\tau_{xz}^2}{2G}\,dA \tag{12.26}$$

where A is the cross-sectional area. The shearing stress τ_{xz} is distributed over the height of the beam in accordance with Eq. (12.20). In the case of a cantilever beam,

$$\tau_{\max} = \frac{3P}{2bh}$$

After substituting Eq. (12.20) into the expression (12.26) for the elastic energy, we have

$$V_s' = \frac{\tau_{\max}^2 b}{2G}\int_{-h/2}^{h/2}\left[1-\left(\frac{2z}{h}\right)^2\right]^2 dz = \frac{4}{15}\frac{\tau_{\max}^2 bh}{G} = \frac{3P^2}{5Gbh}$$

Then, for the entire beam, the strain energy due to shear is

$$V_s = \frac{3P^2L}{5Gbh} = \frac{P^2Lh^2}{20GI}$$

The energy balance condition $W = V_b + V_s$ results in the following formula for the maximum deflection:

$$w_{\max} = \frac{PL^3}{3EI}\left(1 + \frac{3E}{10G}\frac{h^2}{L^2}\right) = \frac{PL^3}{3EI}\left[1 + \frac{3E}{10G}\frac{h^2}{L^2}\right] = \frac{PL^3}{3EI}\left[1 + \frac{3}{5}(1+\nu)\frac{h^2}{L^2}\right] \tag{12.27}$$

where the second term in the parentheses gives the correction due to shear. For a metal beam with $\nu \cong \frac{1}{3}$, this term is about $0.8\ (h/L)^2$. Thus, for a very short beam (e.g., one with $h = L$), the total deflection is 1.8 times greater than the deflection due to bending. On the other hand, if $L = 10h$, then the deflection due to shear is less than 1 percent. This problem will be discussed in greater detail in the second volume, in connection with two- and three-point bending of specimens of rectangular or circular cross section.

12.6 BEAMS ON ELASTIC FOUNDATIONS

Relatively long dual-in-line packages (DIP), i.e., components terminating in two straight rows of pins of lead wires, can be treated as beams on elastic foundation if the wires are sufficiently compliant, their number is large, and the longitudinal distances between the two adjacent wires are small. Stress models based on the theory of beams on elastic foundation can also be used, for instance, in the analyses of the mechanical behavior of surface-mounted devices with compliant external leads and of coated optical fibers. In the latter case, the elastic foundation for the glass fiber is provided by the compliant coating material(s).

The term "elastic foundation" is used to indicate that the beam is supported by some load-bearing medium that responds elastically to beam deformations by developing a resisting load distributed in direct proportion to beam deflections. The problem of finding a complete and rigorous solution for the theory-of-elasticity equations for a beam lying on a semi-infinite elastic body is one of extraordinary analytical difficulty. This is due primarily to the fact that, in such a contact problem, the deformations of this body at the given point depend on the pressures not only at this point but at the adjacent points as well. A simplified engineering theory, however, assumes that a semi-infinite elastic body can be substituted in an approximate analysis by an infinite number of independent linear springs or rods and, therefore, the displacement at the given point is not affected by the pressures at the adjacent points. This assumption, known as *Winkler's hypothesis*, enables us to evaluate the distributed reactions of the foundation by the formula

$$q(x) = -Kw(x) \tag{12.28}$$

where K is the *spring constant* of the foundation, which is assumed to be constant and, as follows from Eq. (12.28), has the unit force per length squared. Obviously, Winkler's hypothesis is especially well suited for surface-mounted devices with separate compliant leads.

Using the equation of beam bending (12.14) and adding the reaction forces (12.28) to the external lateral load $q(x)$, we obtain the differential equation of bending of a beam on an elastic foundation in the form

$$w^{IV}(x) + 4\alpha^4 w(x) = \frac{q(x)}{EI} \tag{12.29}$$

where

$$\alpha = \sqrt[4]{\frac{K}{4EI}} \tag{12.30}$$

The characteristic equation of the homogeneous equation

$$w^{IV}(x) + 4\alpha^4 w(x) = 0 \tag{12.31}$$

has four roots $\pm\alpha(1 \pm i)$ and, therefore, its solution is

$$w(x) = C_1 e^{\alpha(1+i)x} + C_2 e^{-\alpha(1+i)x} + C_3 e^{\alpha(1-i)x} + C_4 e^{-\alpha(1-i)x}$$

where $i = \sqrt{-1}$ is the imaginary unity. Using the Euler formulas

$$\left.\begin{aligned} \cos \alpha x &= \frac{e^{i\alpha x} + e^{-i\alpha x}}{2} & \sin \alpha x &= \frac{e^{i\alpha x} - e^{-i\alpha x}}{2i} \\ \cosh \alpha x &= \frac{e^{\alpha x} + e^{-\alpha x}}{2} & \sinh \alpha x &= \frac{e^{\alpha x} - e^{-\alpha x}}{2} \end{aligned}\right\}$$

to get rid of the imaginaries, we rewrite the preceding solution for the deflection function as

$$w(x) = e^{\alpha x}(A \cos \alpha x + B \sin \alpha x) + e^{-\alpha x}(C \cos \alpha x + D \sin \alpha x) \tag{12.32}$$

In some cases, it is more convenient to present the obtained solution in the form

$$w(x) = A_0 V_0(\alpha x) + A_1 V_1(\alpha x) + A_2 V_2(\alpha x) + A_3 V_3(\alpha x) \tag{12.33}$$

where the functions $V(\alpha x)$ are expressed as follows:

$$\left.\begin{aligned} V_0(\alpha x) &= \cosh \alpha x \cos \alpha x \\ V_1(\alpha x) &= \frac{1}{\sqrt{2}}(\cosh \alpha x \sin \alpha x + \sinh \alpha x \cos \alpha x) \\ V_2(\alpha x) &= \sinh \alpha x \sin \alpha x \\ V_3(\alpha x) &= \frac{1}{\sqrt{2}}(\cosh \alpha x \sin \alpha x - \sinh \alpha x \cos \alpha x) \end{aligned}\right\} \tag{12.34}$$

These functions obey the following simple rule of differentiation, which makes their utilization convenient:

$$\left.\begin{aligned} V'_1(\alpha x) &= \alpha\sqrt{2}\,V_0(\alpha x) \\ V'_2(\alpha x) &= \alpha\sqrt{2}\,V_1(\alpha x) \\ V'_3(\alpha x) &= \alpha\sqrt{2}\,V_2(\alpha x) \\ V'_0(\alpha x) &= -\alpha\sqrt{2}\,V_3(\alpha x) \end{aligned}\right\} \tag{12.35}$$

The particular solution to Eq. (12.29) depends on the function $q(x)$. Obviously, since it is a fourth-order equation, this solution can be presented as

$$w(x) = \frac{q(x)}{K} \tag{12.36}$$

if the function $q(x)$ is expressed by a polynomial not higher than of a third power.

As an example, examine a semi-infinite beam, loaded at its end by a concentrated force N_0 and a bending moment M_0 (Fig. 12.13). The elastic

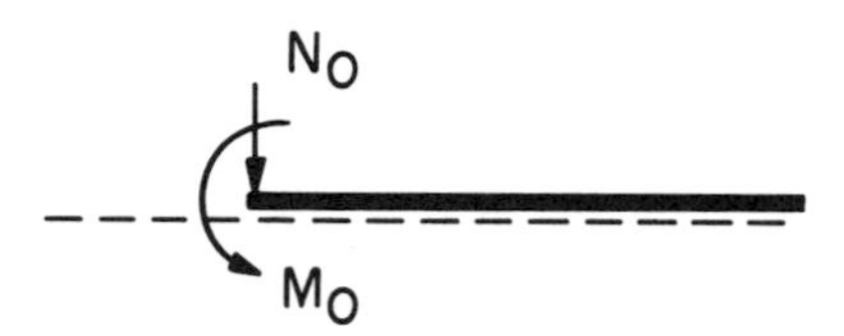

FIGURE 12.13 Semi-infinite beam on an elastic foundation under end force and moment.

curve can be sought in the form (12.32). Since the deflections $w(x)$ must tend to zero, with an increase in the distance x from the end, we have to put the constants A and B equal to zero in order to make the solution physically meaningful. Then we have the following equations for the deflection function $w(x)$ and its derivatives:

$$\left.\begin{aligned} w(x) &= e^{-\alpha x}(C\cos\alpha x + D\sin\alpha x) \\ w'(x) &= -\alpha e^{-\alpha x}[(C-D)\cos\alpha x + (C+D)\sin\alpha x] \\ w''(x) &= 2\alpha^2 e^{-\alpha x}(C\sin\alpha x - D\cos\alpha x) \\ w'''(x) &= 2\alpha^3 e^{-\alpha x}[(C+D)\cos\alpha x - (C-D)\sin\alpha x] \end{aligned}\right\} \tag{12.37}$$

By substituting the last two equations in the boundary conditions

$$w''(0) = \frac{M_0}{EI} \qquad w'''(0) = \frac{N_0}{EI}$$

we obtain the following equations for the constants C and D:

$$-2\alpha^2 EID = M_0 \qquad 2\alpha^3 EI(C+D) = N_0$$

so that

$$C = \frac{N_0 + \alpha M_0}{2\alpha^3 EI} \qquad D = -\frac{M_0}{2\alpha^2 EI}$$

Then Eqs. (12.37) result in the following final formulas for the deflection, angle of rotation, bending moment, and shearing force:

$$\left.\begin{aligned} w(x) &= \frac{e^{-\alpha x}}{2\alpha^3 EI}[N_0\cos\alpha x + \alpha M_0(\cos\alpha x - \sin\alpha x)] \\ w'(x) &= -\frac{e^{-\alpha x}}{2\alpha^2 EI}[N_0(\cos\alpha x + \sin\alpha x) + \alpha M_0\cos\alpha x] \\ M(x) &= EIw''(x) = \frac{e^{-\alpha x}}{\alpha}[N_0\sin\alpha x + \alpha M_0(\cos\alpha x - \sin\alpha x)] \\ N(x) &= EIw'''(x) = e^{-\alpha x}[N_0(\cos\alpha x - \sin\alpha x) - 2\alpha M_0\sin\alpha x] \end{aligned}\right\} \tag{12.38}$$

The diagrams for the functions $w(x)$, $M(x)$, and $N(x)$ are shown in Fig. 12.14 for the case $M_0 = 0$. The obtained formulas will be used in the applied part

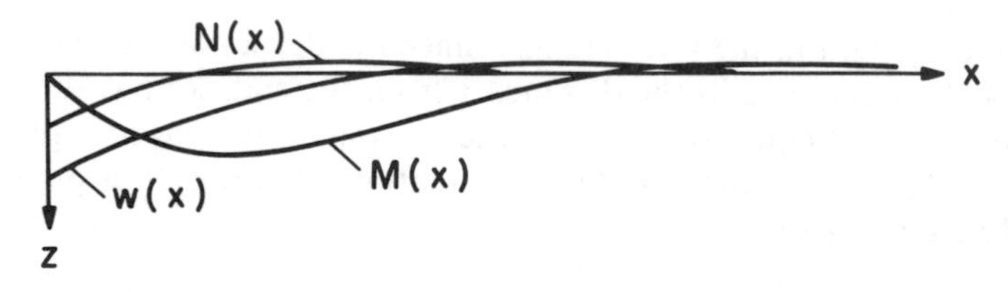

FIGURE 12.14 Distribution of deflections, lateral forces, and bending moments in a semi-infinite beam loaded by an end force.

of the book to evaluate stresses and deflections in clamped optical fibers located within termination fixtures. The solution in the form (12.33) will be applied in the second volume to the case of bending of a surface-mounted device with compliant external leads.

12.7 EFFECT OF SHEAR ON THE BENDING OF BEAMS ON ELASTIC FOUNDATIONS

Consideration of shear might be important, for instance, for surface-mounted components with compliant external leads if their length-to-height ratio is smaller than, say, eight.

In this analysis, we use the following approximate formula for the angle of rotation due to shear

$$\gamma = w_s' = \frac{\tau}{G} \tag{12.39}$$

where w_s is the deflection associated with shear, τ the mean value of the shearing stress, and G the shear modulus of the material. Assuming that the shear angle γ is positive when the tangent to the elastic curve rotates clockwise, we define stress τ by the formula

$$\tau = -\frac{N}{A}$$

so that a positive lateral force N would result in a negative shear angle. In the latter formula, A is the cross-sectional area of the beam. Thus, we have

$$w_s' = \frac{N}{GA}$$

Integrating this equation and considering the second differential relationship in Eq. (12.10), we obtain the following formula:

$$w_s = -\frac{M}{GA} + C \tag{12.40}$$

where C is a constant of integration.

As far as the deflection w_b due to bending is concerned, it can be found from Eq. (12.7), which yields

$$EIw_b'' = M \tag{12.41}$$

After differentiating this equation twice and having in mind that the lateral load acting on the beam consists of an external load $q(x)$ and the reaction $K(w_b + w_s)$ of the elastic foundation, we obtain

$$EIw_b^{\mathrm{IV}} = q(x) - K(w_b + w_s) \tag{12.42}$$

From Eqs. (12.40) and (12.41), we have

$$w_s = -\frac{EI}{GA} w_b'' + C$$

Introducing this formula in Eq. (12.42), we obtain the following equation for bending deflections:

$$EIw_b^{\mathrm{IV}} - \frac{EI}{GA} Kw_b'' + Kw_b = q(x) - KC \tag{12.43}$$

The particular solution to this equation corresponding to the term KC is $w_b = -C$. As we can see using Eq. (12.40), the total deflection $w_b + w_s$ is C-independent, and so are the angle of rotation, bending moment, and lateral force. Therefore, we can put C equal to zero without sacrificing the generality of the solution to the problem. Thus, the differential equation of bending for a beam on elastic foundation with consideration of shear is as follows:

$$EIw_b^{\mathrm{IV}} - \frac{EI}{GA} Kw_b'' + Kw_b = q(x) \tag{12.44}$$

The characteristic equation for the corresponding homogeneous equation

$$EIw_b^{\mathrm{IV}} - \frac{EI}{GA} Kw_b'' + Kw_b = 0 \tag{12.45}$$

is

$$\beta^4 - \frac{K}{GA}\beta^2 + \frac{K}{EI} = 0 \tag{12.46}$$

This equation has the following roots:

$$\beta = \pm\sqrt{\frac{K}{2GA} \pm \sqrt{\left(\frac{K}{2GA}\right)^2 - \frac{K}{EI}}} = \pm\sqrt{\frac{K}{2GA}}\sqrt{1 \pm \sqrt{1 - \left(\frac{2GA}{\sqrt{KEI}}\right)^2}}$$

Examine four special cases characterized by different values of the parameter

$$\eta = \frac{\sqrt{KEI}}{2GA} \tag{12.47}$$

1. η is small. Then the second term in Eq. (12.45) is also small, and shear may not be taken into account. The solution is expressed, in this case, by Eq. (12.32) or (12.33).
2. $\eta < 1$. Then the biquadratic equation (12.46) has two couples of conjugate complex roots:

   $$\beta = \pm\gamma_1 \pm i\gamma_2$$

   where

   $$\gamma_1 = \alpha\sqrt{1+\eta} \qquad \gamma_2 = \alpha\sqrt{1-\eta}$$

   and the α value is given by Eq. (12.30). The integral of Eq. (12.45) is, in this case, as follows:

   $$\begin{aligned} w_b(x) = {} & C_1 \cosh \gamma_1 x \cos \gamma_2 x + C_2 \cosh \gamma_1 x \sin \gamma_2 x \\ & + C_3 \sinh \gamma_1 x \cos \gamma_2 x + C_4 \sinh \gamma_1 x \sin \gamma_2 x \end{aligned} \tag{12.48}$$

3. $\eta > 1$. In this case, the biquadratic equation (12.46) has four real roots:

   $$\beta_{1,2} = \pm\gamma_1 \qquad \beta_{3,4} = \pm\gamma_2$$

   where

   $$\gamma_1 = \alpha\sqrt{2(\eta + \sqrt{\eta^2 - 1})} \qquad \gamma_2 = \alpha\sqrt{2(\eta - \sqrt{\eta^2 - 1})}$$

The integral of Eq. (12.45) is then expressed as

$$w_b(x) = C_1 \cosh \gamma_1 x + C_2 \sinh \gamma_1 x + C_3 \cosh \gamma_2 x + C_4 \sinh \gamma_2 x \tag{12.49}$$

4. $\eta = 1$. In this case,

$$\beta_1 = \beta_2 = \gamma_1 = \sqrt{\frac{K}{2GA}} \qquad \beta_3 = \beta_4 = -\gamma_2 = -\sqrt{\frac{K}{2GA}}$$

and the integral of Eq. (12.45) is

$$w_b(x) = (C_1 + C_2\gamma_1 x) \sinh \gamma_1 x + (C_3 + C_4\gamma_2 x) \cosh \gamma_2 x \tag{12.50}$$

As an example, examine a beam loaded by a uniformly distributed load (Fig. 12.15). Let the parameter η be greater than unity. Then the elastic curve of this beam is expressed as

$$w_b = C_1 \cosh \gamma_1 x + C_3 \cosh \gamma_2 x + \frac{q}{K}$$

where the symmetry of this curve is taken into account. The boundary conditions

$$w_b\left(\pm\frac{L}{2}\right) = 0 \qquad w_b''\left(\pm\frac{L}{2}\right) = 0$$

result in the equations:

$$\left.\begin{aligned} C_1 \cosh u + C_3 \cosh v &= -\frac{q}{K} \\ C_1\gamma_1^2 \cosh u + C_3\gamma_2^2 \cosh v &= 0 \end{aligned}\right\}$$

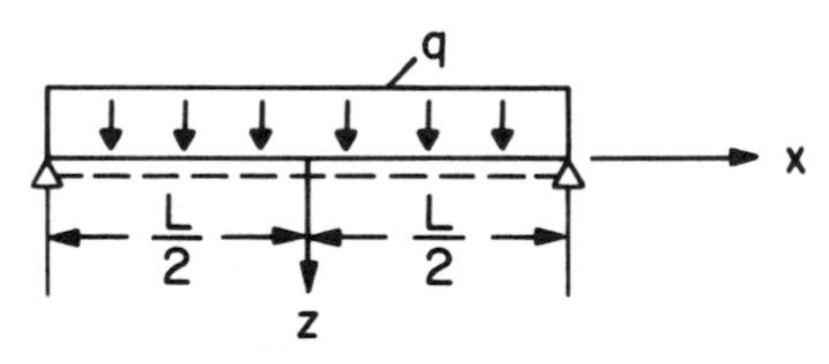

FIGURE 12.15 Single-span beam on an elastic foundation under uniformly distributed lateral load.

where $u = \gamma_1(L/2)$, $v = \gamma_2(L/2)$. Then we have

$$C_1 = \frac{q}{K} \frac{\gamma_2^2}{(\gamma_1^2 - \gamma_2^2)\cosh u} \qquad C_3 = -\frac{q}{K} \frac{\gamma_1^2}{(\gamma_1^2 - \gamma_2^2)\cosh v}$$

and the deflections due to bending are

$$w_b = \frac{q}{K}\left[1 + \frac{1}{\gamma_1^2 - \gamma_2^2}\left(\gamma_2^2 \frac{\cosh \gamma_1 x}{\cosh u} - \gamma_1^2 \frac{\cosh \gamma_2 x}{\cosh v}\right)\right]$$

The elastic curve due to shear is

$$w_s = -\frac{EI}{GA} w_b'' = -\frac{q}{K} \frac{EI}{GA} \frac{\gamma_1^2 \gamma_2^2}{\gamma_1^2 - \gamma_2^2}\left(\frac{\cosh \gamma_1 x}{\cosh u} - \frac{\cosh \gamma_2 x}{\cosh v}\right)$$

Hence, the total deflections are

$$\begin{aligned} w &= w_b + w_s \\ &= \frac{q}{K}\left\{1 - \frac{\gamma_2^2[1 - (EI/GA)\gamma_1^2]\cosh \gamma_1 x}{(\gamma_1^2 - \gamma_2^2)\cosh u} - \frac{\gamma_1^2[1 - (EI/GA)\gamma_2^2]\cosh \gamma_2 x}{(\gamma_1^2 - \gamma_2^2)\cosh v}\right\} \end{aligned} \tag{12.51}$$

The obtained results can be used to evaluate the stresses and deflections in a small-aspect-ratio dual-in-line package mounted on a printed circuit board, with consideration of shear.

12.8 BEAMS UNDER COMBINED ACTION OF LATERAL AND AXIAL LOADS

Now, examine a beam that, in addition to lateral loading, experiences axial tension or compression caused by the given force T (Fig. 12.16). If the beam were absolutely straight, this force would cause only tension or compression of the beam. However, if a beam experiences bending due to lateral loading,

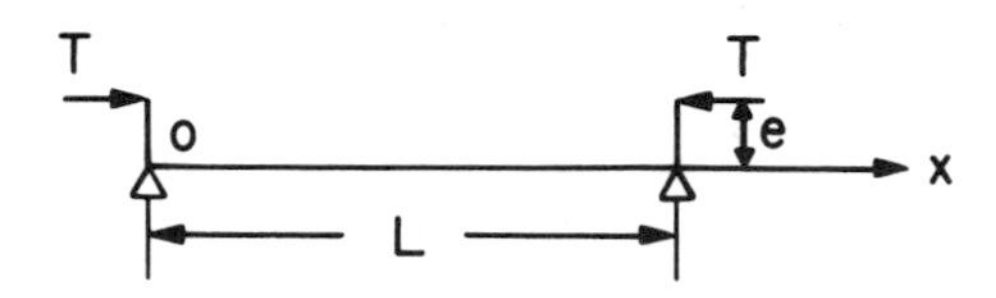

FIGURE 12.16 Beam experiencing both bending and compression.

axial loads will result in an additional bending moment $Tw(x)$, and the differential equation (12.7) yields

$$EIw''(x) - Tw(x) = M(x) \tag{12.52}$$

where $M(x)$ is the bending moment due to lateral loading. The force T is considered positive when it is tensile. Equation (12.52) can be replaced by an equivalent equation,

$$EIw^{IV}(x) - Tw''(x) = q(x) \tag{12.53}$$

where $q(x) = [d^2M(x)]/dx^2$ is the lateral load per unit beam length.

The homogeneous equation

$$EIw^{IV}(x) - Tw''(x) = 0 \tag{12.54}$$

has the following solution:

$$w(x) = A_1 + A_2kx + A_3 \cosh kx + A_4 \sinh kx \tag{12.55}$$

where

$$k = \sqrt{\frac{T}{EI}} \tag{12.56}$$

If the lateral load is distributed linearly,

$$q(x) = q_0 + mx$$

then the particular solution to Eq. (12.53) is as follows:

$$w(x) = -\frac{q_0x^2}{2T} - \frac{mx^3}{6T}$$

If the load T is compressive, then, instead of Eq. (12.55), we have the following solution:

$$w(x) = B_1 + B_2kx + B_3 \cos kx + B_4 \sin kx \tag{12.57}$$

The constants of integration A and B in the solutions (12.56) and (12.57) should be determined from the boundary conditions.

As an example, examine a simply supported beam subjected to compressive forces T applied eccentrically (Fig. 12.16). The boundary conditions

$$w(0) = w(L) = 0 \qquad w''(0) = w''(L) = \frac{Te}{EI} = k^2 e$$

result in the following equations for the constants of integration in the solution (12.57):

$$B_1 + B_3 = 0 \qquad B_1 + B_2 kL + B_3 \cos kL + B_4 \sin kL = 0$$
$$-k^2 B_3 = k^2 e \qquad -k^2(B_3 \cos kL + B_4 \sin kL) = k^2 e$$

Then, we have

$$B_1 = e \qquad B_2 = 0 \qquad B_3 = -e \qquad B_4 = -e\frac{1 - \cos kL}{\sin kL} = -e \tan\frac{kL}{2}$$

and Eq. (12.57) yields

$$w(x) = e\left(1 - \cos kx - \tan\frac{kL}{2}\sin kx\right) \tag{12.58}$$

The maximum deflection is in the mid–cross section of the beam and is

$$w_{\max} = w\left(\frac{L}{2}\right) = e\left(1 - \frac{1}{\cos(kL/2)}\right) \tag{12.59}$$

When the force T is small, the parameter k is also small, and so is the maximum deflection $w_{\max}$. These deflections become, however, infinitely large when the force T approaches a critical value,

$$T_e = \frac{\pi^2 EI}{L^2} \tag{12.60}$$

since, in this case, $\cos(kL/2)$ values are close to zero. This result has to do with the phenomenon of elastic stability, which will be examined in Chapter 16.

When the forces T are tensile, the elastic curve is expressed through hyperbolic functions as follows:

$$w(x) = -e\left(1 - \cosh kx + \tanh \frac{kL}{2} \sinh kx\right)$$

The maximum deflection is

$$w_{\max} = w\left(\frac{L}{2}\right) = -e\left(1 - \frac{1}{\cosh (kL/2)}\right) \tag{12.61}$$

As evident from this formula, the maximum deflection, unlike the case of compressive forces, is always finite and never exceeds the eccentricity e value.

12.9 NONLINEAR BENDING UNDER COMBINED ACTION OF LATERAL AND AXIAL LOADS

In the previous developments, we have tacitly assumed that the beam supports can draw closer to one another during bending and that, therefore, no axial reactive loads can occur. However, if the supports are fixed and cannot move closer, reactive axial forces appear at the supports (Fig. 12.17a).

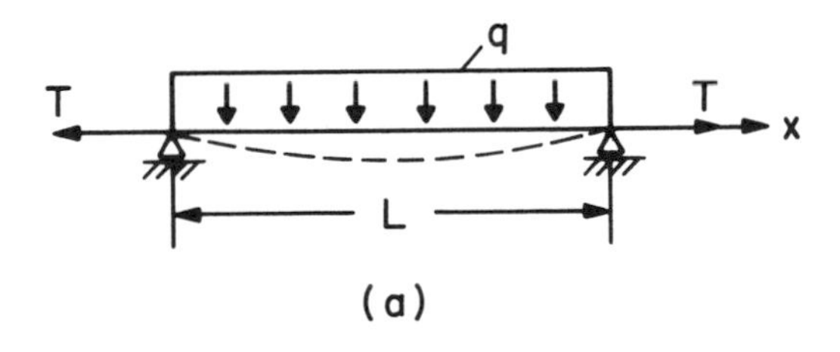

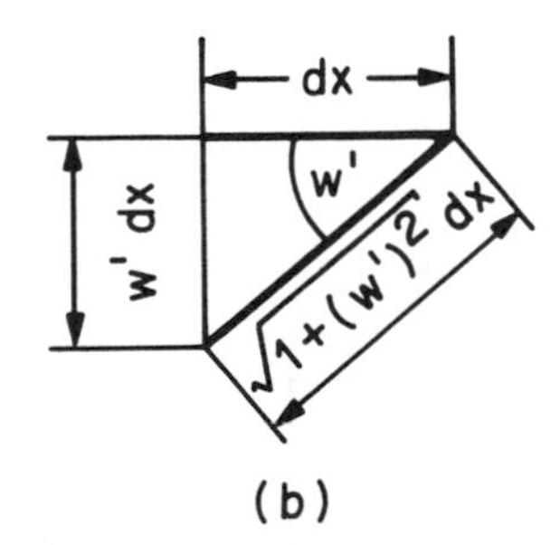

FIGURE 12.17 Beam under combined action of lateral and tensile loads.

The length of the deflected beam can be evaluated by the formula (Fig. 12.17b):

$$s = \int_0^L \sqrt{1 + [w'(x)]^2}\, dx$$

Since the angles of rotation $w'(x)$ are small compared to unity, we can replace the integrand in this formula with the two first terms of the binomial series. Then we have

$$s \cong \int_0^L (1 + \tfrac{1}{2}[w'(x)]^2)\, dx = L + \frac{1}{2}\int_0^L [w'(x)]^2\, dx$$

Hence, the fact that the distance L between the beam supports remains unchanged during bending results in the axial elongation

$$\Delta L = s - L = \frac{1}{2}\int_0^L [w'(x)]^2\, dx$$

The force T needed to produce such elongation can be calculated, in accordance with Hooke's law, as

$$T = EA\frac{\Delta L}{L} = \frac{EA}{2L}\int_0^L [w'(x)]^2\, dx$$

With this formula for the axial force, Eq. (12.53) becomes nonlinear and cannot be solved as simply as in the case of a deflection-independent axial load.

An approximate solution to Eq. (12.53) in the case in question can be obtained assuming a certain longitudinal distribution of the beam deflections. For instance, for a simply supported beam subjected to a uniformly distributed load q, we can take

$$w(x) = w_{\max} \sin\frac{\pi x}{L} \tag{12.62}$$

and replace the load q with the first term in its Fourier series expansion, which is $(4/\pi)q \sin(\pi x/L)$. Then Eq. (12.53) and the formula (12.62) lead to the following nonlinear algebraic equation for the maximum deflection $w_{\max}$:

$$w_{\max} + \frac{A}{4I} w_{\max}^3 = \frac{4L^4}{\pi^5 EI} q \tag{12.63}$$

Obtaining a solution to this cubic equation does not encounter any difficulties. The coefficient $A/4I$ is equal to $3/h^2$ for a rectangular cross section of height h and to $1/r_0^2$ for a circular cross section of radius r_0. Therefore, consideration of the nonlinear effect due to reactive axial forces is more important for thin (flexible) beams (small h or r_0 values) than for deep (stiff) ones.

When the maximum deflection is determined, the maximum bending moment can be evaluated as

$$M_{\max} = EIw''_{\max} = \frac{\pi^2 EI}{L^2} w_{\max}$$

The total stress is due to the bending stress caused by the bending moment $M_{\max}$ and is equal to

$$\sigma_b = \frac{M_{\max}}{I} z_{\max} = \frac{\pi^2 E}{L^2} z_{\max} w_{\max} \tag{12.64}$$

and to the axial ("membrane") stress caused by the axial force

$$T = \frac{\pi^2 EA}{4L^2} w_{\max}^2$$

and equal to

$$\sigma_0 = \frac{T}{A} = \frac{\pi^2 E}{4L^2} w_{\max}^2 \tag{12.65}$$

The total stress is, therefore,

$$\sigma_t = \sigma_b + \sigma_0 = \sigma_b\left(1 + \frac{w_{\max}}{4z_{\max}}\right) \tag{12.66}$$

This formula indicates that the role of the axial stresses reflected by the second term in the parentheses increases with an increase in the beam deflection and decreases in the section modulus $SM = I/z_{\max}$ of the cross section. Note that these two factors are, generally speaking, not independent since small section moduli lead to elevated deflections.

As evident from Eq. (12.63), consideration of the nonlinear effects due to the axial forces results in smaller deflections so that, from the standpoint of the maximum deflections, the linear approach, which ignores the axial forces, is always conservative. This is also true for the bending stresses that are proportional to the deflections. This might not be true, however, for the total stresses, in the case of large deflections, because calculations based on a linear approach and not taking into account the axial stresses may underestimate the total stress, despite the fact that the bending stresses are always overestimated.

12.10 LARGE DEFLECTIONS OF BENT BEAMS (THE ELASTICA)

In the problem examined in the preceding section, we dealt with relatively small nonlinearities whereas, in many practical problems in microelectronics and fiber optics (such as, for instance, problems associated with the mechanical behavior of elastic contactors or optical fibers), we may encounter large elastic deflections. When such deflections occur, an approximate presentation $w''(x)$ for the curvature cannot be used anymore, and the solution for the elastic curve must be based on an exact differential equation of bending. The shape of the elastic curve determined from such an equation is called the *elastica.*

As an example, examine a cantilever beam of very small flexural rigidity EI subjected to a force P, retaining its direction during bending of the cantilever (Fig. 12.18). The exact differential equation of bending is

$$EI \frac{d\theta}{ds} = P(x_B - x) \tag{12.67}$$

Differentiating this equation with respect to the coordinate s and using

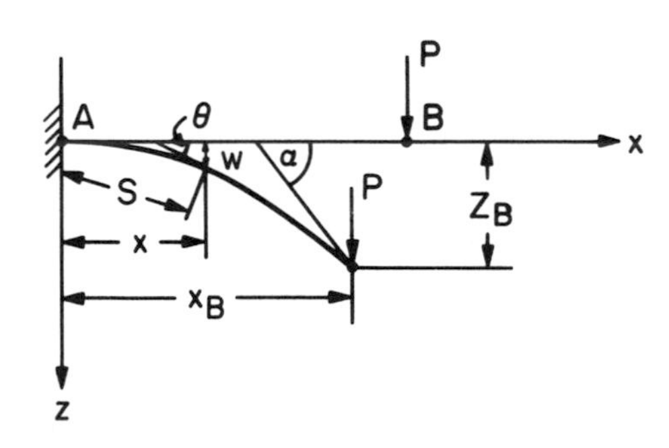

FIGURE 12.18 Large deflections of a cantilever beam under an end force.

the relation

$$\frac{dx}{ds} = \cos\theta$$

we have

$$\frac{d^2\theta}{ds^2} + k^2 \cos\theta = 0 \tag{12.68}$$

where

$$k = \sqrt{\frac{P}{EI}}$$

Equation (12.68) can be rewritten as

$$\frac{d}{ds}\left[\frac{1}{2}\left(\frac{d\theta}{ds}\right)^2 + k^2 \sin\theta\right] = 0$$

and, therefore,

$$\frac{1}{2}\left(\frac{d\theta}{ds}\right)^2 + k^2 \sin\theta = C$$

where C is a constant of integration. This constant can be found from the condition at the end of the cantilever, where $d\theta/ds = 0$, since there is no bending moment at $x = x_B$, and also $\theta = \alpha$. Then we obtain

$$C = k^2 \sin\alpha$$

so that

$$\frac{d\theta}{ds} = k\sqrt{2}\sqrt{\sin\alpha - \sin\theta} \tag{12.69}$$

or

$$L = \frac{1}{k\sqrt{2}} \int_0^\alpha \frac{d\theta}{\sqrt{\sin\alpha - \sin\theta}} \tag{12.70}$$

Introduce a new variable ϕ in such a manner that

$$\sin\theta = \sin\alpha \sin^2\phi$$

It is seen from this relation that when θ varies from zero to α, the angle ϕ varies from zero to $\pi/2$. With the new variable ϕ, the integral (12.70) can be presented as follows:

$$L = \frac{\sqrt{2\sin\alpha}}{k}\int_0^{\pi/2} \frac{\sin\phi\, d\phi}{\sqrt{1 - \sin^2\alpha \sin^4\phi}} = \frac{1}{k}\Phi(p) \tag{12.71}$$

where $p = \sin\alpha$ and the function $\Phi(p)$ is

$$\Phi(p) = \sqrt{2p}\int_0^{\pi/2} \frac{\sin\phi\, d\phi}{\sqrt{1 - p^2\sin^4\phi}} \tag{12.72}$$

The value of the function $\Phi(p)$ depends only on p and can be easily tabulated. With such a table available, we can find the value of p, i.e., the value of the angle α at the end of the cantilever, for any value of k, i.e., for any value of the load P.

When the deflections of the beam are small, α is also a small value, and so is $p = \sin\alpha$, which can be assumed equal to α. In addition, the term $p^2\sin^4\phi$ can be neglected in comparison to unity, and Eq. (12.72) yields $\Phi(p) = \sqrt{2\alpha}$. Then, from Eq. (12.71), we find

$$L = \frac{\sqrt{2\alpha}}{k} = \sqrt{\frac{2\alpha EI}{P}}$$

or

$$P = \frac{2\alpha EI}{L^2} \tag{12.73}$$

This is a well-known result for a linear cantilever beam. Indeed, the differential equation of bending for a cantilever beam (Fig. 12.12) is

$$EIw''(x) = Px$$

By integrating this equation and using a boundary condition $w'(L) = 0$, we have the following formula for the angles of rotation:

$$w'(x) = \frac{P}{2EI}(L^2 - x^2)$$

Then, for $\alpha = w'(0)$, we have

$$\alpha = \frac{PL^2}{2EI}$$

which results in Eq. (12.73).

In calculating deflections, we note that

$$dz = \sin\theta\, ds = \frac{\sin\theta\, d\theta}{k\sqrt{2}\sqrt{\sin\alpha - \sin\theta}}$$

Then the deflection of the beam's end in the vertical direction is

$$z_B = \frac{1}{k\sqrt{2}} \int_0^{\alpha} \frac{\sin\theta\, d\theta}{\sqrt{\sin\alpha - \sin\theta}}$$

or, using the variable ϕ,

$$z_B = \frac{1}{k}\Phi_1(p) \tag{12.74}$$

where

$$\Phi_1(p) = p\sqrt{2p} \int_0^{\pi/2} \frac{\sin^3\phi\, d\phi}{\sqrt{1 - p^2\sin^4\phi}} \tag{12.75}$$

For small deflections, when p can be replaced by α and the second term under the square root in the integrand can be neglected, we have

$$z_B = \frac{\alpha\sqrt{2\alpha}}{k} \int_0^{\pi/2} \sin^3\phi\, d\phi = \frac{2\alpha\sqrt{2\alpha}}{3k} \tag{12.76}$$

or, with consideration of Eq. (12.73),

$$z_B = \frac{PL^3}{3EI} \tag{12.77}$$

This is a well-known formula for the maximum linear deflection of a cantilever beam (see, for instance, Appendix A).

The horizontal displacements x_B of the beam's end can be determined on the basis of the formula

$$dx = \cos\theta\, ds = \frac{\cos\theta\, d\theta}{k\sqrt{2}\sqrt{\sin\alpha - \sin\theta}}$$

Then we have

$$x_B = \frac{1}{k\sqrt{2}}\int_0^{\alpha}\frac{\cos\theta\, d\theta}{\sqrt{\sin\alpha - \sin\theta}} = \frac{\sqrt{2P}}{k} \qquad (12.78)$$

In the case of small deflections, this formula leads to an obvious result: $x_B = L$.

Calculated data (Fig. 12.19) indicate that formulas based on the linear approach can be used if the angles of rotation of the cantilever end do not

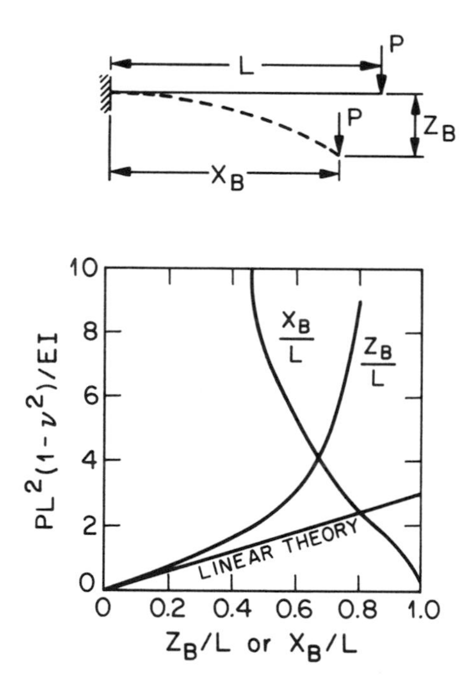

FIGURE 12.19 Linear and nonlinear deflections of a cantilever beam.

exceed 30°. Using Eq. (12.73), we can present this criterion as follows:

$$\frac{PL^2}{EI} \leq 1 \tag{12.79}$$

12.11 THERMAL BENDING OF BEAMS

Beam elements in electronic components often experience bending stresses because of the nonuniform distribution of temperature. We illustrate this phenomenon using an example of a beam whose cross sections are symmetric with respect to the xz plane, and so is the distribution of the temperature.

In accordance with the hypothesis of the plane cross sections, the displacement of an arbitrary point of the given beam cross section along the x axis can be evaluated as

$$u(x) = u_0(x) + zw'(x)$$

where $u_0(x)$ is the displacement of the point located at the x axis, and $w'(x)$ is the angle of rotation of the cross section. The only nonzero strain is ε_x. This can be determined as

$$\varepsilon_x = \frac{\partial u}{\partial x} = u_0'(x) + zw''(x)$$

Supplementing this strain by the thermal strain $\alpha \Delta T(x, z)$, we can evaluate the normal stress in the x direction by the formula

$$\sigma_x = E(\varepsilon_x - \alpha \Delta T) = E[u_0'(x) + zw''(x) - \alpha \Delta T(x, z)] \tag{12.80}$$

where E is Young's modulus. Introducing this equation into the conditions of equilibrium

$$\int_A \sigma_x \, dA = 0 \qquad \int_A \sigma_x z \, dA = M_y$$

we have

$$\left.\begin{aligned} u_0'(x) \int_A E \, dA + w''(x) \int_A Ez \, dA - N_T = 0 \\ u_0'(x) \int_A Ez^2 \, dA - M_T = M_y \end{aligned}\right\} \tag{12.81}$$

Here M_y is the bending moment due to the external forces, and

$$N_T = \int_A E\alpha\Delta T\,dA \qquad M_T = \int_A E\alpha z\Delta T\,dA \tag{12.82}$$

are the thermally induced force and moment. The preceding formulas take into account that Young's modulus E and the coefficient of thermal expansion α might be temperature-dependent and, therefore, are different for different points of the beam.

Using the condition

$$\int_A Ez\,dA = 0$$

to locate an effective centroid of the cross section, we obtain the following formulas for the strains $u_0'(x)$ and $w''(x)$:

$$u_0'(x) = \frac{N_T}{\int_A E\,dA} \qquad w''(x) = \frac{M_y + M_T}{\int_A E\,dA}$$

After introducing these formulas into Eq. (12.80), we obtain the following expression for the total stress:

$$\sigma_x = E\frac{M_y z}{\int_A Ez^2\,dA} + E\left(\frac{N_T}{\int_A E\,dA} + \frac{zM_T}{\int_A Ez^2\,dA} - \alpha\Delta T\right) \tag{12.83}$$

The first term in this expression is due to external loading and the second term to thermally induced loads. Obviously, when the temperature distribution is uniform, no thermal stresses occur. In such a case, the stress σ_x is due to the external loading only and is expressed by a simple formula:

$$\sigma_x = \frac{M_y z}{I_y}$$

where

$$I_y = \int_A z^2\,dA$$

is the cross-sectional moment of inertia. Note that if the thermal strain $\alpha\Delta T$ is distributed linearly over the height of the cross section, no thermal stresses occur either.

We illustrate the above-mentioned theory by examining the problem of a simply supported beam subjected to heating in such a way that its top surface is maintained at temperature T_0 and the temperature of the bottom surface is zero (Fig. 12.20). Hence, the temperature increase throughout the beam is

$$\Delta T = \frac{T_0}{L}\left(z + \frac{h}{2}\right)$$

Since the temperature is distributed symmetrically with respect to the beam's centerline, the force N_T is zero, and the moment M_T is as follows:

$$M_T = \int_A E\alpha z \Delta T \, dA = E\alpha \frac{T_0}{h} b \int_{-(h/2)}^{(h/2)} z\left(z + \frac{h}{2}\right) dz = \frac{1}{12} E\alpha T_0 bh^2$$

The equation of bending

$$EIw''(x) = M_T$$

results in the following formula for the deflection curve:

$$w(x) = -w_{\max}\left[1 - \left(\frac{2x}{L}\right)^2\right] \tag{12.84}$$

where

$$w_{\max} = \frac{M_T L^2}{8EI} = \frac{\alpha T_0 L^2}{8h}$$

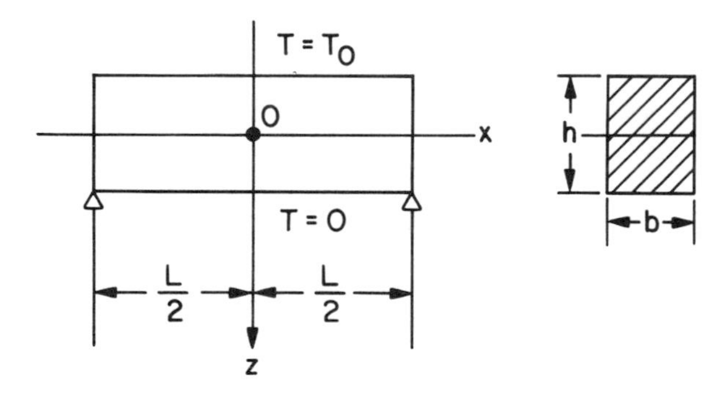

FIGURE 12.20 Simply supported beam experiencing nonuniform distribution of temperature over its height.

is the maximum deflection in the mid–cross section. The solution (12.84) satisfies the boundary condition $w[\pm(L/2)] = 0$. The maximum thermal stress, in accordance with Eq. (12.83), is

$$\sigma_{\max} = \sigma_x|_{z=(h/2)} = \frac{M_T}{I}\frac{h}{2} - E\alpha T_0 = -\frac{1}{2}E\alpha T_0 \qquad (12.85)$$

13

The Variational and Energy Methods, and Some General Principles of Structural Analysis

13.1 VARIATIONAL METHODS

In the previous analyses, we have continually relied upon the differential equations with appropriate boundary conditions as the basic mathematical equipment to obtain solutions to various stress-strain problems. In many cases, however, these equations turn out to be so complicated that it becomes very difficult not only to solve them analytically but even to obtain satisfactory numerical solutions. One of the possible alternative ways to get around these difficulties is to proceed from the energy considerations, as we did, for instance, in Section 12.5, when a law of conservation of energy was used to evaluate the shear deformations in beam bending. Much more general and powerful methods for solving structural problems exist, however, than simply equating the external work to the strain energy of the system. These are *variational methods*, and it is these methods that will be addressed in this chapter. The variational methods do not use differential equations and have as their basis some of the most fundamental principles of mechanics. Although, in the majority of cases, they provide only approximate solutions, in effect, any desirable accuracy can be achieved.

The reason why variational methods can be used alternatively to differential equation–based approaches is that the requirements of structural behavior, such as equilibrium conditions, strain compatibility conditions, and even boundary conditions, can be formulated and satisfied not only by differential equations but by variational equations as well. These methods are based on a condition that a certain *functional* (which is a function of a function) have a stationary value, usually a minimum or a maximum. Such a condition is totally equivalent to the differential equations that describe the mechanical behavior of the system. As an example, the position of equilibrium of a conservative system can be defined as a position for which a function representing the work of all the forces for the given displacements has its minimum value. Variational approach enables us to form this function and to evaluate its minimum, using so-called *direct variational methods*. The fundamentals of direct variational methods in application to structural analysis problems were set down by Rayleigh and Ritz.

13.2 VARIATIONAL PRINCIPLES OF STRUCTURAL ANALYSIS

13.2.1 Generalized Displacements and Generalized Forces

When variational methods are used, it is sometimes more convenient to utilize generalized displacements and forces rather than the "regular" displacements and forces applied in the previous analyses.

A set of independent variables required to specify the configuration of a system is called *generalized coordinates*, and a change in a generalized coordinate is called a *generalized displacement*. The least number of generalized coordinates corresponds to the number of degrees of freedom of the system. The generalized coordinates should be chosen in such a way that their changes are consistent with the constraints. If, for instance, a rigid body is rotating about an immovable axis, the angle of rotation can be taken as a generalized displacement. In an elastic body, the number of degrees of freedom is infinitely large.

For instance, the elastic curve of a simply supported beam of length L can be presented as a series

$$w(x) = \sum_{k=1}^{\infty} q_k \sin \frac{k\pi x}{L} \tag{13.1}$$

Coefficients q_k of this series can be treated as generalized coordinates. Indeed, with an arbitrary change in these coefficients, no constraints due to the supports will be violated since the modes $\sin k\pi x/L$ of the generalized displacements will still satisfy the zero boundary conditions at the beam ends. These modes are called *coordinate functions*. Coordinate functions must be continuous functions of the coordinates of the points of the elastic system and must satisfy the *kinematic boundary conditions*. These are the conditions imposed on the generalized displacements at points on the boundaries. They are called this because the descriptions of the deformation and even motion of a system need not involve forces and are thus based on purely kinematic considerations. The requirement that the coordinate functions be consistent with the constraints means that if these constraints do not allow any displacements (deflections at rigid supports, angles of rotation at rigidly clamped cross sections, and so on), then the corresponding coordinate functions or their derivatives must be zero at such constraints. On the other hand, at least one of these functions must have nonzero values at the points and cross sections at which displacements are possible.

The general expression for the work of all the external forces at the displacement δq_k can be presented as

$$\delta U_k = Q_k \delta q_k \tag{13.2}$$

The coefficient Q_k in this expression is the *generalized force*, corresponding to the displacement δq_k. If, for instance, a simply supported beam is subjected to a uniformly distributed load q, a concentrated force P, and a moment M applied at its mid–cross section, then the external work due to these loads is

$$\begin{aligned}\delta U_k &= \int_0^L \left(q \sin \frac{k\pi x}{L}\, dx + P \sin \frac{k\pi}{L}\frac{L}{2} + M \frac{k\pi}{L} \cos \frac{k\pi}{L}\frac{L}{2} \right) \delta q_k \\ &= \left[\frac{qL}{k\pi}(1 - \cos k\pi) + P \sin \frac{k\pi}{2} + M \frac{k\pi}{L} \cos \frac{k\pi}{2} \right] \delta q_k \qquad k = 1, 2, 3, \ldots\end{aligned}$$

The expression in brackets is the sought generalized force, which is

$$Q_k = \begin{cases} \dfrac{2qL}{k\pi} + P & \text{for } k = 1, 5, 9, \ldots \\[2ex] \dfrac{2qL}{k\pi} - P & \text{for } k = 3, 7, 11, \ldots \\[2ex] -M \dfrac{k\pi}{L} & \text{for } k = 2, 6, 10, \ldots \\[2ex] M \dfrac{k\pi}{L} & \text{for } k = 4, 8, 12, \ldots \end{cases}$$

13.2.2 The Principle of Virtual Displacements

In accordance with the *law of conservation of energy*, an elastic system undergoing deformation receives an amount of energy that is equal to the work of the external forces participating in such a deformation:

$$U = W \tag{13.3}$$

Here U is the total strain energy of the system, including the energy accumulated in the elastic supports and elastically clamped cross sections, and W is the work of the external forces.

Let the elastic system be in equilibrium and the external forces assume infinitely small increments. Then the internal elastic forces will also assume infinitely small increments. Since the work of infinitely small increments of the external and internal forces at an infinitely small displacement of the system is a small quantity of a higher order than the work of the forces themselves, then, when calculating the increments of the work at an infinitely small displacement of the system from the equilibrium position, we may consider that both the external and internal forces remain unchanged during deformation. Therefore, we conclude that the increment of the total strain energy is simply equal to the increment of the external work:

$$\delta U = \delta W \tag{13.4}$$

This result, which follows from the law of conservation of energy, is known as the *principle of virtual displacements*; it can be formulated as follows: A deformable system is in equilibrium if the total external virtual work is equal to the total internal virtual work for every virtual displacement consistent with the constraints. The word "virtual" simply indicates that these works and displacements exist in essence or effect but not necessarily in fact. It is used to distinguish "real work," which is done by true forces moving through true displacements, from the work in which true forces move through imaginary displacements (or vice versa).

It can be shown that both the equilibrium conditions for an elementary tetrahedron, which were used in Part 1, Section 2.4 to establish the boundary conditions at the body surface, and the equilibrium conditions for an elementary parallelepiped (Part 1, Section 2.5), which are the basic equilibrium equations in the three-dimensional elasticity theory, can be obtained from the principle of virtual displacements. Therefore, this principle is considered the most general principle of statics of a deformable body, regardless whether its material obeys Hooke's law or not. It can also be shown that if the functions used to approximate the displacements are continuous and differentiable within the volume of the body and, in addition, satisfy all the kinematic constraints imposed on the body, the conditions of the continuity (compatibility) are satisfied identically within the body and on that portion of its surface that is free of the external forces.

13.2.3 The Lagrange Theorem

If all the virtual displacements are expressed in terms of the generalized displacements, then the sum of the works of all the external forces resulting in the system's displacement from the equilibrium position can be presented as

$$\delta W - \sum_k Q_k \delta q_k \tag{13.5}$$

On the other hand, the strain energy of an ideally elastic body is a single-valued function of the displacement components and, hence, a single-valued function of the generalized displacements. Then the variation of the strain (potential) energy can be presented as

$$\delta U = \sum_k \frac{\partial U}{\partial q_k} \delta q_k \tag{13.6}$$

Introducing Eqs. (13.5) and (13.6) into Eq. (13.4) and equating in the resulting equations the coefficients at the variations δq_k, we obtain the following mathematical formulation of the *Lagrange theorem*:

$$\frac{\partial U}{\partial q_k} = Q_k \tag{13.7}$$

This equation states that if the body is in an equilibrium position, the change in the strain (potential) energy caused by a small unit change in the generalized displacement equals the corresponding generalized force.

13.2.4 Reciprocal Theorem for Forces

In linearly deformable systems, a linear relationship should exist between the generalized forces and generalized displacements:

$$Q_k = \sum_i \beta_{ki} q_i \qquad k = 1, 2, \ldots \tag{13.8}$$

where

$$\beta_{ki} = \frac{\partial Q_k}{\partial q_i} \tag{13.9}$$

Using the Lagrange theorem (13.7), we have

$$Q_k = \frac{\partial U}{\partial q_k} \qquad Q_i = \frac{\partial U}{\partial q_i}$$

Since the second derivative is independent of the order of differentiation, we obtain

$$\frac{\partial Q_k}{\partial q_i} = \frac{\partial Q_i}{\partial q_k} = \frac{\partial^2 U}{\partial q_k \partial q_i}$$

or, considering Eq. (13.9),

$$\beta_{ki} = \beta_{ik} \tag{13.10}$$

This equation is known as the *reciprocal theorem for forces*. It states that, in a linearly elastic system, the kth generalized force for a unit ith generalized displacement is equal to the ith generalized force for a unit kth generalized displacement. This theorem is illustrated in Fig. 13.1a and b. In the first case, the force P_1 is applied at point 1 and causes a unit displacement of point 2. In the second case, the force P_2 is applied at point 2 and causes a unit displacement of point 1. According to the reciprocal theorem for forces, $P_1 = P_2$. Indeed, the elastic curve of a simply supported beam loaded by a concentrated force P (Fig. 13.1c) is expressed as

$$w(x) = \frac{PL^3}{6EI}\left[\frac{c_1}{L}\frac{x}{L}\left(1 - \frac{c_1^2}{L^2} - \frac{x^2}{L^2}\right) + \Big\|_c \left(\frac{x}{L} - \frac{c}{L}\right)^3\right]$$

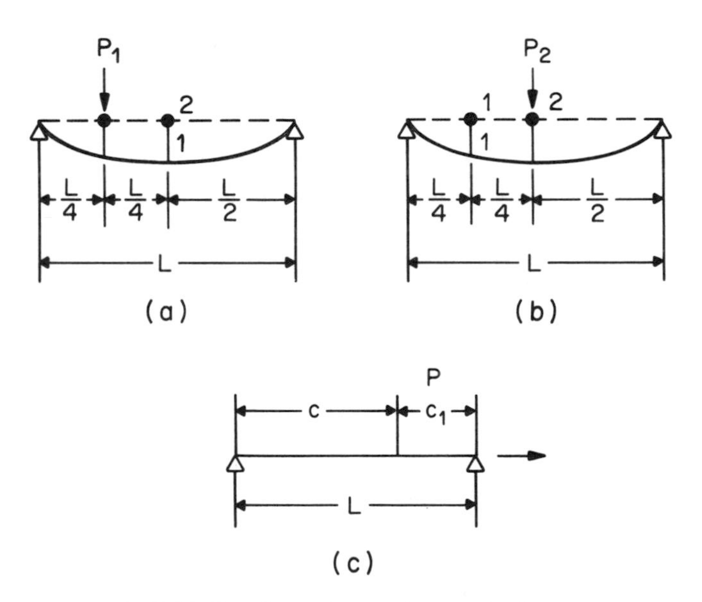

FIGURE 13.1 The reciprocal theorem for forces.

Assuming first that $P = P_1$, $c = L/4$, $c_1 = 3L/4$, and $x = L/2$, we find $w_2 = 11/768\ P_1L^3/EI$. Assuming, then, that $P = P_2$, $c = L/2$, $c_1 = L/2$, and $x = L/4$, we obtain $w_1 = 11/768\ P_2L^3/EI$. With $w_1 = w_2 = 1$, we have

$$P_1 = P_2$$

13.2.5 Klapeiron's Theorem

In a linearly deformable system, the change in the strain (potential) energy caused by a small unit change in the generalized displacement is a linear function of the generalized displacements. This simply follows from Eqs. (13.7) and (13.8). Then the strain energy itself is a homogeneous quadratic function of the generalized displacements. In accordance with the Euler theorem of homogeneous functions, we have

$$\sum_k \left(\frac{\partial U}{\partial q_k}\right) q_k = 2V$$

This equation, with consideration of the Lagrange theorem (13.7), yields

$$U = \frac{1}{2} \sum_k Q_k q_k \tag{13.11}$$

This formula expresses *Klapeiron's theorem*: The strain energy of a linearly deformable system is twice as small as the sum of the products of the generalized forces and the generalized displacements corresponding to the equilibrium position of the system. Note that this result was obtained earlier for a special case of a cantilever beam in Section 12.5.

13.2.6 Castigliano's Theorem

The strain energy can be expressed not only through the generalized displacements, as it was assumed in the Lagrange theorem, but through the generalized forces as well. If we treat the relationships (13.8) as a system of linear algebraic equations for the unknowns q_i, then the solutions to these equations can be presented as follows:

$$q_k = \sum_{i=1} \alpha_{ki} Q_i \tag{13.12}$$

As noted earlier, the strain energy of a linearly deformable system is a homogeneous quadratic function of the generalized displacements. Therefore, when the generalized displacements are replaced by the generalized forces, the strain energy becomes a homogeneous quadratic function of the generalized forces as well. Using the Euler theorem for the homogeneous functions, we have

$$U = \frac{1}{2}\sum_k \frac{\partial U}{\partial Q_k} Q_k \tag{13.13}$$

Comparing Eqs. (13.11) and (13.13), we conclude that

$$\frac{\partial U}{\partial Q_k} = q_k \tag{13.14}$$

This relationship expresses *Castigliano's theorem:* In a linearly deformable system, the change in the strain energy, expressed in terms of the generalized forces and caused by a small unit change in one of these forces, equals the generalized displacement corresponding to this force. We would like to remind the reader that the generalized displacements q_k corresponding to the generalized forces Q_k are those whose increments are related to the increment of the work produced by the force Q_k by the expression (13.12). We also note that a generalized displacement corresponding to an external concentrated force is the linear displacement of the point of the application of this force in the direction of the force. A generalized displacement corresponding to an external concentrated bending moment is the angle of rotation of the cross section to which the moment is applied in the direction of the bending moment.

Castigliano's theorem enables us to evaluate a generalized displacement in a linearly deformable system if it is possible to form an expression for the strain energy as a function of the external load and the generalized force corresponding to the generalized displacement. If there is no generalized force corresponding to the sought displacement, then a "fictitious" force can be introduced. This force should be put equal to zero after the displacement is determined.

Let us illustrate the application of Castigliano's theorem using an example of a cantilever beam (Fig. 12.12). The stored energy for such a beam is given by Eq. (12.25) and is

$$U = \frac{P^2 L^3}{6EI}$$

Since the energy is expressed in terms of the load, it is suitable for the application of Castigliano's theorem. The vertical deflection under the force P is

$$w = \frac{\partial U}{\partial P} = \frac{PL^3}{3EI}$$

If a beam is subjected to the combined action of bending moments $M(x)$ and axial forces $T(x)$, then its total strain energy, with consideration of shear, can be evaluated by the formula

$$U = \int_0^L \left[\frac{M^2(x)}{2EI} + \frac{T^2(x)}{2EA} + \frac{N^2(x)}{2GA} \right] dx \tag{13.15}$$

where $M(x)$ is the bending moment, $T(x)$ the axial force, $N(x)$ the lateral force, E Young's modulus, G the shear modulus, I the moment of inertia of the cross section, and A the cross-sectional area of the beam. The first term in this formula is the same as in Eq. (12.24). The third term can be obtained from Eq. (12.26) by integrating this expression over the beam length. The second term can be determined from Eqs. (5.1), assuming that all the stress components except $\sigma_x = T/A$ are zero and integrating the obtained expression over the beam length.

Let us now demonstrate the procedure to be used for calculating displacements of points at which no forces are applied. Consider, for instance, a cantilever in Fig. 13.2, and suppose that we wish to find the rotation of its end cross section. Since P is the only applied force acting on the beam, we must introduce a "fictitious" (dummy) moment M_f corresponding to the sought generalized displacements. The strain energy is

$$U = \int_0^L \frac{M^2(x)\,dx}{2EI} = \int_0^L \frac{(M_f + Px)^2}{2EI}\,dx = \frac{1}{2EI}\left(M_f^2 L + M_f PL^2 + \tfrac{1}{3}P^2L^3\right)$$

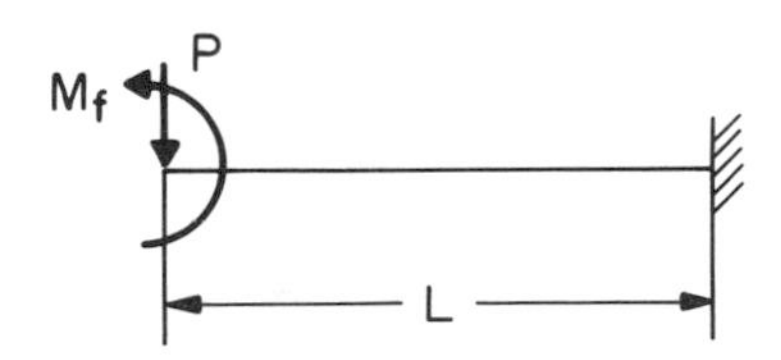

FIGURE 13.2 Cantilever beam with an end force and moment.

Determining the derivative

$$\frac{\partial U}{\partial M_f} = \frac{1}{2EI}(2M_f L + PL^2)$$

and equating the dummy bending moment M_f to zero, we find that the angle of rotation of the beam's end is

$$\alpha = \left.\frac{\partial U}{\partial M_f}\right|_{M_f=0} = \frac{PL^2}{2EI}$$

13.2.7 Theorem of Least Work

In the preceding examples, Castigliano's theorem was applied to finding displacements in a statically determinate system. It can also be used, however, to solve statically indeterminate problems. The general approach in this case is such that the redundant forces are being included in the external loads, so that the system becomes statically determinate, although with some unknown forces. Then, forming an expression for the strain energy, which includes both the external and the redundant forces, and equating the derivative of this equation with respect to the unknown redundant forces to the corresponding displacements, we obtain equations for these unknowns.

As an example, examine a beam shown in Fig. 13.3a. This is a two times statically indeterminate system. Using an imaginary cut, we divide the beam into two parts (Fig. 13.3b) and substitute the actions of the removed parts by the forces N and moments M. Let U_1 be the total strain energy of the left part of the beam, and U_2 the total energy of the right part. Then, in accordance with Castigliano's theorem, we have the following relationships for the evaluation of the generalized displacements (Fig. 13.3c):

$$q_1 = \frac{\partial U_1}{\partial N} \qquad q_2 = \frac{\partial U_2}{\partial N} \qquad \alpha_1 = \frac{\partial U_1}{\partial M} \qquad \alpha_2 = \frac{\partial U_2}{\partial M}$$

The conditions of the compatibility of the displacements require that $q_1 = -q_2$ and $\alpha_1 = -\alpha_2$. Then we have

$$\frac{\partial U}{\partial N} = 0 \qquad \frac{\partial U}{\partial M} = 0 \tag{13.16}$$

where $U = U_1 + U_2$ is the total strain energy of the system, determined with consideration of both the external and statically indeterminate forces and moments. Since the strain energy is always positive and can increase

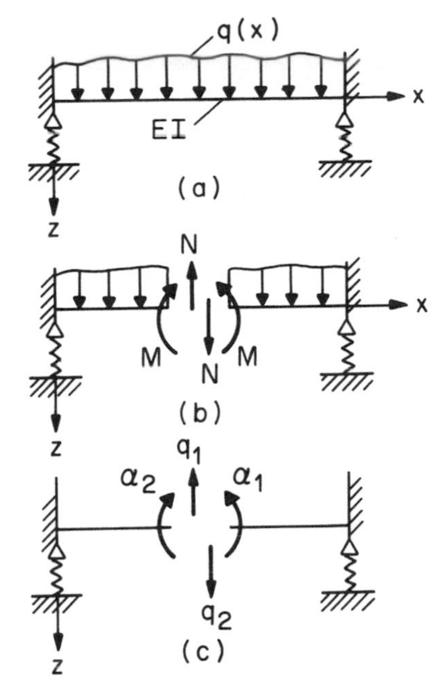

FIGURE 13.3 Two times statically indeterminate beam.

only with an increase in the loading, then Eqs. (13.16) are the conditions of the minimum of the potential energy of deformation. These conditions are usually referred to as the *theorem of least work*. It can be formulated as follows: All the possible sets of values of the redundants in a linear statically indeterminate structural system which satisfy the equilibrium and the compatibility conditions make the strain energy a minimum, provided the prescribed displacements of these redundants are zero.

As an example, examine a cantilever beam shown in Fig. 13.4a. By removing the support at the left end and replacing it with a redundant unknown force Q, we obtain a statically determinate system as shown in Fig. 13.4b. The expression for the bending moment is

$$M(x) = Qx - \frac{qx^2}{2}$$

Then the strain energy of the beam is

$$U = \int_0^L \frac{M^2(x)}{2EI}\,dx = \int_0^L \frac{1}{2EI}\left(Qx - \frac{qx^2}{2}\right)^2 dx = \frac{Q^2L^3}{6EI} - \frac{QqL^4}{8EI} + \frac{q^2L^5}{20EI}$$

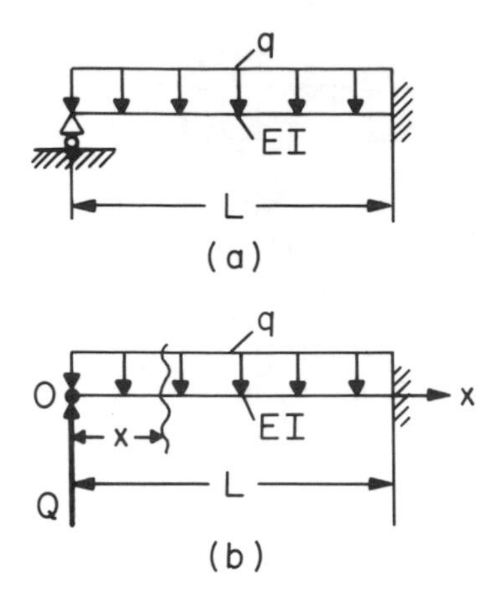

FIGURE 13.4 The theorem of least work.

In accordance with the theorem of least work, we have

$$\frac{\partial U}{\partial Q} = \frac{QL^3}{3EI} - \frac{qL^4}{8EI} = 0$$

so that

$$Q = \tfrac{3}{8}qL \tag{13.17}$$

Although the theorem of least work was derived here from Castigliano's theorem, its applications are much more far-reaching than simply evaluating deflections or finding redundant reactions in statically indeterminate systems. The theorem of least work is most useful in complicated cases in which an approximate solution is sought.

13.2.8 Reciprocal Property of Displacements

From Eq. (13.14), expressing Castigliano's theorem, and Eq. (13.12), we have

$$\frac{\partial U}{\partial Q_k} = \sum_{i=1} \alpha_{ki} Q_i$$

Simply by changing the indices, we have

$$\frac{\partial U}{\partial Q_i} = \sum_{k=1} \alpha_{ik} Q_k$$

Since the value of a mixed derivative is independent of the order of differentiation, we have

$$\frac{\partial^2 U}{\partial Q_k \partial Q_i} = \frac{\partial^2 U}{\partial Q_i \partial Q_k}$$

so that

$$\alpha_{ki} = \alpha_{ik} \tag{13.18}$$

As follows from Eq. (13.12), α_{ki} is the kth generalized displacement occurring as a result of a unit ith generalized force, and α_{ik} is the ith generalized displacement occurring as a result of a unit kth generalized force. The relationship (13.18) is usually referred to as *Maxwell's theorem of reciprocity of displacements*: For a linearly elastic body, the displacement of point i due to a force Q_k at point k is equal to the displacement of point k due to a force Q_i at point i.

13.2.9 Reciprocal Theorem for Works

Let a linearly elastic body be in equilibrium under the action of two different systems of external forces, $Q_1, Q_2, \ldots, Q_m$ and $R_1, R_2, \ldots, R_n$, applied consequently. The generalized displacements corresponding to the forces of the second system and caused by the forces of the first system can be determined as

$$r_i = \sum_{k=1}^{m} \alpha_{ik} Q_k \qquad i = 1, 2, \ldots, m \tag{13.19}$$

Similarly, the generalized displacements corresponding to the forces of the first system and caused by the forces of the second system can be determined as

$$q_k = \sum_{i=1}^{n} \alpha_{ki} R_i \qquad k = 1, 2, \ldots, m \tag{13.20}$$

The *reciprocal theorem for works*, also known as the *Rayleigh–Betti theorem*, states: If a linearly elastic system is subjected to two different force systems, the work that would be done by the first system of forces in moving through the displacements produced by the second system of forces is equal to the work that would be done by the second system of forces in moving through the displacements produced by the first system of forces, i.e.,

$$\sum_{k=1}^{m} Q_k q_k = \sum_{i=1}^{n} R_i r_i \tag{13.21}$$

In order to prove this equation, we present the left part of it, using Eq. (13.20), as follows:

$$\sum_{k=1}^{m} Q_k q_k = \sum_{k=1}^{m} Q_k \sum_{i=1}^{n} \alpha_{ki} R_i = \sum_{k=1}^{m} \sum_{i=1}^{n} \alpha_{ki} Q_k R_i$$

The right part of Eq. (13.21) can be presented, using Eq. (13.19), as

$$\sum_{i=1}^{n} R_i r_i = \sum_{i=1}^{n} R_i \sum_{k=1}^{m} \alpha_{ik} Q_k = \sum_{k=1}^{m} \sum_{i=1}^{n} \alpha_{ik} Q_k R_i$$

Since $\alpha_{ki} = \alpha_{ik}$, then the relationship (13.21) is true.

13.3 RAYLEIGH–RITZ AND BUBNOV–GALERKIN METHODS

The *Rayleigh–Ritz* method is the most widespread and perhaps the most important variational method in structural analysis. It was first presented by Lord Rayleigh (John William Strutt) in 1877 and then refined and extended in 1909 by W. Ritz. This approximate method enables us to replace a deformable system having an infinite number of degrees of freedom with systems having a finite number of degrees of freedom. In the Rayleigh–Ritz method, the displacements of the system are presented by functions containing a finite number of independent parameters, which are determined in such a way that the total strain energy of the system is at a minimum.

Using an example of a single-span beam, we present its sought deflection under the given distributed load $q(x)$ in the form of a series

$$w(x) = \sum_{k=1}^{\infty} a_k \Phi_k(x) \tag{13.22}$$

where $\Phi_k(x)$ are coordinate functions (see Section 13.2.1) that must satisfy the kinematic boundary conditions. The parameters a_k behave as generalized coordinates. Using the expansion (13.22), we compute the total strain energy of the system U and the work W of the external forces, and form an *energy functional*

$$\Pi = U - W = \Pi(w) = \Pi(x, a_1, a_2, \ldots)$$

Since

$$\delta w = \sum_k \Phi_k \delta a_k$$

then the variation of the energy functional is

$$\delta\Pi = \delta U - \delta W = \sum_k \frac{\partial\Pi}{\partial a_k} \partial a_k$$

If the system is in equilibrium, then, in accordance with Lagrange's theorem (12.89), $\delta\Pi = 0$, and, therefore,

$$\sum_k \frac{\partial\Pi}{\partial a_k} \delta a_k = 0$$

With arbitrary variations δa_k, this equation can be satisfied only if all the derivatives $\partial\Pi/\partial a_k$ are zero:

$$\frac{\partial\Pi}{\partial a_1} = 0 \qquad \frac{\partial\Pi}{\partial a_2} = 0, \ldots \tag{13.23}$$

These equations represent a system of linearly independent simultaneous equations for the unknown parameters a_k. The more terms are taken in the expansion (13.22), the more accurate is the obtained solution. Once Eqs. (13.23) are solved, the results are introduced into the expansion (13.22). The latter is then used to calculate the displacements and other stress-strain characteristics.

It should be emphasized that the Rayleigh–Ritz method usually leads to fairly accurate expressions for displacements, especially if the coordinate functions represent the actual elastic curve closely. The stresses, however, are evaluated less accurately, since they generally depend on derivatives of the displacements, and the derivatives of approximate functions are less accurate approximations than the approximate functions themselves. Therefore, if a more precise prediction of the stresses is important, we have to retain a greater number of terms in the expansion (13.22). We would also like to note that since the Rayleigh–Ritz method approximates systems having infinitely many degrees of freedom with systems having a finite number of degrees of freedom, the approximate system is stiffer than the actual systems. Therefore, the Rayleigh–Ritz method, when used in elastic stability problems, overestimates the buckling loads and, when used in vibration problems, overestimates the natural frequencies. Finally, it is noteworthy that the conditions of equilibrium are satisfied in this method in an average sense only, through minimization of the total potential energy.

As an illustration of the application of the method, let us examine the deflections of a beam as shown in Fig. 13.5. The potential (strain) energy of this beam, in accordance with Eq. (13.15), is

$$U = \int_0^L \frac{M^2(x)}{2EI}\,dx = \int_0^L \frac{[EIw''(x)]^2}{2EI}\,dx = \frac{1}{2}\int_0^L EI[w''(x)]^2\,dx \tag{13.24}$$

After substituting the expansion (13.22) in this formula, we have

$$U = \frac{1}{2}\int_0^L EI\left[\sum_{k=1}^{\infty} a_k \Phi_k''(x)\right]^2 dx \tag{13.25}$$

The external work is expressed as follows:

$$W = \frac{1}{2}\int_0^L q(x)w(x)\,dx = \frac{1}{2}\int_0^L q(x)\sum_{k=1}^{\infty} a_k \Phi_k(x)\,dx = \sum_{k=1}^{\infty} a_k \Delta_k \tag{13.26}$$

where

$$\Delta_k = \frac{1}{2}\int_0^L q(x)\Phi_k(x)\,dx \tag{13.27}$$

is the load term. Forming an energy functional $\Pi = U - W$ and substituting it into Eqs. (13.23), we obtain the following equations for the unknown

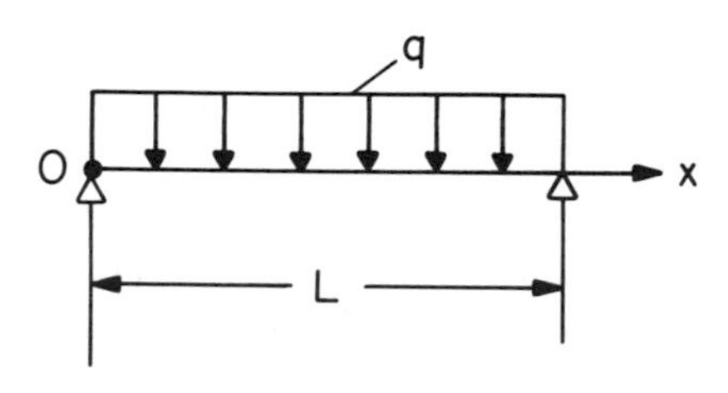

FIGURE 13.5 Simply supported beam under uniformly distributed lateral load.

parameters a_k:

$$\sum_{k=1}^{\infty} \delta_{kj} a_k = \Delta_j \tag{13.28}$$

where

$$\delta_{kj} = \int_0^L EI\Phi_k''(x)\Phi_j''(x)\, dx \tag{13.29}$$

and

$$\Delta_j = \int_0^L q(x)\Phi_j(x)\, dx \tag{13.30}$$

In the case of a concentrated force applied at the cross section $x = c$, the load term Δ_j is expressed as

$$\Delta_j = P\Phi_j(c) \tag{13.31}$$

This formula can be obtained from Eq. (13.30) if the concentrated force P is presented in the form of a distributed load q, using delta function

$$q = P\delta(c)$$

After introducing this formula in Eq. (13.30) and using the following major property of a delta function:

$$\int \delta(c)\Phi_j(x)\, dx = \Phi_j(c)$$

we obtain Eq. (13.31). In the expanded form, Eqs. (13.28) yield

$$\left.\begin{array}{l} \delta_{11}a_1 + \delta_{21}a_2 + \delta_{31}a_3 + \cdots = \Delta_1 \\ \delta_{12}a_1 + \delta_{22}a_2 + \delta_{32}a_3 + \cdots = \Delta_2 \\ \delta_{13}a_1 + \delta_{23}a_2 + \delta_{33}a_3 + \cdots = \Delta_3 \\ \dots \end{array}\right\} \tag{13.32}$$

These equations are called the *canonical equations* of the Rayleigh–Ritz method. In these equations, $\delta_{ik} = \delta_{ki}$.

Equations (13.32) decompose into separate equations if the second derivatives of the coordinate functions $\Phi_k(x)$ form a system of orthogonal functions (see Section 6.5), i.e., if they have the following property:

$$\int_0^L \Phi_k''(x)\Phi_j''(x)\,dx = 0 \quad k \neq j \tag{13.33}$$

In this case, all the coefficients δ_{kj}, for $k \neq j$, are zero, and Eqs. (13.32) yield

$$\delta_{11}a_1 = \Delta_1 \qquad \delta_{22}a_2 = \Delta_2 \qquad \delta_{33}a_3 = \Delta_3, \ldots \tag{13.34}$$

Assuming that the coordinate functions take the form

$$\Phi_k(x) = \sin\frac{k\pi x}{L}$$

we conclude that they satisfy the condition of orthogonality (13.33) and that, therefore, Eqs. (13.34) can be used. In these equations, the coefficients δ_{kk} and the load terms Δ_k are, in accordance with the formulas (13.29) and (13.30), as follows:

$$\delta_{kk} = \int_0^L EI[\Phi_k''(x)]^2\,dx = \int_0^L EI\left(\frac{k\pi}{L}\right)^4 \sin^2\frac{k\pi x}{L}\,dx = EI\left(\frac{k\pi}{L}\right)^4\frac{L}{2}$$

$$\Delta_k = \int_0^L q\Phi_k(x)\,dx = \frac{qL}{k\pi}(1-\cos k\pi) = \begin{cases} \dfrac{2qL}{k\pi} & k = 1, 3, 5, \ldots \\ 0 & k = 2, 4, 6, \ldots \end{cases}$$

Then the parameters a_k are expressed as

$$a_k = \frac{\Delta_k}{\delta_{kk}} = \frac{4qL^4}{\pi^5 EIk^5}$$

and the deflection function (13.22) is as follows:

$$w(x) = \frac{4qL^4}{\pi^5 EI}\sum_{k=1,3,5,\ldots}^{\infty}\frac{1}{k^5}\sin\frac{k\pi x}{L} \tag{13.35}$$

The bending moments can be computed by differentiation:

$$M(x) = EIw''(x) = -\frac{4qL^2}{\pi^3}\sum_{k=1,3,5,\ldots}\frac{1}{k^3}\sin\frac{k\pi x}{L} \tag{13.36}$$

Both series (13.35) and (13.36) converge very well; however, the convergence of the series for the deflections is significantly better than for the bending moments. Retaining only the first terms in these expansions, we obtain the following formulas for the maximum deflection and the maximum moment:

$$w_{\max} = w\left(\frac{L}{2}\right) = \frac{4qL^4}{\pi^5 EI} = 0.011307\,\frac{qL^4}{EI}$$

$$M_{\max} = M\left(\frac{L}{2}\right) = -\frac{4qL^2}{\pi^3} = -0.129qL^2$$

The exact values of these parameters, obtained by solving the differential equation of bending, are

$$w_{\max} = \frac{5}{384}\frac{qL^4}{EI} = 0.01302\,\frac{qL^4}{EI}$$

$$M_{\max} = -0.125qL^2$$

Thus, the approximate solution for the maximum deflection when only one term of the series is used differs from the exact solution by less than 0.4 percent. For the maximum bending moments, the error is about 3.2 percent. With three terms in the expansion (13.36), the maximum bending moment is

$$M(x) = -\frac{4qL^2}{\pi^3}\left(1 - \frac{1}{3^3} + \frac{1}{5^4}\right) = -0.1252qL^2$$

which is in excellent agreement with the exact solution.

As another example, we evaluate the maximum deflection for a cantilever beam, using, in the first approximation, the following coordinate function (Fig. 13.6):

$$\Phi_1(x) = x^2$$

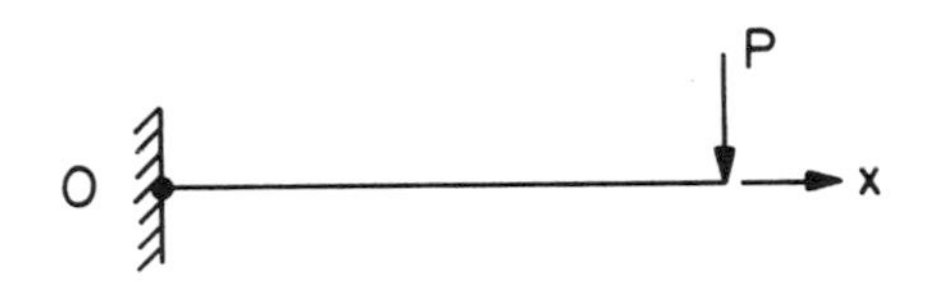

FIGURE 13.6 Cantilever beam under an end force.

which satisfies the conditions $w(0) = 0$, $w'(0) = 0$. The coefficient δ_{11} and the load term Δ_1 are, in accordance with Eqs. (13.29) and (13.31), as follows:

$$\delta_{11} = \int_0^L EI[\Phi_1''(x)]^2 \, dx = 4EIL$$

$$\Delta_1 = P\Phi_1(L) = PL^2$$

so that the parameter a_1 is

$$a_1 = \frac{\Delta_1}{\delta_{11}} = \frac{PL}{4EI}$$

and the deflections of the beam are as follows:

$$w(x) = a_1\Phi_1(x) = \frac{PL}{4EI} x^2$$

The maximum deflection is

$$w_{\max} = \frac{PL^3}{4EI}$$

whereas the exact solution is

$$w_{\max} = \frac{PL^3}{3EI}$$

The error is 25 percent. In the second approximation, we use two coordinate functions

$$\Phi_1(x) = x^2 \qquad \Phi_2(x) = x^3$$

Then Eqs. (13.29) and (13.30) yield

$$\delta_{11} = 4EIL \qquad \delta_{22} = 12EIL^3 \qquad \delta_{12} = \delta_{21} = 6EIL^2$$

$$\Delta_1 = PL^2 \qquad \Delta_2 = PL^3$$

The canonical equations (13.32) yield

$$\left.\begin{aligned}4EILa_1 + 6EIL^2a_2 &= PL^2\\ 6EIL^2a_1 + 12EIL^3a_2 &= PL^3\end{aligned}\right\}$$

so that

$$a_1 = \frac{PL}{2EI} \qquad a_2 = -\frac{PL}{6EI}$$

The deflection function is then expressed as

$$w(x) = a_1\Phi_1(x) + a_2\Phi_2(x) = \frac{PL}{2EI}x^2 - \frac{PL}{6EI}x^3$$

which coincides with the exact solution. The addition of new terms to this equation will not change the final result since the assumed coordinate functions correspond to the exact solution of the problem.

In the *Bubnov–Galerkin method*, the displacements are sought in the same form (13.22) as in the Rayleigh–Ritz method. However, whereas the Rayleigh–Ritz method proceeds from the energy functional, the Bubnov–Galerkin method is based on the differential equation of the problem. The Bubnov–Galerkin method is one of the most powerful methods for obtaining solutions to various differential equations: linear and nonlinear, ordinary and partial derivatives, static and dynamic, etc.

We illustrate the application of this method using, as an example, the beam shown in Fig. 13.5. The differential equation of bending of this beam is

$$EIw^{\mathrm{IV}}(x) - q(x) = 0 \tag{13.37}$$

After substituting the expansion (13.22) into the left part of Eq. (13.37), we obtain a function

$$r(x) = \left[EI \sum_{k=1}^{\infty} a_k\Phi_k^{\mathrm{IV}}(x) - q(x)\right] \tag{13.38}$$

which is, generally speaking, not equal to zero. In the Bubnov–Galerkin method, the parameters a_k are determined from the condition that the function $r(x)$ be orthogonal with respect to all the coordinate functions:

$$\int_0^L r(x)\Phi_j(x)\,dx = 0 \qquad k = 1, 2, \ldots \tag{13.39}$$

Introducing Eq. (13.38) into Eq. (13.39), we have

$$\int_0^L \left[EI \sum_{k=1}^{\infty} a_k \Phi_k^{\mathrm{IV}}(x) - q(x) \right] \Phi_j(x)\, dx = 0 \qquad k = 1, 2, \ldots$$

or

$$\sum_{k=1}^{\infty} \delta_{kj} a_k = \Delta_j \qquad k = 1, 2, \ldots \tag{13.40}$$

where

$$\delta_{kj} = \int_0^L EI \Phi_k^{\mathrm{IV}}(x) \Phi_j(x)\, dx \tag{13.41}$$

$$\Delta_j = \int_0^L q(x) \Phi_j(x)\, dx \tag{13.42}$$

Equations (13.40) are the same as Eqs. (13.28) in the Rayleigh–Ritz method, and the load terms Δ_j also coincide. Although the expression (13.41) for the coefficients δ_{kj} looks different from Eq. (13.29), both expressions result in the same formulas for the coefficients δ_{kj} if the coordinate functions satisfy all the boundary conditions. The Bubnov–Galerkin method is advisable in those cases in which the differential equation of the problem is available but is too complicated to be solved exactly.

13.4 FINITE-ELEMENT METHOD

The Rayleigh–Ritz method can be used as a basis for one of the most powerful and most commonly used computer analysis techniques for calculating stress and deformation in structures: the *method of finite elements*. The main difficulty in applying the Rayleigh–Ritz procedure to bodies of geometrically complex shapes arises in selecting a convenient set of coordinate functions. The finite-element method provides an elegant solution to this problem by breaking the entire domain into many subdomains of a simple geometry. For such subdomains, called *finite elements*, it suffices to use constants or low-order polynomials as coordinate functions. This enables the analyst to apply the Rayleigh–Ritz method to bodies of arbitrary shape. The finite-element method brings together three areas: stress analysis, matrix algebra, and the digital computer.

The first papers on the finite-element method were published in the mid-1950s by J. H. Argyris, by Argyris with S. Kelsey, and by M. J. Turner, R. W. Claugh, H. C. Martin, and J. L. Topp in application to aircraft structures. Since then, numerous papers and books were published on the use of this method in different fields of engineering and science. There are several reasons for the revolutionary effect of the finite-element method, which currently is widely used not only in stress analysis but also in such areas as heat transfer, hydrodynamics, and the physics of plasma:

It utilizes a digital computer and thus handles large amounts of processing in a short time.
It enables the user to assemble the properties of a complex large structure from a number of elements with simple properties.
It is capable of handling static, dynamic, plasticity, creep, viscoelastic, and stability problems for a very broad range of structures.

We illustrate the procedure of the finite-element method using the case of plane stress (see Section 6.1). With $\sigma_z = \tau_{xz} = \tau_{yz} = 0$, Eqs. (7.35) result in the following expression for the strain energy per unit volume:

$$W = \frac{1}{2E}[(\sigma_x^2 + \sigma_y^2 - 2\nu\sigma_x\sigma_y) + 2(1 + \nu)\tau_{xy}^2]$$

The particular subtopic of the finite-element method we deal with here is known as the *stiffness method*, or the *displacement method*. In this method, the displacements of the extreme points of the adjacent elements or the points at which the structure is supported (*nodal points, or nodes*) are the unknowns, and the general solution of the structural equations centers on finding the nodal displacements. With this in mind, we should obtain an expression for the strain energy in terms of the strains. Substituting Eqs. (6.10) for strains into the preceding equation, we have

$$W = \frac{E}{2(1 - \nu^2)}\left(\varepsilon_x^2 + \varepsilon_y^2 + 2\nu\varepsilon_x\varepsilon_y + \frac{1 - \nu}{2}\gamma_{xy}^2\right)$$

Then the energy functional can be expressed as

$$\Pi = \frac{E}{2(1 - \nu^2)}\int_A \left(\varepsilon_x^2 + \varepsilon_y^2 + 2\nu\varepsilon_x\varepsilon_y^2 + \frac{1 - \nu}{2}\gamma_{xy}^2\right) dA$$
$$- \int_A (Xu + Yv)\, dA - \int_C (X_\nu u + Y_\nu v)\, ds$$

where A is the body area, s is its surface, X and Y are body forces, X_ν and Y_ν are surface forces, and u and v are the displacements in the x and y directions, respectively. The last two terms in this formula represent external work due to the body and surface forces.

Let the area A be decomposed into M finite elements. Then the total strain energy can be evaluated as the sum of the strain energies for each element:

$$\Pi = \sum_{m=1}^{M} \Pi_m$$

where the energy of one element is

$$\Pi_m = \frac{E}{2(1-\nu^2)} \int_{A_m} \left(\varepsilon_x^2 + \varepsilon_y^2 + 2\nu\varepsilon_x\varepsilon_y + \frac{1-\nu}{2}\gamma_{xy}^2 \right) dA$$
$$- \int_{A_m} (Xu + Yv)\, dA - \int_{C_m} (X_\nu u + Y_\nu v)\, ds \qquad (13.43)$$

Suppose that the entire area A is partitioned into M triangles, and consider a typical triangle such as that shown in Fig. 13.7a. Because the finite elements—triangles—are small, it is reasonable to assume that the displacements u and v vary linearly over A_m:

$$u(x, y) = a + bx + cy \qquad v(x, y) = d + ex + fy$$

We choose the displacements

$$u_1 = a + bx_1 + cy_1 \qquad u_2 = a + bx_2 + cy_2 \qquad u_3 = a + bx_3 + cy_3$$
$$r_1 = d + ex_1 + fy_1 \qquad r_2 = d + ex_2 + fy_2 \qquad r_3 = d + ex_3 + fy_3$$

at the vertices of the triangles (nodal points) as generalized displacements and solve the preceding equations for the constants a,b,c and d,e,f. For the first set of constants, for instance, we have

$$\left.\begin{aligned} a &= \frac{1}{2\Delta_m}[(x_2y_3 - x_3y_2)u_1 + (x_3y_1 - x_1y_3)u_2 + (x_1y_2 - x_2y_3)u_3] \\ b &= \frac{1}{2\Delta_m}[(y_2 - y_3)u_1 + (y_3 - y_1)u_2 + (y_1 - y_2)u_3] \\ c &= \frac{1}{2\Delta_m}[(x_3 - x_2)u_1 + (x_1 - x_3)u_2 + (x_2 - x_1)u_3] \end{aligned}\right\}$$

where Δ_m is the area of the triangle A_m. Thus, the displacement u of any point within this element can be expressed through the nodal displacements as follows:

$$u(x, y) = u_1\Phi_1(x, y) + u_2\Phi_2(x, y) + u_3\Phi_3(x, y) \qquad (13.44)$$

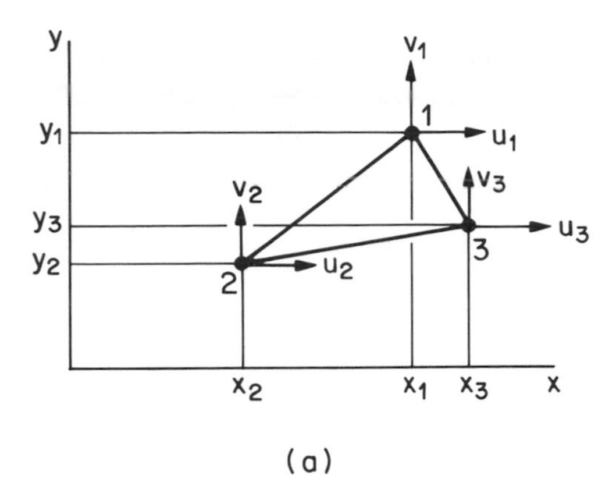

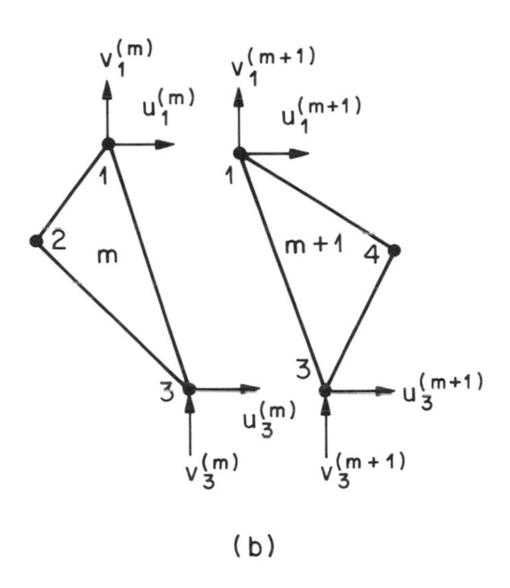

FIGURE 13.7 Triangular finite element.

where the coordinate functions are

$$
\left.
\begin{aligned}
\Phi_1(x, y) &= \frac{1}{2\Delta_m}\,[(x_2 y_3 - x_3 y_2) + (y_2 - y_3)x + (x_3 - x_2)y] \\
\Phi_2(x, y) &= \frac{1}{2\Delta_m}\,[(x_3 y_1 - x_1 y_3) + (y_3 - y_1)x + (x_1 - x_3)y] \\
\Phi_3(x, y) &= \frac{1}{2\Delta_m}\,[(x_1 y_2 - x_2 y_1) + (y_1 - y_2)x + (x_2 - x_1)y]
\end{aligned}
\right\}
$$

Observe that, at nodal points, these functions assume the following values:

$$\begin{array}{lll} \Phi_1(x_1, y_1) = 1 & \Phi_1(x_2, y_2) = 0 & \Phi_1(x_3, y_3) = 0 \\ \Phi_2(x_1, y_1) = 0 & \Phi_2(x_2, y_2) = 1 & \Phi_2(x_3, y_3) = 0 \\ \Phi_3(x_1, y_1) = 0 & \Phi_3(x_2, y_2) = 0 & \Phi_3(x_3, y_3) = 1 \end{array}$$

and that

$$\Phi_1(x, y) + \Phi_2(x, y) + \Phi_3(x, y) = 1$$

Similarly,

$$v(x, y) = v_1\Phi_1(x, y) + v_2\Phi_2(x, y) + v_3\Phi_3(x, y) \tag{13.45}$$

The strains in element A_m are therefore

$$\left.\begin{aligned} \varepsilon_x &= \frac{\partial u}{\partial x} = u_1 \frac{\partial \Phi_1}{\partial x} + u_2 \frac{\partial \Phi_2}{\partial x} + u_3 \frac{\partial \Phi_3}{\partial x} \\ \varepsilon_y &= \frac{\partial v}{\partial y} = v_1 \frac{\partial \Phi_1}{\partial y} + v_2 \frac{\partial \Phi_2}{\partial y} + v_3 \frac{\partial \Phi_3}{\partial y} \\ \gamma_{xy} &= \frac{\partial u}{\partial x} + \frac{\partial v}{\partial y} = u_1 \frac{\partial \Phi_1}{\partial y} + v_1 \frac{\partial \Phi_1}{\partial x} + u_2 \frac{\partial \Phi_2}{\partial y} + v_2 \frac{\partial \Phi_2}{\partial x} + u_3 \frac{\partial \Phi_3}{\partial y} + v_3 \frac{\partial \Phi_3}{\partial x} \end{aligned}\right\} \tag{13.46}$$

By introducing these formulas into Eq. (13.43), we obtain the expression for the energy functional for the element A_m in terms of the generalized displacements $u_1, v_1, u_2, v_2, u_3, v_3$. To obtain this expression in the most compact form, matrix notation can be used.

The displacements u and v can be treated as elements of a matrix

$$\mathbf{u} = \begin{Bmatrix} u(x, y) \\ v(x, y) \end{Bmatrix} = \begin{bmatrix} \Phi_1 & \Phi_2 & \Phi_3 & 0 & 0 & 0 \\ 0 & 0 & 0 & \Phi_1 & \Phi_2 & \Phi_3 \end{bmatrix} \begin{bmatrix} u_1 \\ u_2 \\ u_3 \\ v_1 \\ v_2 \\ v_3 \end{bmatrix} = \mathbf{\Phi}_m(x, y)\mathbf{q}_m$$

where $\mathbf{\Phi}$ is the 2×6 matrix of coordinate functions and $\mathbf{q}_m$ is the 6×1 column matrix of generalized displacements. Then the stresses and strains can be displayed in the form of column matrices,

$$\boldsymbol{\sigma} = \begin{Bmatrix} \sigma_x \\ \sigma_y \\ \tau_{xy} \end{Bmatrix} \qquad \boldsymbol{\varepsilon} = \begin{Bmatrix} \varepsilon_x \\ \varepsilon_y \\ \gamma_{xy} \end{Bmatrix}$$

and the formulas (12.46) for the strains can be presented as

$$\boldsymbol{\varepsilon} = \boldsymbol{\Psi}(x, y)\mathbf{q}_m \tag{13.47}$$

where the matrix $\boldsymbol{\Psi}(x, y)$ is determined by the coordinate functions, and is

$$\boldsymbol{\Psi}(x, y) = \begin{vmatrix} \Phi_{1x} & \Phi_{2x} & \Phi_{3x} & 0 & 0 & 0 \\ 0 & 0 & 0 & \Phi_{1y} & \Phi_{2y} & \Phi_{3y} \\ \Phi_{1y} & \Phi_{2y} & \Phi_{3y} & \Phi_{1x} & \Phi_{2x} & \Phi_{3x} \end{vmatrix}$$

In this formula $\Phi_{1x} = \partial\Phi_1/\partial x$, $\Phi_{2x} = \partial\Phi_2/\partial x$, etc. The stress-strain relation (Hooke's law) can be written as

$$\boldsymbol{\sigma} = \mathbf{E}\boldsymbol{\varepsilon} \tag{13.48}$$

where the matrix of the elastic constants is

$$\mathbf{E} = \frac{E}{1 - v^2}\begin{bmatrix} 1 & v & 0 \\ v & 1 & 0 \\ 0 & 0 & \dfrac{1 - v}{2} \end{bmatrix}$$

Likewise, the strain energy (13.43) of an element can be presented as

$$\Pi_m = \frac{1}{2}\int_A \boldsymbol{\varepsilon}^T\mathbf{E}\boldsymbol{\varepsilon}\, dA - \int_A \mathbf{u}^T\mathbf{X}\, dA - \int_C \mathbf{u}^T\mathbf{S}\, ds \tag{13.49}$$

where

$$\mathbf{X} = \begin{Bmatrix} X \\ Y \end{Bmatrix} \qquad \mathbf{S} = \begin{Bmatrix} X_v \\ Y_v \end{Bmatrix}$$

and T denotes matrix transposition.

Upon substituting Eq. (13.47) into (13.49), we obtain the following matrix formula for the energy functional Π_m of the mth finite element:

$$\Pi_m = \tfrac{1}{2}\mathbf{q}_m^T \mathbf{k}_m \mathbf{q}_m - \mathbf{f}_m^T \mathbf{q}_m$$

where

$$\mathbf{k}_m = \int_{A_m} \boldsymbol{\Psi}^T \mathbf{E} \boldsymbol{\Psi} \, dA$$

is the 6×6 *stiffness matrix* for the element A_m and

$$\mathbf{f}_m = \int_{A_m} \boldsymbol{\Phi}^T \mathbf{X} \, dA + \int_{C_m} \boldsymbol{\Phi}^T \mathbf{S} \, ds$$

is the 6×1 generalized force vector for this element.

We now impose a compatibility condition for the displacements of the adjacent finite elements. These conditions demand that all the elements of the body fit together in such a way that the displacements of points shared by adjacent elements are the same. For example, if adjacent elements m and $m + 1$ are to fit together so that no gaps or discontinuities are to occur in the displacement field, we demand that (Fig. 13.7b):

$$u_1^{(m)} = u_1^{(m+1)} \qquad v_1^{(m)} = v_1^{(m+1)}$$

$$u_3^{(m)} = u_3^{(m+1)} \qquad v_3^{(m)} = v_3^{(m+1)}$$

When these conditions are imposed, there remain only two degrees of freedom per nodal point in the connected collection of finite elements. The energy functional for the entire system is

$$\Pi = \sum_{m=1}^{M} \Pi_m = \frac{1}{2} \sum_{m=1}^{M} (\mathbf{q}^T \mathbf{k}_m \mathbf{q}_m - \mathbf{q}_m^T \mathbf{f}_m) = \frac{1}{2} \boldsymbol{\Delta}^T \mathbf{K} \boldsymbol{\Delta} - \mathbf{F}^T \boldsymbol{\Delta}$$

where $\boldsymbol{\Delta}$ is the column matrix containing the final collection of generalized displacements of the connected system of elements, $\mathbf{K}$ is the *global stiffness matrix*, and $\mathbf{F}$ is the *global generalized force matrix*. Application of Eqs. (13.23) which, in the case in question, should be written as

$$\frac{\partial \Pi}{\partial \boldsymbol{\Delta}_i} = 0$$

results in the following system of equations for the unknown nodal displacements:

$$\mathbf{K}\mathbf{\Delta} = \mathbf{F} \tag{13.50}$$

Thus, in the finite-element method, the structure is mathematically modeled by examining the load deflection behavior of its nodes. All the loading and displacement occur at the nodes, and all the distributed stiffness properties of the structure are "lumped" into the nodes. Lumping the structural stiffness properties at the nodes is accomplished by assembling the structural stiffness matrix. The general entry of a stiffness matrix K_{ij} is, by definition, the force at the ith node that occurs when a unit displacement is given to the jth node while all the other nodes are held fixed. The global (overall) stiffness matrix is affected by the location of the nodes, the size of the finite elements, material properties, and the way the elements are connected (to which nodes). It is extremely fortunate that all these factors can be dealt with by considering the elements one by one and that the relatively complex job of assembling the structural stiffness matrix consists of taking each element's stiffness matrix, transforming this matrix to structural coordinates, and adding the result to the proper location of the structural stiffness matrix.

The displacement field described by Eq. (13.50) contains an arbitrary rigid motion, which should be eliminated by application of appropriate boundary conditions restricting this motion. These conditions eliminate the rows and columns in the matrices **K** and **F**, which correspond to the restricted degrees of freedom. This leads to a reduced system of equations,

$$\hat{\mathbf{K}}\hat{\mathbf{\Delta}} = \hat{\mathbf{F}}$$

from which the displacements can be found:

$$\hat{\mathbf{\Delta}} = \hat{\mathbf{K}}^{-1}\hat{\mathbf{F}}$$

Here $\hat{\mathbf{K}}^{-1}$ is the inverse of the matrix $\hat{\mathbf{K}}$. Once the matrix $\hat{\mathbf{\Delta}}$ is determined, the displacements u_1, u_2, u_3 and v_1, v_2, v_3 become known for each element. Strains and stresses are then computed using Eqs. (13.46) and (13.48), and the analysis is complete.

The only difficult part of a finite-element procedure is to transform the element stiffness matrix written in element coordinates to the global structural coordinate system. It is the element-by-element assembly process that gives the finite-element method its power. If a *library of element types* is available, then more complicated structures can be assembled. The elements in a library are each described completely by their relatively simple stiffness matrices. Here are some of the most common element types: beam elements (take bending in two planes, shear in two planes, and torsion), rod elements (take axial forces and sometimes torsion), shear panels (take in-plane shear), membrane elements (take in-plane shear and normal stresses), plate elements (take in-plane shear and normal stress plus bending stress and plate shear). The solutions are dependent on finding suitable elements with which to build a preprocessing model. A large number of choices are available, and these are widely cataloged in the literature. Note that the choice of element can affect the results significantly.

14

Bending of Frames

14.1 FRAME STRUCTURES, SIMPLEST FRAMES

Some chassis and other structural framework employed in electronic equipment can be treated as frame structures. In this section, we discuss methods for structural analysis of frames, limiting our consideration to plane (two-dimensional) frames only.

A *plane frame* is a beam (rod) system whose members (beams) are rigidly connected at the corners (nodes) in such a way that the angles between the connected beams do not change when the beams are deformed. The external loads acting on the frame are located in its plane and cause bending of the frame beams. It is assumed that no tension occurs and that the axial forces in the beams are small enough to permit neglecting any effects due to these forces compared to the effects caused by bending. The number and location of the beams in frame structures are usually such that the replacement of the rigid connections at the nodes by hinges turns the frame structure into a rod mechanism with one or more degrees of freedom. This is one of the main differences between frames and trusses. In the latter, the rods are joined together by smooth pins, which make the configuration of the truss structure geometrically unchanged. In addition, the external forces acting on a truss are always concentrated at its hinges, so that the truss rods can be treated as two-force bodies experiencing tension or compression but never bending.

Frame structures are, as a rule, statically indeterminate systems and, therefore, the primary purpose of the analysis of such structures is to establish and determine the redundant generalized forces. Once these forces are found, the evaluation of stresses and deflections for the frame beams does not cause any difficulties and can be carried out using methods described previously for single-span beams. Let us show how it can be done using an example of a simplest frame structure in Fig. 14.1a.

In order to make our analysis as simple as possible, we assume that the nodes of the frame are fixed in space. The elastic curves of the particular

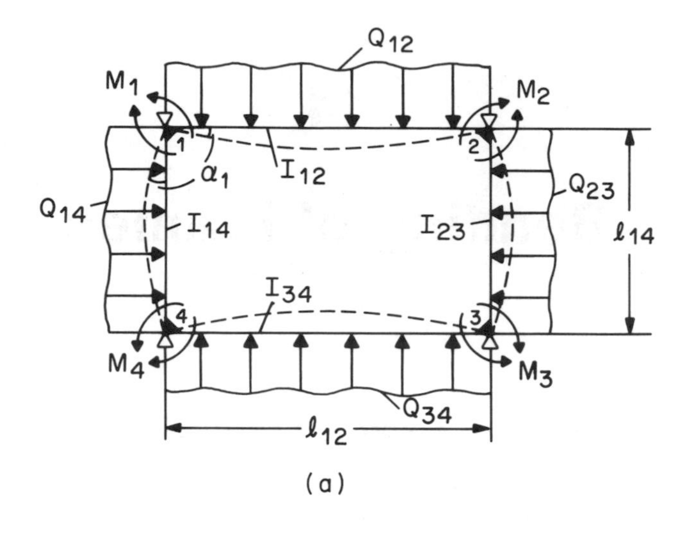

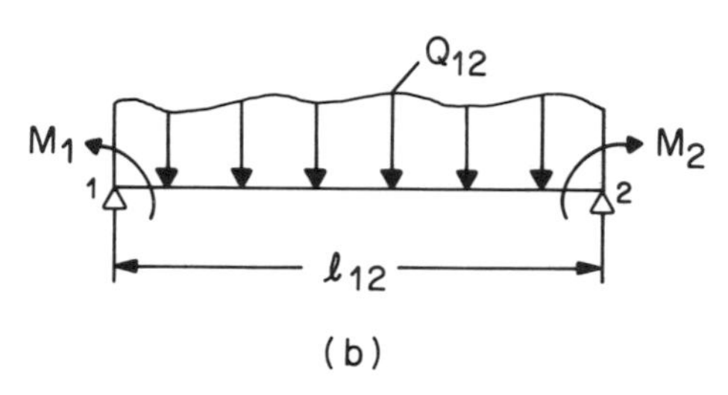

FIGURE 14.1 Simplest frame.

beams are drawn in broken lines. Such a configuration of a deformed frame structure occurs because the initially right angles at the nodes have to remain straight after the beam deformation.

We choose the nodal bending moments as the redundant unknowns and use the condition of the compatibility of angular displacements at the nodes to determine these moments. Examine first the horizontal beam 1–2 (Fig. 14.1b). It is subjected to the given external load Q_{12}, which may include all kinds of distributed loads and concentrated forces and moments, etc. The action of the adjacent beams 1–4 and 2–3, we replace with as yet unknown nodal bending moments M_1 and M_2. The angle of rotation of the beam's cross section at node 1 due to moments M_1 and M_2 is

$$\alpha_M = \frac{M_1 l_{12}}{3(EI)_{12}} + \frac{M_2 l_{12}}{6(EI)_{12}}$$

This formula can be easily obtained by using one of the methods of single-span beam analysis or can just be taken from the beam deflection table in Appendix A. Note that moment M_1 applied to node 1 is twice as effective as the remote moment M_2 and that the moments cause angles of rotation in the same direction. The total angle of rotation at node 1 can be determined by subtracting from angle α_M an angle α_{12} due to the external load Q_{12}:

$$\alpha_1 = \alpha_M - \alpha_{12} = \frac{M_1 l_{12}}{3(EI)_{12}} + \frac{M_2 l_{12}}{6(EI)_{12}} - \alpha_{12}$$

The angle of rotation of the upper end of the beam 1–4 due to the external load and nodal moments acting on this beam can be determined, by analogy, as

$$\alpha_1 = \frac{M_1 l_{14}}{3(EI)_{14}} - \frac{M_4 l_{14}}{6(EI)_{14}} + \alpha_{14}$$

We deliberately use the same notation for both the total angles of rotation since the condition of the compatibility of the angular displacements requires that these angles be equal. This results in the following equation for the unknown moments M_1, M_2, and M_4:

$$\left[\frac{l_{12}}{3(EI)_{12}} + \frac{l_{14}}{3(EI)_{14}}\right] M_1 + \frac{l_{12}}{6(EI)_{12}} M_2 + \frac{l_{14}}{6(EI)_{14}} M_4 = \alpha_{12} + \alpha_{14}$$

The conditions of compatibility applied to nodes 2, 3, and 4 yield

$$\left[\frac{l_{14}}{3(EI)_{23}} + \frac{l_{12}}{3(EI)_{12}}\right] M_2 + \frac{l_{14}}{6(EI)_{23}} M_3 + \frac{l_{12}}{6(EI)_{12}} M_1 = \alpha_{23} + \alpha_{21}$$

$$\left[\frac{l_{12}}{3(EI)_{34}} + \frac{l_{14}}{3(EI)_{23}}\right] M_3 + \frac{l_{12}}{6(EI)_{34}} M_4 + \frac{l_{14}}{6(EI)_{23}} M_2 = \alpha_{34} + \alpha_{32}$$

$$\left[\frac{l_{14}}{3(EI)_{14}} + \frac{l_{12}}{3(EI)_{34}}\right] M_4 + \frac{l_{14}}{6(EI)_{14}} M_1 + \frac{l_{12}}{6(EI)_{34}} M_3 = \alpha_{41} + \alpha_{43}$$

The obtained four equations are sufficient for the evaluation of the four unknown moments M_1, M_2, M_3, and M_4.

If, for instance, all the beams are made of the same material and have identical cross sections, then the flexural rigidity of all the beams is EI.

Assuming further that all the beams but 1–2 are load-free and that the load $Q_{12} = Q$ is evenly distributed along the beam, we have

$$\alpha_{12} = \alpha_{21} = \frac{Qb^2}{24EI}$$

where $b = l_{12}$ is the frame width. Obviously, all the other rotation angles caused by any external load are zero. A further important simplification is due to the geometrical, materials, and loading symmetry of the structure, so that we know beforehand that $M_1 = M_2$ and $M_3 = M_4$. Then we have the following two equations for the nodal moments M_1 and M_3:

$$\left.\begin{aligned} (3b + 2h)M_1 + hM_3 &= \frac{Qb^2}{4} \\ hM_1 + (3b + 2h)M_3 &= 0 \end{aligned}\right\}$$

where h is the height of the frame structure. Then we have

$$M_1 = \frac{Qb^2}{12} \frac{3b + 2h}{3b^2 + 2bh + h^2} \qquad M_3 = -\frac{Qb^2}{12} \frac{h}{3b^2 + 2bh + h^2}$$

For a square frame ($b = h$), we have

$$M_1 = \tfrac{5}{72}Qb \qquad M_3 = -\frac{Qb}{72}$$

The bending moment in the x cross section of the beam 1–2 (the origin is at node 1) is

$$M(x) = M_1 - \frac{Qx}{2} + \frac{Qx^2}{2b}$$

so that the moment in the mid–cross section is

$$M\left(\frac{b}{2}\right) = -\tfrac{1}{18}Qb$$

The distribution of the bending moments along the frame beams is shown in Fig. 14.2. The maximum bending moments occur, at nodes 1 and 2 and are $M_1 = M_2 = (5/72)Qb$. Note that if load Q were applied to a single-span beam

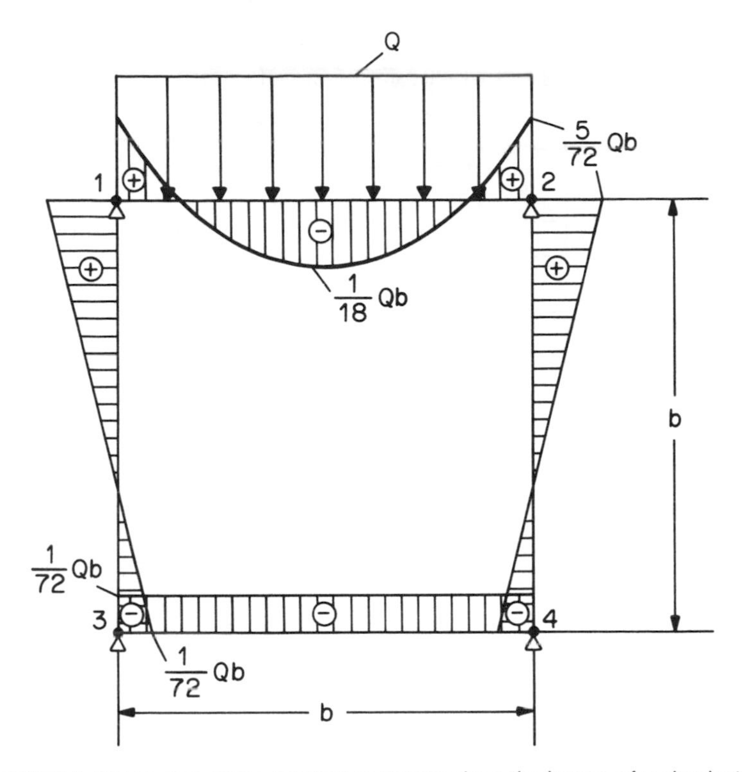

FIGURE 14.2 Distribution of the bending moments along the beams of a simplest frame.

1–2, which would work separately and whose bending would not be relieved by the rest of the frame structure, then the maximum bending moment would occur in the mid–cross section, and its magnitude would be $M_{max} = -(Qb/8)$. This bending moment is greater by a factor of 1.8 than the node moments M_1 and M_2 and greater by a factor of 2.25 than the bending moment in the mid–cross section of a beam that is a member of the frame structure. The maximum bending moments M_1 and M_2 increase with an increase in the b/h ratio and, for a wide and low frame ($h \ll b$), are $M_1 = M_2 = (1/12)Qb$, i.e., greater by a factor of 1.2 than in a square frame of the same width.

It is also noteworthy that geometric and structural symmetry is a very useful property, enabling us to reduce the degree of indeterminacy. If even the loading is asymmetric, a symmetric structure can be presented as a sum of a symmetric and an antisymmetric loading. Then the superposition principle can be used to obtain a solution to the given problem.

We would like to emphasize that a structure of the type shown in Fig.

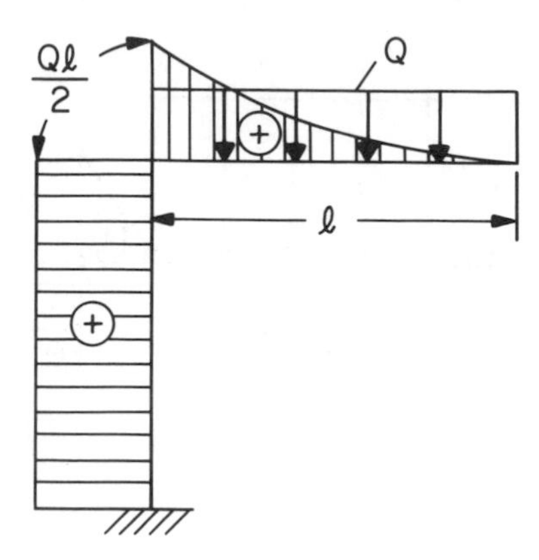

FIGURE 14.3 Broken beam.

14.3 is not a frame, but just a broken beam, and that methods used for its analysis should be, in effect, the same as the methods applied to single-span beams.

14.2 COMPLEX FRAMES

In this section, we examine a case in which a frame has more than one "field." An example of such a frame is shown in Fig. 14.4. Frames with many fields are usually referred to as *complex frames*.

Consider an arbitrary ith node of a complex frame, which is a hub of several rods connecting this node to the adjacent nodes $j, k, \ldots, m$ (Fig. 14.5). It is assumed that the given load may include, in addition to the span loads Q_{ij}, concentrated moments M_i applied directly to the nodes. In the analysis of a simple frame in the preceding section, we used the nodal moments as the redundant unknowns. Obviously, we can use a similar approach in the case of a complex frame. However, if n rods are connected at the given node, then we can form n equations for the unknown nodal moments: an equilibrium equation,

$$M_i = \sum_j M_{ij} \tag{14.1}$$

reflecting the fact that the external moment applied to this node must be equilibrated by the reactive moment and $n - 1$ equations of the compatibility of the angular displacements. Since one of the unknown moments can be expressed through the other moments from the equilibrium equation and then can be excluded from the equations of compatibility, we can consider

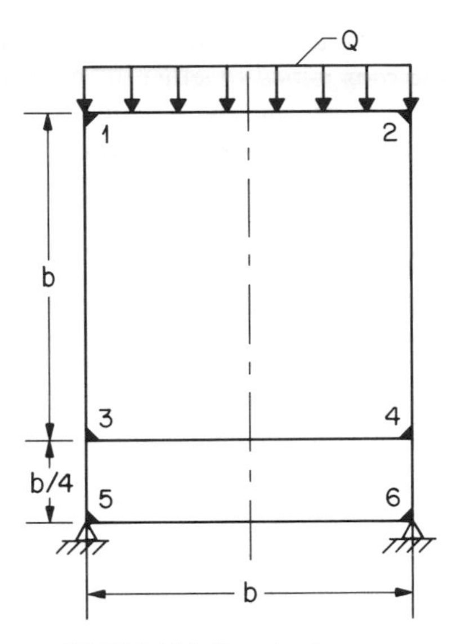

FIGURE 14.4 Complex frame.

that there are $n - 1$ unknowns in each node. Thus, if, in simple frames, there is just one unknown moment in each node at which two rods are connected ($n = 2$), in complex nodes ($n > 2$), there can be many unknown moments. Accordingly, the number of the equations of the compatibility of angular displacements also increases. Therefore, it is advisable to choose as the redundant unknowns those quantities whose number is independent of the number of rods in the nodes of a complex frame and does not exceed the number of nodes. This can be achieved if the angles of rotation of the nodal

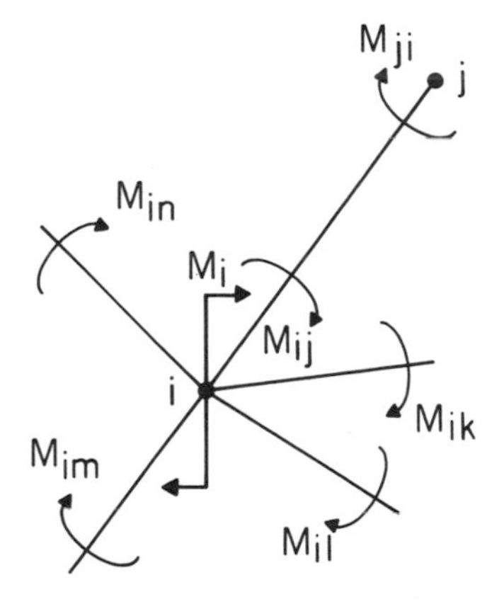

FIGURE 14.5 A node of a complex frame.

cross sections are taken as the redundant unknowns. Indeed, since the end cross sections of all the rods joined together at the given node rotate at the same angle, the latter can be chosen as a redundant unknown, thereby reducing significantly the total number of unknowns. In order to form equations for the unknown nodal rotation angles, we have to express all the bending moments through the external loads and the nodal rotation angles and form the equilibrium conditions for these moments.

The bending moment M_{ij} at the ith node of a beam connecting this node to the jth node and the bending moment M_{ji} at the jth node of this beam can be evaluated, assuming immovable nodes, on the basis of Eq. (13.50). Having in mind that the sign of the moment M_{ji} is opposite to the sign of the moment M_{ji} in this formula, we have the following expression for the rotation angle α_{ij} (Fig. 14.6):

$$\alpha_{ij} = \frac{M_{ij} l_{ij}}{3(EI)_{ij}} - \frac{M_{ji} l_{ij}}{6(EI)_{ij}}$$

Similarly,

$$\alpha_{ji} = \frac{M_{ji} l_{ij}}{3(EI)_{ij}} - \frac{M_{ij} l_{ij}}{6(EI)_{ij}}$$

Solving these equations for the unknown moments M_{ij} and M_{ji} and supplementing the obtained expressions by bending moments $\bar{M}_{ij}$ and $\bar{M}_{ji}$ caused by the loads applied in the span of the beam, we have

$$\left. \begin{aligned} M_{ij} &= \bar{M}_{ij} + \frac{2(EI)_{ij}}{l_{ij}}(2\alpha_{ij} + \alpha_{ji}) \\ M_{ji} &= \bar{M}_{ji} + \frac{2(EI)_{ij}}{l_{ij}}(2\alpha_{ji} + \alpha_{ij}) \end{aligned} \right\} \tag{14.2}$$

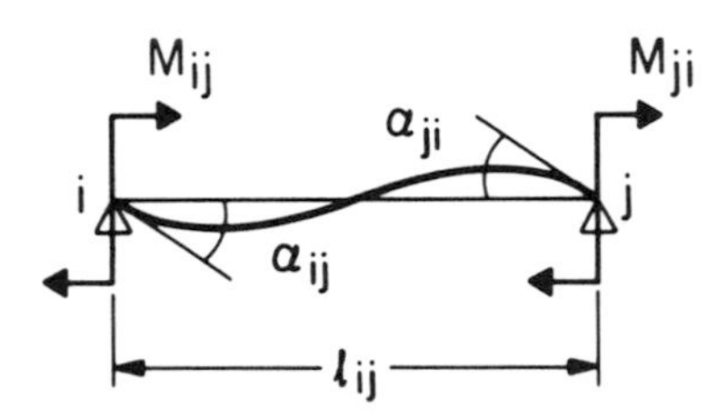

FIGURE 14.6 Simply supported beam under concentrated end moments.

As evident from these formulas, the moments $\bar{M}_{ij}$ and $\bar{M}_{ji}$ should be calculated assuming that the given beam is clamped at both ends. Substituting these formulas into the condition of equilibrium (14.1), we obtain the following equations for the unknown nodal angles of rotation:

$$4\alpha_i \sum_j \frac{(EI)_{ij}}{l_{ij}} + 2\sum_j \alpha_j \frac{(EI)_{ij}}{l_{ij}} = M_i - \sum_j \bar{M}_{ij} \qquad i = 1, 2, \ldots, p$$

where p is the number of nodes that are able to rotate, i the given node, and j the general notation for the nodes adjacent to node i. These equations can be rewritten in a more convenient and compact form as

$$m_i = \sum_j m_j \eta_{ij} = m_i^0 \qquad i = 1, 2, \ldots, p \tag{14.3}$$

where the following notation is used:

$$\left.\begin{array}{lll} m_i = \dfrac{2EI_0}{l_0}\alpha_i & m_i^0 = \dfrac{1}{K_i}\left(M_i - \sum\limits_j \bar{M}_{ij}\right) & \\ K_i = 2\sum\limits_j k_{ij} & k_{ij} = k_{ji} = \dfrac{(EI)_{ij}}{EI_0}\dfrac{l_0}{l_{ij}} & \eta_{ij} = \dfrac{k_{ij}}{K_i} \end{array}\right\} \tag{14.4}$$

Here EI_0 and l_0 are the arbitrarily chosen flexural rigidity and length of one of the frame's beams. The m_i values, directly proportional to the angles α_i of the node rotations, can be called *factors of nodal rotation angles.* The η_{ij} values are *influence coefficients* of the angle of rotation of the node j on the angle of rotation of the node i. Obviously, $\eta_{ij} = \eta_{ji}$. We would like to remind the reader that, in this section, all the moments are considered positive if they are directed clockwise and that the angles of rotation are positive if the cross sections rotate clockwise. After Eqs. (14.3) are solved, then the bending moments at the nodes can be evaluated by Eqs. (14.2) which, using the notations (14.4), can be presented as

$$\left.\begin{array}{l} M_{ij} = \bar{M}_{ij} + k_{ij}(2m_i + m_j) \\ M_{ji} = \bar{M}_{ji} + k_{ij}(2m_j + m_i) \end{array}\right\} \tag{14.5}$$

After these moments are evaluated, they should be included in the loads acting on the beams, which are already statically determinate and freely supported at the nodes.

In a symmetric frame subjected to a symmetric load, the angles of rotation of symmetric nodes i and j are related as $\alpha_i = -\alpha_j$, so that $m_i = -m_j$. If this frame is subjected to an antisymmetric load, then $\alpha_i = \alpha_j$ and $m_i = m_j$. This enables us to reduce the number of the unknowns in Eqs. (14.3) from p to $\frac{1}{2}(p - p_1)$ for a symmetric load and to $\frac{1}{2}(p + p_1)$ for an antisymmetric load, where p is the total number of nodes, and p_1 is the number of nodes located on the axis of the frame's symmetry. For complex frames having a symmetry axis, it is feasible to break down an arbitrary asymmetric load into symmetric and antisymmetric components, to carry out the calculations for these components separately, and then to sum up the final results.

As an example, examine the frame in Fig. 14.4, which is similar to the frame in Fig. 14.2 but has additional "fields" 3–4–6–5. Let all the beams have the same flexural rigidity EI. Since both the frame and the loading are symmetric, then $\alpha_1 = -\alpha_2$, $\alpha_3 = -\alpha_4$, $\alpha_5 = -\alpha_6$ and $m_1 = -m_2$, $m_3 = -m_4$, $m_5 = -m_6$, so that there are only three unknown angles of rotation. Putting $l_0 = b$, we have

$$k_{12} = k_{13} = k_{34} = k_{56} = 1 \qquad k_{35} = 4$$

$$K_1 = 2(k_{12} + k_{13}) = 4 \qquad K_3 = 2(k_{13} + k_{34} + k_{35}) = 12$$

$$K_5 = 2(k_{35} + k_{56}) = 10$$

$$\eta_{12} = \frac{k_{12}}{K_1} = \frac{1}{4} \qquad \eta_{13} = \frac{k_{13}}{K_1} = \frac{1}{4} \qquad \eta_{31} = \frac{k_{31}}{K_3} = \frac{1}{12}$$

$$\eta_{34} = \frac{k_{34}}{K_3} = \frac{1}{12} \qquad \eta_{35} = \frac{k_{35}}{K_3} = \frac{1}{3} \qquad \eta_{53} = \frac{k_{53}}{K_5} = \frac{2}{5}$$

$$\eta_{56} = \frac{k_{56}}{K_5} = \frac{1}{10}$$

Using the beam deflection tables in Appendix A for clamped-clamped beams and considering our sign convention, we have

$$\bar{M}_{12} = -\frac{Qb}{12}$$

Then,

$$m_1^0 = \frac{1}{K_1}(-\bar{M}_{12}) = \frac{1}{4}\frac{Qb}{12} = \frac{Qb}{48}$$

Obviously, $m_3^0 = m_5^0 = 0$. Then Eqs. (14.3) yield

$$\left.\begin{aligned} m_1 + m_2\eta_{12} + m_3\eta_{13} &= m_1^0 \\ m_3 + m_1\eta_{31} + m_4\eta_{34} + m_5\eta_{35} &= 0 \\ m_5 + m_3\eta_{53} + m_6\eta_{56} &= 0 \end{aligned}\right\}$$

or, considering $m_2 = -m_1, m_4 = -m_3, m_6 = -m_5$ and the calculated values of the coefficients η_{ij} and the load term m_1^0, we have

$$\left.\begin{aligned} 3m_1 + m_3 &= Qb/12 \\ m_1 + 11m_3 + 4m_5 &= 0 \\ 4m_3 + 9m_5 &= 0 \end{aligned}\right\}$$

These equations have the following solutions:

$$m_1 = 0.028819Qb \qquad m_3 = -0.003127Qb \qquad m_5 = 0.001390Qb$$

Then the nodal bending moments, in accordance with Eqs. (14.5), are

$$M_{12} = \bar{M}_{12} + k_{12}(2m_1 + m_2) = -\frac{Qb}{12} + k_{12}m_1 = -0.05451Qb$$

$$M_{13} = k_{13}(2m_1 + m_3) = 0.05451Qb$$

$$M_{31} = k_{13}(2m_3 + m_1) = 0.02257Qb$$

$$M_{34} = k_{34}(2m_3 + m_4) = k_{34}m_3 = -0.00312Qb$$

$$M_{35} = k_{35}(2m_3 + m_5) = -0.01945Qb$$

$$M_{53} = k_{35}(2m_5 + m_3) = -0.00139Qb$$

$$M_{56} = k_{56}(2m_5 + m_6) = k_{56}m_5 = 0.00139Qb$$

The distribution of the bending moments along the beams is diagrammed in Fig. 14.7 where, in order to compare the results with the Fig. 14.2 data, we use the same sign convention for the moments that was used in Section 14.1. Comparing the obtained numerical data to the calculated bending moments for a single-field frame in Fig. 14.2, we conclude that the addition of the field 3–4–6–5 brought down the maximum bending moment in the frame from $0.06944Qb$ to $0.05451Qb$, i.e., by 21.5 percent.

For frames with many nodes, the system of algebraic equations (14.2) for the moment factors m_i should be solved using a computer. The effect of the possible linear displacements of the nodes is not considered here, nor is the analysis of three-dimensional frames.

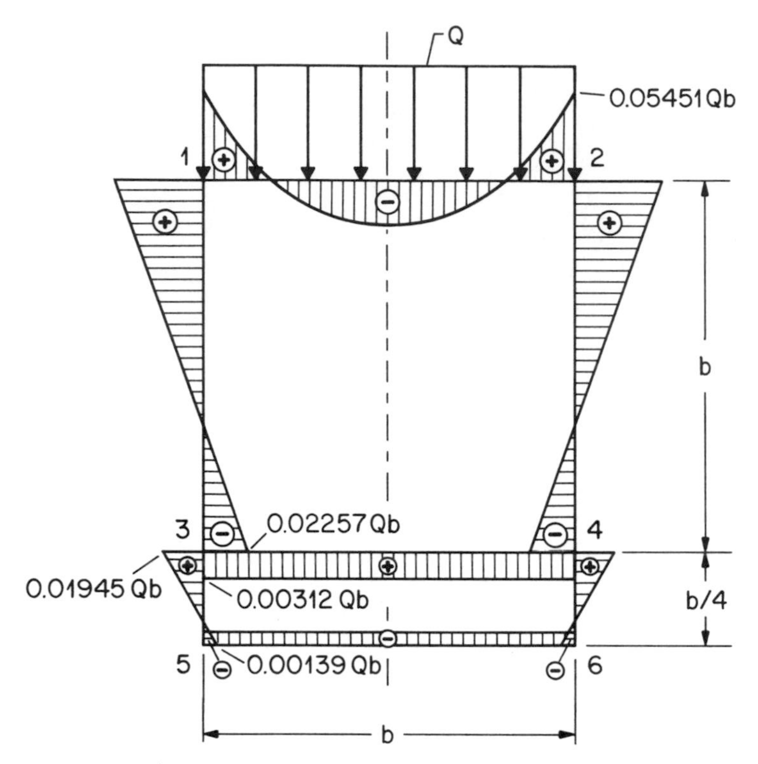

FIGURE 14.7 Distribution of the bending moments along the beams of a complex frame.

14.3 FORCE METHOD AND DEFORMATION METHOD

The approach used in the analysis of a single-field frame, examined in Section 14.1, was such that the bending moments at the frame nodes were chosen as the principal unknowns, and the equations for these unknowns were obtained on the basis of the condition of the compatibility of angular displacements of the end cross sections of the beams connected at the given node. Such an approach is called the *force method.* On the other hand, in the analysis of multifield frames, we chose the angular displacements at the frame nodes as the principal unknowns and formed the equations for the evaluation of these displacements using the conditions of equilibrium of the frame nodes. This is an example of a *deformation (displacement) method.* Generally, when the force method is used, we choose the bending moments and/or the forces at the end cross sections of the beams as the principal redundant unknowns, which are determined first. After these factors are found, all the other quantities of interest are evaluated by examining each beam separately. When the deformation method is used, it is the angular and/or linear displacements (deformations) at the end cross sections (nodes) that are sought first. In each case, we should apply a method that leads to a system of equations with the smallest number of unknowns. If the degree of the static indeterminacy of a system is greater than the number of the principal unknown displacements which should be determined for the further analysis of each beam or a part of a system separately, then the deformation method should be used. Otherwise, the force method is advisable.

We would like to point out that similar approaches underlie the basic theory-of-elasticity equations, where the Beltrami–Mitchell equations reflect the major idea of the force method, and the Lamé equations use the concept of the deformation (displacement) method. Indeed, the Beltrami–Mitchell equations contain stress components as the principal unknowns and are obtained on the basis of Saint-Venant's compatibility conditions for the strains. The Lamé equations contain the displacements as the principal unknowns and are based on the equilibrium equations of an elementary parallelepiped. In both cases, Hooke's law is used to express the stresses through the strains, and the equilibrium equations for an elementary tetrahedron are used to formulate the boundary conditions. Since the elements in an elastic body of an arbitrary form are infinitely small and their number is infinitely large, all the equations in the theory of elasticity are differential and, since the unknowns, whether displacements, strains, or stresses, depend on more than one argument, these differential equations are in partial derivatives. In the theory of beam (rod) systems, all the basic equations are algebraic equations.

Obviously, there are cases in which a combination of the force and deformation methods should be used. If such a mixed method is applied, one chooses force factors as the redundant unknowns in some cross sections (nodes), and deformations (displacements) as the principal unknowns in other cross sections. When this method is used, we should ensure the fulfillment of the equilibrium conditions when the force factors are introduced and the fulfillment of the displacement compatibility conditions when the unknown displacement factors are introduced. Then, where the force factors have been chosen as the principal unknowns, we form the equations of the compatibility of displacements for the nodes. In these equations, the displacements should be expressed through the forces and the displacements chosen as the principal unknowns. In the nodes, where the displacements are chosen as the principal unknowns, we should form the equilibrium equations, in which the forces and moments are expressed through the displacements and forces (moments) chosen as the principal unknowns.

15

Bending of Plates

15.1 MAJOR DEFINITIONS AND ASSUMPTIONS

A *plate* is a structural element that, initially, has the form of a straight prism whose height is significantly smaller than its base. In other words, a plate is a thin, flat body, symmetric with respect to a plane dividing the *thickness* (height) of the body into equal halves. This plane is called the *midplane* of the plate (Fig. 15.1). The particles of the plate are identified by coordinates (x, y, z), with $-(h/2) \leq z \leq (h/2)$, where $h = h(x, y)$ is the thickness of the plate. In the following analyses, we will always assume h to be a constant value throughout the plate. Structural elements that can be treated as plates are often encountered in microelectronic systems. Examples are printed circuit boards, silicon wafers, various substrates, and even chips themselves.

Plates are usually classified in two groups, *thin plates* and *thick plates*, depending on the ratio of the thickness to the length of the smaller span. If this ratio is greater than 0.2, the plate is considered thick, and its analysis should take into account the effects of shear on the deflections. Plate elements encountered in microelectronics have relatively small thicknesses and can therefore be evaluated on the basis of a simpler theory of thin plates. Indeed, even for a small-sized chip, whose side is, say, 3 mm, the above-mentioned ratio is about 0.17. Plates of extremely small thickness are called *membranes*. The *flexural rigidity* of plates is defined as

$$D = \frac{Eh^3}{12(1 - \nu^2)}$$

In membranes, this value is so small that such plates scarcely resist bending and experience in-plane tensile stresses only. An example is a nonreinforced thin film (i.e., a film separated from its substrate).

A flat plate experiences lateral (transverse) loads and, in a sense, can be treated as a two-dimensional generalization of a beam. As in a beam, these

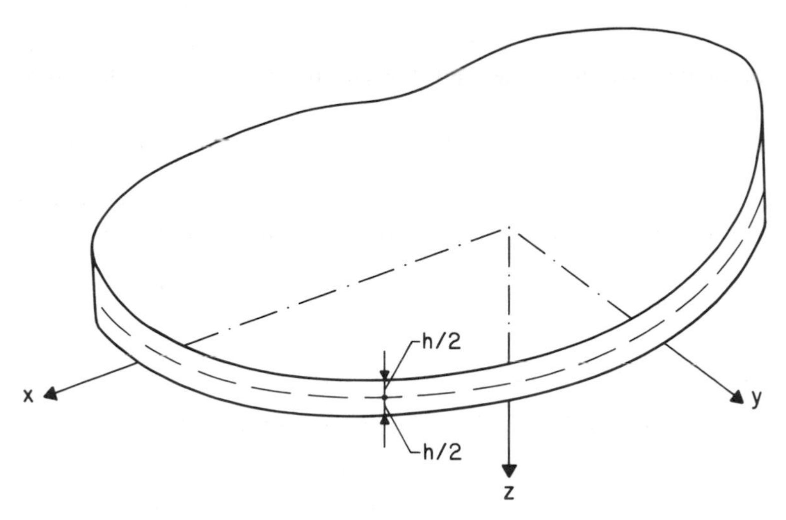

FIGURE 15.1 Flat plate.

loads cause distributed bending moments. The distribution of the bending moments over the cross sections of the plate is, however, two-dimensional and is usually accompanied by torsion. Therefore, a torsional moment should also be considered in the bending of plates.

The classical theory of bending of thin plates, which will be studied here, was developed principally by J. L. Lagrange, S. Germain, H. Navier, G. R. Kirchhoff, M. Lévy, and A. E. Love. This theory is based on the following basic assumptions adapted directly from the elementary theory of bending of beams:

The material of the plate is elastic, homogeneous, continuous, and isotropic.

The bending deflections are small compared to the thickness of the plate and do not alter its geometry. The slope of the deflected surface is therefore also small, so that the square of this slope is a negligible quantity compared to unity. Tentatively, a maximum deflection of one fifth of the plate's thickness is considered as the limit for the classical small-deflection theory. This limitation can also be stated in terms of length: the maximum deflection should be less than one twenty-fifth of the smaller span length.

The deformations are such that the straight lines normal to the midplane before bending remain straight lines and normal to the middle surface after bending. This means that plane sections initially normal to the midplane remain plane and normal to that surface after bending. Therefore, the vertical shearing strains γ_{xz} and γ_{yz} are negligible.

The stresses σ_z normal to the middle surface of the plate are small and can be neglected.

The deflections of the plate are due to the displacements of points of the middle surface in the direction normal to the initial plane.

These assumptions are known as *Kirchhoff–Love hypotheses.*

15.2 BENDING OF RECTANGULAR PLATES

15.2.1 Governing Differential Equation for Deflections of Plates

Pure bending of a rectangular plate subjected to bending moments uniformly distributed over the plate's edges (Fig. 4.4) was examined in Section 4.6 as an elementary problem of the theory of elasticity. Now, examine a case in which a rectangular plate experiences bending due to a lateral load of

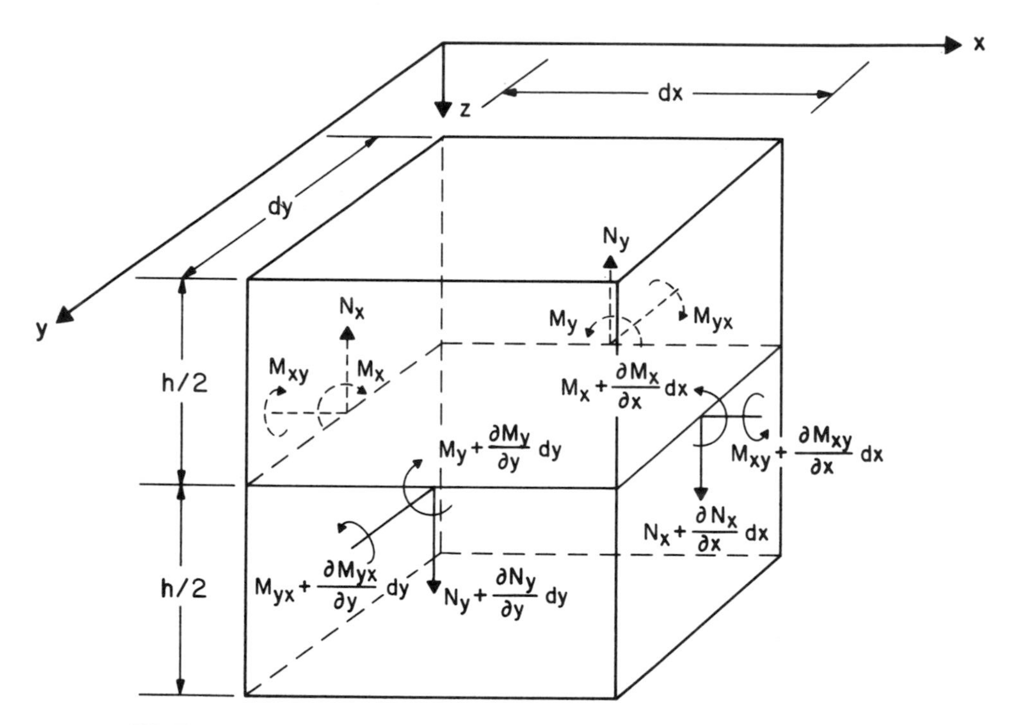

FIGURE 15.2 Variations in moments and forces over an element of a flat plate.

intensity q distributed over its surface. The load q may vary along the surface of the plate, i.e., may be a function of the coordinates x and y.

The forces and moments applied to a plate element cut from the plate by two pairs of planes parallel to the xz and yz planes are shown in Fig. 15.2. This element is subjected to bending moments

$$M_x = \int_{-h/2}^{h/2} \sigma_x z \, dz \qquad M_y = \int_{-h/2}^{h/2} \sigma_y z \, dz \tag{15.1}$$

to twisting moments

$$M_{xy} = -\int_{-h/2}^{h/2} \tau_{xy} z \, dz \qquad M_{yx} = \int_{-h/2}^{h/2} \tau_{yx} z \, dz \tag{15.2}$$

and to shearing forces

$$N_x = \int_{-h/2}^{h/2} \tau_{xz} \, dz \qquad N_y = \int_{-h/2}^{h/2} \tau_{yz} \, dz \tag{15.3}$$

The negative sign is used in the formula for the twisting moment M_{xy} because a positive shearing stress τ_{xy} produces a negative moment when z is positive.

The projections of all the forces on the z axis result in the equation

$$\frac{\partial N_x}{\partial x} dx\, dy + \frac{\partial N_y}{\partial y} dy\, dx + q\, dx\, dy = 0$$

or, after simplification,

$$\frac{\partial N_x}{\partial x} + \frac{\partial N_y}{\partial y} + q = 0 \tag{15.4}$$

Taking all the moments acting with respect to the x axis, we have the following equilibrium equation:

$$\frac{\partial M_{xy}}{\partial x} dx\, dy - \frac{\partial M_y}{\partial y} dy\, dx + N_y\, dx\, dy = 0$$

The moment due to the load q and the moment due to the change in the force N_y are not included in this equation since they are small quantities of a higher order of magnitude than the other terms. We encountered a similar situation in Section 2.5 when deriving an equilibrium equation for an elementary parallelepiped in the theory of elasticity. After simplifying the last equation, we have

$$\frac{\partial M_{xy}}{\partial x} - \frac{\partial M_y}{\partial y} + N_y = 0 \tag{15.5}$$

Similarly, the condition of equilibrium of all the moments with respect to the y axis yields

$$\frac{\partial M_{yx}}{\partial y} + \frac{\partial M_x}{\partial x} - N_x = 0 \tag{15.6}$$

Determining N_x from Eq. (15.6) and N_y from Eq. (15.5), and substituting the obtained expressions into Eq. (15.4), we obtain

$$\frac{\partial^2 M_x}{\partial x^2} + \frac{\partial^2 M_{yx}}{\partial x \partial y} + \frac{\partial^2 M_y}{\partial y^2} - \frac{\partial^2 M_{xy}}{\partial x \partial y} = -q$$

Since, as can be seen from Eq. (15.2), the twisting moments M_{xy} and M_{yx}, with $\tau_{xy} = \tau_{yx}$, are related as $M_{xy} = -M_{yx}$, we rewrite the last equation as

$$\frac{\partial^2 M_x}{\partial x^2} - 2\frac{\partial^2 M_{xy}}{\partial x \partial y} + \frac{\partial^2 M_y}{\partial y^2} = -q \tag{15.7}$$

Let us now express the obtained equation in terms of the plate's deflections $w(x, y)$. When the problem of pure bending of a rectangular plate was examined in Section 4.6, the relationship between the moments M_1 and M_2, uniformly distributed over the edges of the plate, and the deflection function $w(x, y)$ was obtained in the form (4.29). Since, in the case of pure bending, the bending moments do not change along and across the plate, the relationships (4.29) are true for the moments acting in the span of the plate as well:

$$M_x = -D\left(\frac{\partial^2 w}{\partial x^2} + \nu\frac{\partial^2 w}{\partial y^2}\right) \qquad M_y = -D\left(\frac{\partial^2 w}{\partial y^2} + \nu\frac{\partial^2 w}{\partial x^2}\right) \tag{15.8}$$

Although these formulas are, strictly speaking, valid for the case of pure bending only, they are used as general formulas in the technical theory of plates. Indeed, as long as any straight line normal to the middle surface of the plate remains straight and normal to the middle surface after bending as well, this surface does not experience any deformation during such bending and is therefore a stress-free *neutral surface*. With the curvatures $1/\rho_x$ and $1/\rho_y$ of the surface in sections parallel to the xz and yz planes, respectively, we have the following formulas for the normal stresses ε_x and ε_y:

$$\varepsilon_x = \frac{z}{\rho_x} \qquad \varepsilon_y = \frac{z}{\rho_y} \tag{15.9}$$

Then, using the two-dimensional Hooke's law equations (6.10),

$$\sigma_x = \frac{E}{1 - \nu^2}(\varepsilon_x + \nu\varepsilon_y) \qquad \sigma_y = \frac{E}{1 - \nu^2}(\varepsilon_y + \nu\varepsilon_x)$$

we obtain the following relationships between the normal stresses and the curvatures:

$$\sigma_x = \frac{Ez}{1 - \nu^2}\left(\frac{1}{\rho_x} + \frac{\nu}{\rho_y}\right) \qquad \sigma_y = \frac{Ez}{1 - \nu^2}\left(\frac{1}{\rho_y} + \frac{\nu}{\rho_x}\right) \tag{15.10}$$

Since, for small deflections,

$$\frac{1}{\rho_x} \cong -\frac{\partial^2 w}{\partial x^2} \qquad \frac{1}{\rho_y} \cong -\frac{\partial^2 w}{\partial y^2} \tag{15.11}$$

then the first formula in Eq. (15.1) yields

$$\begin{aligned} M_x &= \int_{-h/2}^{h/2} \frac{Ez^2}{1 - \nu^2}\left(-\frac{\partial^2 w}{\partial x^2} - \nu\frac{\partial^2 w}{\partial y^2}\right) dz = -\frac{E}{1 - \nu^2}\left(\frac{\partial^2 w}{\partial x^2} + \nu\frac{\partial^2 w}{\partial y^2}\right) \\ &\quad \times \int_{-h/2}^{h/2} z^2\, dz = -\frac{Eh^3}{12(1 - \nu^2)}\left(\frac{\partial^2 w}{\partial x^2} + \nu\frac{\partial^2 w}{\partial y^2}\right) \\ &= -D\left(\frac{\partial^2 w}{\partial x^2} + \nu\frac{\partial^2 w}{\partial y^2}\right) \end{aligned}$$

A similar equation can be obtained for the moment M_y.

In order to express the twisting moment M_{xy} through the deflection function $w(x, y)$, we use the formulas (15.9) for the normal strains. Considering Eq. (15.11), we have

$$\varepsilon_x = -z\frac{\partial^2 w}{\partial x^2} \qquad \varepsilon_y = -z\frac{\partial^2 w}{\partial y^2} \tag{15.12}$$

The linear displacements $u(x, y)$ and $v(x, y)$ in the directions x and y are related to the normal strains ε_x and ε_y by the Cauchy formulas (6.3):

$$\varepsilon_x = \frac{\partial u}{\partial x} \qquad \varepsilon_y = \frac{\partial v}{\partial y}$$

Introducing these formulas into Eqs. (15.12) and integrating the obtained equations, we have

$$u = -z\frac{\partial w}{\partial x} \qquad v = -z\frac{\partial w}{\partial y}$$

The shearing strain is therefore

$$\gamma_{xy} = \frac{\partial u}{\partial y} + \frac{\partial v}{\partial x} = -2z\frac{\partial^2 w}{\partial x \partial y}$$

and the shearing stress is

$$\tau_{xy} = G\gamma_{xy} = -2Gz\frac{\partial^2 w}{\partial x \partial y} = -\frac{E}{1+\nu}z\frac{\partial^2 w}{\partial x \partial y}$$

Introducing this formula into the first integral in Eq. (15.2), we obtain the following equation for the twisting moment:

$$M_{xy} = \frac{E}{1+\nu}\frac{\partial^2 w}{\partial x \partial y}\int_{-h/2}^{h/2} z^2\,dz = \frac{Eh^3}{12(1+\nu)}\frac{\partial^2 w}{\partial x \partial y} = D(1-\nu)\frac{\partial^2 w}{\partial x \partial y} \qquad (15.13)$$

Substituting Eqs. (15.8) for the bending moments and Eq. (15.13) for the twisting moment into Eq. (15.7), we obtain the following *basic equation of bending of a rectangular plate*:

$$\nabla^4 w \equiv \frac{\partial^4 w}{\partial x^4} + 2\frac{\partial^4 w}{\partial x^2 \partial y^2} + \frac{\partial^4 w}{\partial y^4} = \frac{q}{D} \qquad (15.14)$$

This equation was first obtained by S. Germain who, however, made an error in the derivation. This error was corrected by J. L. Lagrange, whose name is usually associated with the preceding governing differential equation for deflection of thin plates. The operator ∇^4 is called a biharmonic operator, and the equation

$$\nabla^4 w \equiv \frac{\partial^4 w}{\partial x^4} + 2\frac{\partial^4 w}{\partial x^2 \partial y^2} + \frac{\partial^4 w}{\partial y^4} = 0 \tag{15.15}$$

is called a biharmonic equation. We have encountered such an equation before in Section 6.3.

If the solution to Eq. (15.14) is found, the bending and the twisting moments can then be evaluated by the formulas (15.8) and (15.13), and the lateral shearing forces can be obtained from the formulas (15.5) and (15.6) which, considering Eqs. (15.8) and (15.13), can be presented as

$$\left.\begin{aligned} N_x &= -D\frac{\partial}{\partial x}\left(\frac{\partial^2 w}{\partial x^2} + \frac{\partial^2 w}{\partial y^2}\right) = -D\frac{\partial}{\partial x}(\Delta w) \\ N_y &= -D\frac{\partial}{\partial y}\left(\frac{\partial^2 w}{\partial x^2} + \frac{\partial^2 w}{\partial y^2}\right) = -D\frac{\partial}{\partial y}(\Delta w) \end{aligned}\right\} \tag{15.16}$$

The operator

$$\Delta = \nabla^2 = \frac{\partial^2}{\partial x^2} + \frac{\partial^2}{\partial y^2}$$

is called a harmonic, or Laplacean, operator. We have encountered this operator before, in Section 6.2. The normal stresses σ_x and σ_y can be evaluated using the formulas (15.10) which, taking into account Eqs. (15.11), can be presented as

$$\left.\begin{aligned} \sigma_x &= -\frac{Ez}{1-\nu^2}\left(\frac{\partial^2 w}{\partial x^2} + \nu\frac{\partial^2 w}{\partial y^2}\right) = \frac{Ez}{1-\nu^2}\frac{M_x}{D} = \frac{12M_x}{h^3}z \\ \sigma_y &= -\frac{Ez}{1-\nu^2}\left(\frac{\partial^2 w}{\partial y^2} + \nu\frac{\partial^2 w}{\partial x^2}\right) = \frac{Ez}{1-\nu^2}\frac{M_y}{D} = \frac{12M_y}{h^3}z \end{aligned}\right\} \tag{15.17}$$

The maximum stresses occur at $z = \pm(h/2)$ and are as follows:

$$(\sigma_x)_{\max} = \frac{6M_x}{h^2} \qquad (\sigma_y)_{\max} = \frac{6M_y}{h^2} \tag{15.18}$$

When these stresses have opposite signs, the maximum shearing stress τ_{xy} acts in the plane bisecting the angle between the xz and yz planes and, in accordance with Eqs. (2.26), is

$$\tau_{max} = \frac{(\sigma_x)_{max} - (\sigma_y)_{max}}{2} = \frac{3}{h^2}(M_x - M_y) \tag{15.19}$$

If the normal stresses (15.18) have the same sign, the maximum shearing stress acts in the plane bisecting the angle between the xy and xz planes, or in the plane bisecting the angle between the xy and yz planes, and is equal to $\frac{1}{2}(\sigma_y)_{max}$ or $\frac{1}{2}(\sigma_x)_{max}$, depending on which of the two normal stresses, $(\sigma_y)_{max}$ or $(\sigma_x)_{max}$, is greater. The shearing stress (15.19) is, however, substantially smaller than the maximum shearing stress acting in the through-thickness direction. The latter can be evaluated assuming, as it was assumed in beam theory (see Section 12.5), that the shearing forces N_x and N_y are distributed over the plate's height in accordance with a parabolic law. This leads to the following formulas for the maximum shearing stresses:

$$(\tau_{xz})_{max} = \frac{3}{2h}(N_x)_{max} \qquad (\tau_{yz})_{max} = \frac{3}{2h}(N_y)_{max} \tag{15.20}$$

Thus, once the deflection function $w(x, y)$ is determined, evaluation of all the maximum stresses does not cause essential difficulties. Of course, the function $w(x, y)$ should satisfy not only the basic governing equation (15.14) but the appropriate boundary conditions as well. These will be discussed in the next subsection.

In conclusion, we examine several important special cases.

In a special case, $M_y = 0$, Eqs. (15.18) yield

$$M_x = -D(1 - \nu^2)\frac{\partial^2 w}{\partial x^2} = -\frac{Eh^3}{12}\frac{\partial^2 w}{\partial x^2} = -EI_y\frac{\partial^2 w}{\partial x^2}$$

This is the equation of bending of a beam of a unit width.

In another special case, $\partial^2 w/\partial y^2 = 0$, the bending moments are expressed as follows:

$$M_x = -D\frac{\partial^2 w}{\partial x^2} \qquad M_y = -\nu D\frac{\partial^2 w}{\partial x^2} = \nu M_x \tag{15.21}$$

This case is usually referred to as *cylindrical bending* of a plate and leads also to a situation in which an equivalent beam ("*strip beam*") can be examined.

Finally, in a special case, $M_x = M_y = M$, the curvature of the plate is the same in all directions and is

$$\frac{1}{\rho} = \frac{M}{D(1 + \nu)} \tag{15.22}$$

This case is referred to as *spherical bending*.

15.2.2 Boundary Conditions

Let us consider the most common boundary conditions encountered in engineering practice and, particularly, in microelectronic and fiber-optic structures.

1. *Simply supported edge*. If an edge is simply supported, the deflection along this edge must be zero. In addition, since a simply supported edge is free to rotate, then the bending moment along this edge must be zero as well. If, for instance, the edge $y = 0$ is simply supported, then the boundary conditions at this edge are

$$w = 0 \qquad \frac{\partial^2 w}{\partial y^2} + \nu \frac{\partial^2 w}{\partial x^2} = 0 \qquad \text{at } y = 0$$

   Since the curvature $\partial^2 w/\partial y^2$ is zero along the edge $y = 0$, the preceding conditions can be expressed in the following equivalent form:

$$w = 0 \qquad \frac{\partial^2 w}{\partial x^2} = 0 \qquad \text{at } y = 0 \tag{15.23}$$

2. *Built-in (clamped) edge*. If an edge is clamped, both the deflection and the angle of rotation of the plate's cross section at this edge are zero. If, for instance, the edge coinciding with the x axis is clamped, then the boundary conditions along this edge are

$$w = 0 \qquad \frac{\partial w}{\partial y} = 0 \qquad \text{at } y = 0 \tag{15.24}$$

3. *Free edge*. In the case of a free edge, it is natural to assume that the bending moment, the twisting moment, and the shearing force are zero at this edge. If, for instance, the edge $x = a$ is free, then

$$M_x = 0 \qquad M_{xy} = 0 \qquad N_x = 0 \qquad \text{at } x = a$$

The second and the third conditions, however, are not independent and can be replaced by one boundary condition for the deflection function.

To determine this condition, we proceed from the following expression for the work produced by the transverse force and the twisting moment for a free edge subjected to bending (Fig. 15.3):

$$W = \int_{y_1}^{y_2} \left(N_x w + M_{xy} \frac{\partial w}{\partial y} \right) dy \tag{15.25}$$

This formula presents the work produced for a segment of the free edge located between points y_1 and y_2. The transverse force N_x produces work at the displacement w, whereas the twisting moment M_{xy} produces work at the twist angle $\partial w/\partial y$. Let us introduce an "equivalent" transverse force

$$\tilde{N}_x = N_x - \frac{\partial M_{xy}}{\partial y}$$

so that

$$N_x = \tilde{N}_x + \frac{\partial M_{xy}}{\partial y}$$

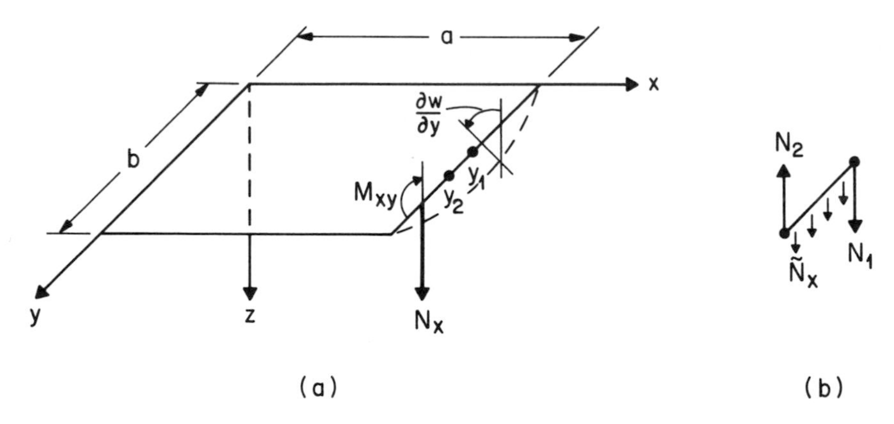

FIGURE 15.3 Evaluation of the boundary conditions for a free edge.

Substituting this formula into Eq. (15.25), we have

$$W = \int_{y_1}^{y_2} \left[\tilde{N}_x w + \frac{\partial}{\partial y}(M_{xy} w) \right] dy = \int_{y_1}^{y_2} \tilde{N}_x w \, dy + [M_{xy} w]_{y_1}^{y_2}$$

$$= \int_{y_1}^{y_2} \tilde{N}_x w \, dy + N_2 w_2 - N_1 w_1$$

where

$$N_1 = M_{xy}(a, y_1) \qquad w_1 = w(a, y_1)$$

$$N_2 = M_{xy}(a, y_2) \qquad w_2 = w(a, y_2)$$

Thus, the combined action of the twisting moment M_{xy} and the shearing force N_x, applied to the free edge, is equivalent to the action of the distributed forces $\tilde{N}_x$ and two concentrated forces N_1 and N_2 applied at points (a, y_1) and (a, y_2). The segment (y_1, y_2) can be, of course, of any length and, particularly, can be equal to an elementary length dy. Therefore, we conclude that the distribution of twisting moments M_{xy} is statically equivalent to the distribution of shearing forces of the intensity $-(\partial M_{xy}/\partial y)$. Hence, the joint requirement regarding twisting moments and shearing forces acting along the free edge $x = a$ will be fulfilled if we put

$$\tilde{N}_x = N_x - \frac{\partial M_{xy}}{\partial y} = 0 \qquad \text{at } x = a$$

Using Eqs. (15.16) and (15.13), we obtain

$$\frac{\partial^3 w}{\partial x^3} + (2 - \nu) \frac{\partial^3 w}{\partial x \partial y^2} = \qquad \text{at } x = a \tag{15.26}$$

4. *Sliding edge.* In this case, the edge is free to move vertically, but its rotation is prevented. Then we have

$$\frac{\partial w}{\partial x} = 0 \qquad \frac{\partial^3 w}{\partial x^3} + (2 - \nu) \frac{\partial^3 w}{\partial x \partial y^2} = 0 \qquad \text{at } x = a \tag{15.27}$$

5. *Edge loaded by the given distributed lateral load and bending moments.* If the edge $x = a$ is loaded by the given distributed load $q(y)$ and the

given distributed bending moments $M_0(y)$, then the boundary conditions are

$$\left.\begin{aligned} D\left(\frac{\partial^2 w}{\partial x^2} + \nu \frac{\partial^2 w}{\partial y^2}\right) &= -M_0(y) \\ D\left[\frac{\partial^3 w}{\partial x^3} + (2-\nu)\frac{\partial^3 w}{\partial x \partial y^2}\right] &= -q(y) \end{aligned}\right\} \tag{15.28}$$

at $x = a$.

Similarly, we can form the boundary conditions for other types of supports. We would like to point out that the boundary conditions describing end constraints pertaining to deflections or slopes are called *geometric* or *kinematic conditions*, whereas the boundary conditions equating the internal forces and moments at the plate's edges to the given external forces are called *static conditions*. Accordingly, both the conditions (15.24) are kinematic, both the conditions (15.28) are static, and conditions (15.23) are *mixed conditions*.

15.2.3 Navier's Solution for Simply Supported Plates

Let a rectangular plate be simply supported along all its edges (Fig. 15.4) and subjected to a distributed load $q(x, y)$. This problem dates back to H. Navier in 1820, who suggested that the deflection function be sought in the form of the double Fourier series

$$w = \sum_{m=1}^{\infty} \sum_{n=1}^{\infty} A_{mn} \sin \alpha_m x \sin \beta_n y \tag{15.29}$$

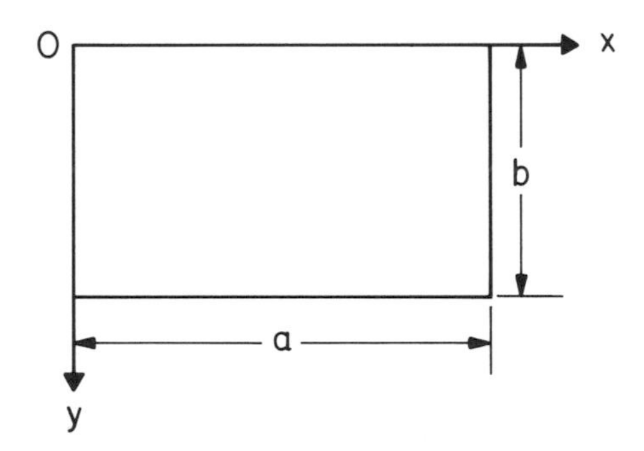

FIGURE 15.4 Rectangular plate.

where

$$\alpha_m = \frac{m\pi}{a} \qquad \beta_n = \frac{n\pi}{b}$$

and A_{mn} are Fourier coefficients to be determined. As we can easily see, the boundary conditions (15.23) are fulfilled automatically.

In accordance with Navier's method, the distributed load q should also be presented as the double Fourier series

$$q(x, y) = \sum_{m=1}^{\infty} \sum_{n=1}^{\infty} q_{mn} \sin \alpha_m x \sin \beta_n y \tag{15.30}$$

where, as we know from the theory of Fourier series,

$$q_{mn} = \frac{4}{ab} \int_0^a \int_0^b q(x, y) \sin \alpha_m x \sin \beta_n y \, dx \, dy \tag{15.31}$$

Note that this formula can be obtained from Eq. (15.30) by multiplying both parts of Eq. (15.30) by $\sin \alpha_i x \sin \beta_j y$, integrating over the surface of the plate and considering the fact that sine is an orthogonal function, so that

$$\int_0^a \sin \alpha_m x \sin \alpha_i x \, dx = \begin{cases} 0 & m \neq i \\ \dfrac{a}{2} & m = i \end{cases}$$

Substituting Eqs. (15.29) and (15.30) in the basic equation (15.14), we have

$$\sum_{m=1}^{\infty} \sum_{n=1}^{\infty} \left[A_{mn}(\alpha_m^2 + \beta_n^2)^2 - \frac{q_{mn}}{D} \right] \sin \alpha_m x \sin \beta_n y = 0$$

Since this equation must be valid for all values of the variables x and y, then we should require that the expression in brackets is zero. This leads to the following formula for the coefficients A_{mn}:

$$A_{mn} = \frac{q_{mn}}{D(\alpha_m^2 + \beta_n^2)^2} \tag{15.32}$$

The double series (15.29), with the coefficients A_{mn} expressed by Eq. (15.32), is convergent and can be used for practical evaluations.

In the case of a uniform load, the formula (15.31) yields

$$q_{mn} = \frac{4q}{ab} \int_0^a \int_0^b \sin \alpha_m x \sin \beta_n y \, dx \, dy = \frac{16q}{\pi^2 mn}$$

where m and n are odd numbers. Then the expansion (15.29) is

$$w = \frac{16}{\pi^2 D} \sum_m^\infty \sum_n^\infty \frac{\sin \alpha_m x \sin \beta_n y}{mn(\alpha_m^2 + \beta_n^2)^2} \qquad m, n = 1, 3, 5, \ldots \qquad (15.33)$$

The maximum deflection takes place at the center of the plate and is

$$w_{\max} = \frac{16}{\pi^2 D} \sum_m^\infty \sum_n^\infty \frac{(-1)^{(m+n)/2-1}}{mn(\alpha_m^2 + \beta_n^2)^2} \qquad m, n = 1, 3, 5, \ldots \qquad (15.34)$$

The bending moments, in accordance with Eqs. (15.8), are as follows:

$$\left. \begin{aligned} M_x &= \frac{16q}{\pi^4} \sum_m^\infty \sum_n^\infty \frac{(m/a)^2 + \nu(n/b)^2}{mn[(m/a)^2 + (n/b)^2]^2} \sin \alpha_m x \sin \beta_n y \\ M_y &= \frac{16q}{\pi^4} \sum_m^\infty \sum_n^\infty \frac{\nu(m/a)^2 + (n/b)^2}{mn[(m/a)^2 + (n/b)^2]^2} \sin \alpha_m x \sin \beta_n y \end{aligned} \right\}$$

We can observe that these moments are zero at the edges, whereas the twisting moment does not vanish at the edges and at the corners of the plate.

For a square plate ($a = b$), assuming $\nu = 0.3$ and taking the first four terms in the expansion (15.34), i.e., using terms $n = m = 1$; $m = 1$, $n = 3$; $n = 1$, $m = 3$; $m = n = 3$, we have

$$w_{\max} = 0.0443 \frac{qa^4}{Eh^3}$$

As another example, we examine the case in which the plate is subjected to a load P uniformly distributed over the shaded subregion in Fig. 15.5. The difference between this case and the case of a load distributed over the entire surface of the plate is that the integration in the formula (15.31) should be performed within the limits $x_0 - (a_0/2) \leq x \leq x_0 + (a_0/2)$ and $y_0 - (b_0/2) \leq y \leq y_0 + (b_0/2)$. Then we have

$$q_{mn} = \frac{16P}{\pi^2 mna_0 b_0} \sin \alpha_m x_0 \sin \beta_n y_0 \sin \alpha_m \frac{a_0}{2} \sin \beta_n \frac{b_0}{2}$$

When $a_0 = b_0$, i.e., when the area of the application of the load is located

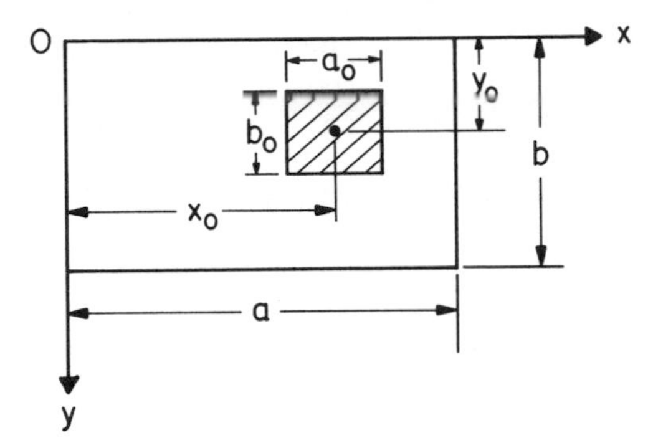

FIGURE 15.5 Rectangular plate under a load distributed over a finite subregion.

in the center of the plate,

$$q_{mn} = \frac{16P}{\pi^2 mna_0^2} \sin \alpha_m x_0 \sin \beta_n y_0 \sin \alpha_m \frac{a_0}{2} \sin \beta_n \frac{a_0}{2}$$

In the case of a concentrated force, we should make a_0 approach zero. Since

$$\lim_{a_0/2 \to 0} \frac{\sin \alpha_m(a_0/2)}{a_0} = \frac{\alpha_m}{2} \lim_{a_0/2 \to 0} \frac{\sin \alpha_m(a_0/2)}{\alpha_m(a_0/2)} = \frac{\alpha_m}{2}$$

and, similarly,

$$\lim_{a_0/2 \to 0} \frac{\sin \beta_n(a_0/2)}{a_0} = \frac{\beta_n}{2}$$

then we have

$$q_{mn} = \frac{4P}{ab} \sin \alpha_m x_0 \sin \beta_n y_0$$

The expression for the deflection function can be obtained by inserting the q_{mn} values into the formula (15.32) and then into the expansion (15.29). For a concentrated force, for instance, we obtain

$$w = \frac{4P}{\pi^4 Dab} \sum_m^\infty \sum_n^\infty \frac{\sin \alpha_m x_0 \sin \beta_n y_0}{[(m/a)^2 + (n/b)^2]^2} \sin \alpha_m x \sin \beta_n y$$

The obtained results can be used, for example, to evaluate deflections of printed wire boards subjected to the weights of heavy mounted components.

The Navier method, despite its simplicity, is not used very often in engineering practice because the convergence of the double Fourier series is not very good, especially for bending moments and shearing forces. As will be shown in the next subsection, much better convergence can be achieved if ordinary trigonometric series are applied.

15.2.4 The M. Lévy Solution

The solution developed by M. Lévy in 1900 not only results in a significantly better convergence of the series but can also be applied in cases in which only two opposite edges of the plate are simply supported while the other two edges can have arbitrary boundary conditions. An additional advantage of the Lévy solution is that it is based on single (ordinary) series, which lend themselves to much easier calculations than the double series.

Let the edges $x = 0$ and $x = a$ of a rectangular plate be simply supported (Fig. 15.6). The edges $y = \pm(b/2)$ may have arbitrary conditions of support. The deflection function w is sought in the M. Lévy method in the form

$$w = \sum_{m=1}^{\infty} f_m(y) \sin \alpha_m x \qquad \alpha_m = \frac{m\pi}{a} \tag{15.35}$$

where the functions $f_m(y)$ are to be determined from the boundary conditions at the edges $y = \pm(b/2)$. As to the boundary conditions at the simply supported edges $x = 0$ and $x = a$, these are fulfilled automatically as long as the distribution of the deflections and the bending moments along the x axis is expressed by the functions $\sin \alpha_m x$. If the plate is subjected to a lateral load $q = q(x, y)$, then the latter should also be presented in the form of a series:

$$q = \sum_{m=1}^{\infty} q_m(y) \sin \alpha_m x \tag{15.36}$$

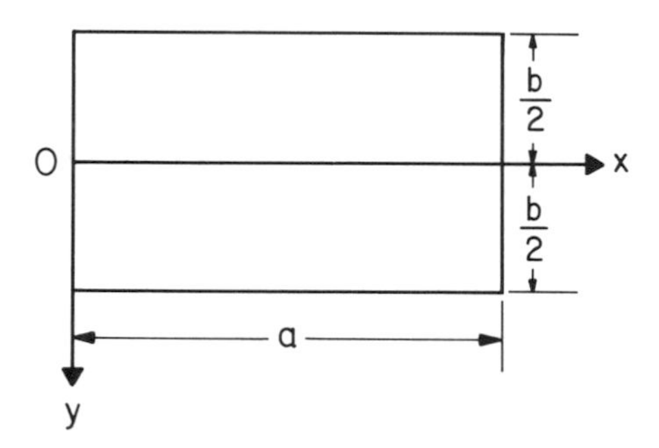

FIGURE 15.6 Rectangular plate.

where the functions $q_m(y)$ are

$$q_m(y) = \frac{2}{a}\int_0^a q(x, y) \sin \alpha_m x \, dx \tag{15.37}$$

This formula can be obtained in the same way as the similar formula in the Navier method, i.e., by multiplying both parts of Eq. (15.36) by $\sin \alpha_j x$, and integrating over the plate's length a.

After substituting the expansions (15.35) and (15.36) into the governing equation (15.14), we obtain the following ordinary differential equation for the unknown functions $f_m(y)$:

$$f_m^{\mathrm{IV}}(y) - 2\alpha_m^2 f_m''(y) + \alpha_m^4 f_m(y) = \frac{q_m(y)}{D} \tag{15.38}$$

The general solution to this equation can be presented as the sum of the general solution to the homogeneous equation

$$f_m^{\mathrm{IV}}(y) - 2\alpha_m^2 f_m''(y) + \alpha_m^4 f_m(y) = 0 \tag{15.39}$$

and the particular solution to the nonhomogeneous equation (15.38). This particular solution depends on the type of the functions $q_m(y)$, i.e., on the loading q.

The solution to the homogeneous equation (15.39) is (see Section 6.5)

$$f_m(y) = (A_m + B_m y) \cosh \alpha_m y + (C_m + D_m y) \sinh \alpha_m y \tag{15.40}$$

As an example, let us examine a plate simply supported along all the edges and loaded by a uniformly distributed load q. With $q = \text{const}$, the formula (15.37) yields

$$q_m(y) = \frac{2}{a} q \int_0^a \sin \alpha_m x \, dx = \frac{4q}{m\pi} \qquad m = 1, 3, 5, \ldots$$

Then the particular solution of Eq. (15.38) can be sought as a constant value and is

$$f_m = \frac{4qa^4}{D(m\pi)^5} \qquad m = 1, 3, 5, \ldots$$

Placing the origin in the middle of the edge $x = 0$ (Fig. 15.8) and noticing that, because of the symmetry of bending, we should assume $B_m = C_m = 0$, so that only the even functions would be retained in the formula for $f_m(y)$, we seek the functions $f_m(y)$ in the form

$$f_m(y) = A_m \cosh \alpha_m y + D_m y \sinh \alpha_m y + \frac{4qa^4}{D(m\pi)^5} \qquad m = 1, 3, 5, \ldots \tag{15.41}$$

The boundary conditions for the functions $f_m(y)$ at the edges $y = \pm(b/2)$ can be obtained by substituting the expansion (15.35) into the conditions (15.23). This yields

$$f_m\left(\pm\frac{b}{2}\right) = 0 \qquad f''_m\left(\pm\frac{b}{2}\right) = 0$$

Introducing (15.41) into these conditions, we obtain the following system of algebraic equations for the unknown constants A_m and D_m:

$$\left.\begin{aligned} A_m \cosh u_m + \frac{b}{2} D_m \sinh u_m &= -\frac{4qa^4}{D(m\pi)^5} \\ \alpha_m A_m \cosh u_m + D_m(2\cosh u_m + u_m \sinh u_m) &= 0 \end{aligned}\right\}$$

where

$$u_m = \frac{b}{2}\alpha_m = \frac{m\pi}{2}\frac{b}{a} \tag{15.42}$$

Solving these equations, we find

$$\left.\begin{aligned} A_m &= -\frac{4qa^4}{D(m\pi)^5}\frac{1 + (u_m/2)\tanh u_m}{\cosh u_m} \\ D_m &= \frac{4qa^4}{Db(m\pi)^5}\frac{u_m}{\cosh u_m} \end{aligned}\right\}$$

With these values of the constants of integration, the functions $f_m(y)$ are expressed as follows:

$$f_m(y) = \frac{4qa^4}{D(m\pi)^5}\left(1 - \frac{1 + (u_m/2)\tanh u_m}{\cosh u_m}\cosh \alpha_m y + \frac{\alpha_m}{2\cosh u_m} y \sinh \alpha_m y\right)$$

and the deflection function is

$$w = \frac{4qa^4}{D\pi^5} \sum_{m=1,3,\ldots}^{\infty} \frac{1}{m^5} \times \left[1 - \frac{1 + (u_m/2)\tanh u_m}{\cosh u_m} \cosh \alpha_m y + \frac{\alpha_m}{2\cosh u_m} y \sinh \alpha_m y\right] \sin \alpha_m x \tag{15.43}$$

This series converges quite rapidly. The convergence can be made even better, however, if the following approach is applied.

Let the length of the plate in the direction of the y axis be significantly larger than the other dimension. Then the u_m value is also large, and Eq. (15.43) yields

$$w_\infty = w|_{b\to\infty} = \frac{4qa^4}{D\pi^5} \sum_{m=1,3,\ldots}^{\infty} \frac{1}{m^5} \sin \alpha_m x \tag{15.44}$$

On the other hand, we can analyze the bending of a plate elongated in the y direction using the equation

$$Dw_\infty^{\mathrm{IV}}(x) = q \tag{15.45}$$

for a "strip beam." This equation simply follows from the first formula in Eq. (15.21), where $q = -(d^2M_x/dx^2)$. The solution to Eq. (15.45) can be obtained in a closed form:

$$w_\infty = \frac{qa^4}{24D}\left[\frac{x}{a} - 2\left(\frac{x}{a}\right)^3 + \left(\frac{x}{a}\right)^4\right] \tag{15.46}$$

As we can see from Eqs. (15.43) and (15.44), the deflection function (15.43) can be presented as follows:

$$w = w_\infty - \frac{4qa^4}{D\pi^5} \sum_{m=1,3,\ldots}^{\infty} \frac{1}{m^5} \times \left(\frac{1 + (u_m/2)\tanh u_m}{\cosh u_m} \cosh \alpha_m y - \frac{\alpha_m}{2\cosh u_m} y \sinh \alpha_m y\right) \sin \alpha_m x \tag{15.47}$$

The term containing the series is, in effect, a correction taking into account the finite aspect ratio of the plate. The term w_∞ determines the deflections of a long and narrow strip and can be evaluated by Eq. (15.46).

The maximum deflection takes place in the center of the plate, i.e., at point $x = a/2$, $y = 0$, and is as follows:

$$w_{\max} = \frac{5qa^4}{384D} - \frac{4qa^4}{D\pi^5} \sum_{m=1,3,\ldots} (-1)^{(m-1)/2} \frac{1}{m^5} \frac{1 + (u_m/2)\tanh u_m}{\cosh u_m} \tag{15.48}$$

since

$$\sin\frac{m\pi}{2} = (-1)^{(m-1)/2} \qquad m = 1, 3, \ldots$$

The first term in this equation is the deflection in the middle of a uniformly loaded, simply supported strip, and the second term is the correction accounting for the actual aspect ratio of the plate. The convergence of the series is very rapid, and even one term gives satisfactory accuracy. For instance, for a square plate ($a = b$, $u_m = m\pi/2$), the maximum deflection is

$$w_{\max} = \frac{5qa^4}{384D} - \frac{4qa^4}{D\pi^5} \times 0.6856 = 0.00406\frac{qa^4}{D}$$

Similarly, the bending moments, the shearing forces, the maximum reactions of the support contour, and the concentrated reactions at the corner of the plate can be evaluated for plates of any aspect ratio.

Note that when the M. Lévy solution is used, $3 \to 5$ terms are usually sufficient to calculate the deflections, and about $10 \to 13$ terms are sufficient to calculate the bending moments with an accuracy of about 1 percent. All the calculations can be easily computerized and carried out on mini-computers and even on manual calculators.

15.2.5 M. Lévy's Method for a Plate Clamped Along Two Opposite Edges

Let the edges $y = \pm(b/2)$ of the plate be clamped, while the edges $x = 0$ and $x = a$ are simply supported (Fig. 15.6). The functions $f_m(y)$ entering Eq. (15.35) for the deflection function can still be sought in the form (15.41); however, the boundary conditions in this case are

$$f_m\left(\pm\frac{b}{2}\right) = 0 \qquad f'_m\left(\pm\frac{b}{2}\right) = 0$$

These conditions are due to the fact that the deflections and the angles of rotation are zero at the clamped edges. These conditions lead to the following equations for the constants A_m and D_m:

$$\left.\begin{aligned} A_m \cosh u_m + \frac{b}{2} D_m \sinh u_m &= -\frac{4qa^4}{D(m\pi)^5} \\ \alpha_m A_m \sinh u_m + D_m(\sinh u_m + u_m \cosh u_m) &= 0 \end{aligned}\right\}$$

where the parameter u_m is given by (15.42). Then we have

$$\left.\begin{aligned} A_m &= -\frac{8qa^4}{D(m\pi)^5} \frac{\sinh u_m + u_m \cosh u_m}{\sinh 2u_m + 2u_m} \\ D_m &= \frac{8qa^3}{D(m\pi)^4} \frac{\sinh u_m}{\sinh 2u_m + 2u_m} \end{aligned}\right\}$$

so that the functions $f_m(y)$ are expressed as

$$f_m(y) = \frac{4qa^4}{D(m\pi)^5}\left(1 - 2\frac{\sinh u_m + u_m \cosh u_m}{\sinh 2u_m + 2u_m} \cosh \alpha_m y + 2\frac{\alpha_m \sinh u_m}{\sinh 2u_m + 2u_m} y \sinh \alpha_m y\right)$$

and the deflection function is

$$w = \frac{4qa^4}{D\pi^5} \sum_{m=1,3,\ldots}^{\infty} \frac{1}{m^5}\left(1 - \frac{\sinh u_m + u_m \cosh u_m}{\sinh 2u_m + 2u_m} \cosh \alpha_m y + 2\frac{\alpha_m \sinh u_m}{\sinh 2u_m + 2u_m} y \sinh \alpha_m y\right) \sin \alpha_m x \tag{15.49}$$

Although an approach similar to the one used in the preceding section can be applied to enhance the convergence of this series, this convergence is good enough to be used for practical calculations. With Eq. (15.49), we can easily calculate all the elements of bending.

The maximum deflection occurs in the center of the plate ($x = a/2$, $y = 0$) and is

$$w_{\max} = \frac{4qa^4}{D\pi^5} \sum_{m=1,3,\ldots}^{\infty} (-1)^{(m-1)/2} \frac{1}{m^5}\left(1 - 2\frac{\sinh u_m + u_m \cosh u_m}{\sinh 2u_m + 2u_m}\right) \tag{15.50}$$

In the case of a square plate ($a = b$, $u_m = m\pi/2$), retaining just one term in the series, we obtain

$$w_{\max} = \frac{4qa^4}{D\pi^5} \times 0.1501 = 0.001962 \frac{qa^4}{D}$$

If, for instance, $\nu = 0.3$, then this formula yields

$$w_{\max} = 0.02142 \frac{qa^4}{Eh^3}$$

The M. Lévy method requires that at least two opposite edges of the plate be simply supported and, therefore, generally speaking, this method cannot be used directly for arbitrary boundary conditions at all the four edges. However, in this case, it can still be used as a constituent part of some more general and more complicated methods and approaches.

A natural generalization of the M. Lévy method is to try to find the solution to the homogeneous equation (15.15) in the form

$$w = \sum_{m=1}^{\infty} f_m(y) \sin \alpha_m x + \sum_{n=1}^{\infty} g_n(x) \sin \beta_n y$$

where

$$\alpha_m = \frac{m\pi}{a} \qquad \beta_n = \frac{n\pi}{b}$$

and the functions $f_m(y)$ and $g_n(x)$ are determined from the boundary conditions at the edges of the plate. After substituting the preceding expression for the elastic surface in Eq. (15.15), we obtain two equations of the type (15.39). The solutions to these two equations are of the type (15.40), i.e.,

$$\left.\begin{aligned} f_m(y) &= (A_m + B_m y) \cosh \alpha_m y + (C_m + D_m y) \sinh \alpha_m y \\ g_n(x) &= (A'_m + B_n x) \cosh \beta_n x + (C'_n + D'_n x) \sinh \beta_n x \end{aligned}\right\}$$

These solutions contain eight constants of integration, which can be determined from the eight boundary conditions at the edges (two conditions at each edge). As far as the particular solution to the nonhomogeneous equation (15.14) is concerned, it can be found by using any suitable method, depending on the particular loading. For instance, Navier's method or a solution in polynomials can be applied. The resulting expressions turn out, however, to be quite cumbersome and, therefore, this method cannot be recommended for practical calculations.

The most widely used approach in the theory of plates with arbitrary boundary conditions at the edges is the *method of superposition.* First, we solve a relatively simple problem of bending of a simply supported plate subjected to thus far undefined bending moments distributed in an arbitrary way along the edges of the plate. This can be done using the M. Lévy method. After the elastic curve of the plate is determined, the angles of rotation at the edges are evaluated. These angles turn out, of course, to be functions of the edge moments. As the next step, we use a solution for the same plate subjected to the given lateral load and evaluate the angles of rotation at the edges. If we demand that the angles of rotation due to the external lateral load be equal in magnitude and opposite in sign to the angles of rotation caused by the edge moments, then the edges of the plate will not undergo any rotation at all, so that the boundary conditions for a clamped plate will be fulfilled. The requirement that the angles of rotation due to the bending moments and to the external load be equal in magnitude is, in effect, a displacement compatibility condition, similar to the condition used previously in the theory of frames. Solving the equations of compatibility for the unknown edge moments and substituting the obtained formulas into the solution for the deflection surface of the plate subjected to distributed moments, we obtain the equation for the elastic surface of a plate whose rotation angles at the edges are expressed through the external lateral load and are of the same magnitude as, but opposite in sign to, the simply supported plate under the lateral load. By superposing the solutions for these two cases, we can obtain the final solution to the problem in question. This approach was widely used by A. E. H. Love, I. G. Bubnov, and S. P. Timoshenko. Before we show how it is used in practical problems, let us examine an important auxiliary problem of bending of a rectangular plate under distributed edge moments.

15.2.6 A Plate Under Distributed Edge Moments

Consider a simply supported rectangular plate (Fig. 15.7) subjected to edge moments $M_0(x)$ symmetrically distributed along the edges $y = \pm(b/2)$. These moments can be presented as a Fourier sine series as follows:

$$M_0(x) = \sum_{m=1}^{\infty} \mu_m \sin \alpha_m x \tag{15.51}$$

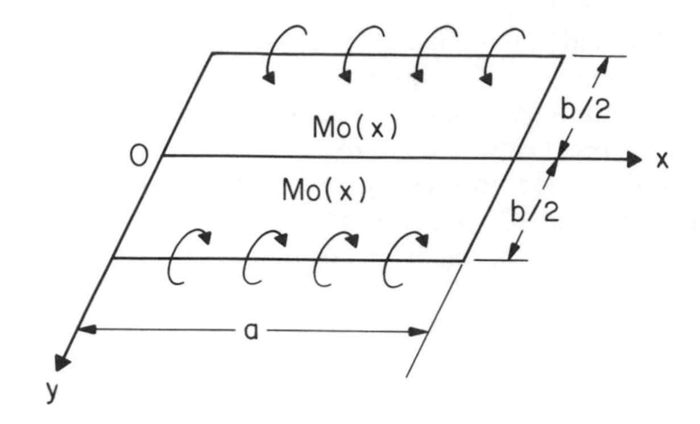

FIGURE 15.7 Bending of a plate subjected to distributed edge moments.

where the expansion coefficients μ_m are

$$\mu_m = \frac{2}{a}\int_0^a M_0(x) \sin \alpha_m x \, dx \tag{15.52}$$

Since the deflection surface must be symmetric with respect to the x axis, it can be sought in the form

$$w = \sum_{m=1,3,\ldots}^{\infty} (A_m \cosh \alpha_m y + D_m y \sinh \alpha_m y) \sin \alpha_m x \tag{15.53}$$

The deflections w must be zero at all the edges. The solution in the form (15.53) automatically satisfies this condition at the edges $x = 0$ and $x = a$. The condition $w[x, \pm(b/2)] = 0$ results in the equation

$$A_m \cosh u_m + D_m \frac{b}{2} \sinh u_m = 0$$

where

$$u_m = \alpha_m \frac{b}{2} = \frac{m\pi}{2}\frac{b}{a}$$

Hence,

$$A_m = -\frac{b}{2} D_m \tanh u_m$$

and the formula (15.53) yields

$$w = \sum_{m=1,3,\ldots}^{\infty} D_m\left(y \sinh \alpha_m y - \frac{b}{2} \tanh u_m \cosh \alpha_m y\right) \sin \alpha_m x$$

The static boundary conditions for the bending moments require that

$$\frac{\partial^2 w}{\partial x^2} = 0 \qquad \text{at } x = 0 \qquad x = a$$

and

$$\frac{\partial^2 w}{\partial y^2} = -\frac{M_0(x)}{D} \qquad \text{at } y = \pm\frac{b}{2}$$

With the formula (15.54) for the deflections, the zero boundary conditions for the curvature at the edges $x = 0$ and $x = a$ are fulfilled automatically. The conditions at $y = \pm(b/2)$ yield

$$\left.\frac{\partial^2 w}{\partial y^2}\right|_{y=\pm(b/2)} = 2 \sum_{m=1,3,\ldots}^{\infty} \alpha_m D_m \cosh u_m \sin \alpha_m x = -\frac{M_0(x)}{D}$$

Then, using the expansion (15.51), we find that the constant D_m is

$$D_m = -\frac{\mu_m}{2D\alpha_m \cosh u_m}$$

so that the deflection function of the plate is as follows:

$$w = \frac{b^2}{8D} \sum_{m=1,3,\ldots}^{\infty} \frac{\mu_m}{u_m \cosh u_m}\left(\tanh u_m \cosh \alpha_m y - \frac{2y}{b} \sinh \alpha_m y\right) \sin \alpha_m x \tag{15.54}$$

In the case of uniformly distributed moments $M_0(x) = M_0$, the formula (15.52) yields

$$\mu_m = \frac{2M_0}{a} \int_0^a \sin \alpha_m x \, dx = \frac{2M_0}{m\pi}(1 - \cos m\pi) = \frac{4M_0}{m\pi} \qquad m = 1, 3, \ldots$$

and the equation for the deflection function becomes

$$w = \frac{abM_0}{D\pi^2} \sum_{m=1,3,\ldots}^{\infty} \frac{1}{m^2 \cosh u_m} \left(\tanh u_m \cosh \alpha_m y - \frac{2y}{b} \sinh \alpha_m y \right) \sin \alpha_m x \tag{15.55}$$

For a square plate ($a = b$), the maximum deflection at the center ($x = a/2$, $y = 0$) is

$$w_{\max} = \frac{M_0 a^2}{D\pi^2} \sum_{m=1,3,\ldots}^{\infty} (-1)^{(m-1)/2} \frac{\sinh m\pi/2}{m^2 \cosh^2 m\pi/2}$$

Retaining two terms in this series, we obtain

$$w_{\max} \cong \frac{M_0 a^2}{D\pi^2} \left(\frac{\sinh \pi/2}{\cosh^2 \pi/2} - \frac{\sinh 3\pi/2}{9 \cosh^2 3\pi/2} \right)$$

$$= \frac{M_0 a^2}{D\pi^2} (0.3655 - 0.0020) = 0.0368 \frac{M_0 a^2}{D}$$

In another special case, when the aspect ratio a/b is very large, the u_m value is small, so that we can assume

$$\cosh u_m \cong 1 \qquad \tanh u_m \cong u_m$$

and Eq. (15.55) results in the following formula for the displacements at the points located on the x axis ($y = 0$):

$$w_{\max} = \frac{b^2 M_0}{2D\pi} \sum_{m=1,3,\ldots}^{\infty} \frac{1}{m} \sin \alpha_m x$$

Multiplying both parts of this equation by $\sin \alpha_j x$, integrating the obtained formula over the length a, and using the fact that sine is an orthogonal function, we have

$$w_{\max} = \frac{M_0 b^2}{8D}$$

This is a formula for the maximum deflection of a strip of length b subjected to pure bending under the action of the moments M_0 applied to its ends.

Once the deflection function is known, all the other characteristics of bending can easily be obtained. The angles of rotation at the edges $x = 0$ and $y = b/2$, particularly, can be evaluated using Eq. (15.54) and are as follows:

$$\left.\begin{aligned}\alpha(y) &= \left.\frac{\partial w}{\partial x}\right|_{x=0} = \frac{b}{4D}\sum_{m=1,3,\ldots}^{\infty}\frac{\mu_m}{\cosh u_m}\left(\tanh u_m \cosh \alpha_m y - \frac{2y}{b}\sinh \alpha_m y\right)\\ \beta(x) &= \left.\frac{\partial w}{\partial y}\right|_{y=b/2} = \frac{b}{8D}\sum_{m=1,3,\ldots}\mu_m \frac{\sinh 2u_m + 2u_m}{u_m \cosh^2 u_m}\sin \alpha_m x\end{aligned}\right\} \tag{15.56}$$

The angles of rotation of the edges $x = 0$ and $x = a$ change from zero at the corners to

$$\alpha_{\max} = \frac{b}{4D}\sum_{m=1,3,\ldots}^{\infty}\mu_m \frac{\sinh u_m}{\cosh^2 u_m}$$

in the middle of the edges. The angles of rotation of the edges $y = \pm b/2$ change from zero at the corners to

$$\beta_{\max} = \frac{b}{8D}\sum_{m=1,3,\ldots}^{\infty}(-1)^{(m-1)/2}\mu_m \frac{\sinh 2u_m + 2u_m}{u_m \cosh^2 u_m}$$

in the middle of the edges. In the case of moments uniformly distributed along the edges, when $\mu_m = 4M_0/m\pi$, we have

$$\left.\begin{aligned}\alpha_{\max} &= \frac{bM_0}{D\pi}\sum_{m=1,3,\ldots}^{\infty}\frac{1}{m}\frac{\sinh u_m}{\cosh^2 u_m}\\ \beta_{\max} &= \frac{aM_0}{D\pi^2}\sum_{m=1,3,\ldots}^{\infty}(-1)^{(m-1)/2}\frac{1}{m^2}\frac{\sinh 2u_m + 2u_m}{\cosh^2 u_m}\end{aligned}\right\}$$

Note that the convergence of the series for the rotation angles of the edges where the moments M_0 are applied is essentially better than for load-free edges.

15.2.7 Method of Superposition for a Plate Clamped Along Two Opposite Edges

Now, examine the problem of bending of a plate whose edges $x = 0$ and $x = a$ are simply supported, and the edges $y = \pm(b/2)$ are clamped (Fig. 15.6). The plate is subjected to a uniform load q. This problem was examined previously by direct application of M. Lévy's method. Here we will obtain a solution to this problem by superposing the solutions obtained for a simply supported plate subjected to a uniform load and for a plate subjected to edge moments.

The deflection function for a simply supported plate subjected to a uniform load is expressed by Eq. (15.47). The angles of rotation of the edges $y = \pm(b/2)$ can be evaluated as

$$\left.\frac{\partial w}{\partial y}\right|_{y=b/2} = -\frac{qa^3}{D\pi^4}\sum_{m=1,3,\ldots}^{\infty}\frac{1}{m^4}\frac{\sinh 2u_m - 2u_m}{\cosh^2 u_m}$$

The angles of rotation of the edges $y = \pm(b/2)$ are zero, if the bending moments applied to these edges are such that the angles given by these formulas are equal in magnitude and opposite in sign to the angles determined by the second formula in Eqs. (15.56). This results in the following equation for the coefficients μ_m in Eqs. (15.56):

$$\mu_m = \frac{4qa^2}{(m\pi)^3}\frac{\sinh 2u_m - 2u_m}{\sinh 2u_m + 2u_m}$$

With this expression for the coefficients μ_m, the formula (15.54) for the deflection function yields

$$w = \frac{qa^3b}{D\pi^4}\sum_{m=1,3,\ldots}^{\infty}\frac{1}{m^4}\frac{\sinh 2u_m - 2u_m}{(\sinh 2u_m + 2u_m)\cosh u_m} \times \left(\tanh u_m \cosh \alpha_m y - \frac{2y}{b}\sinh \alpha_m y\right)\sin \alpha_m x \qquad (15.57)$$

The sought deflection surface of the plate clamped at the edges $y = \pm(b/a)$, simply supported at the edges $x = 0$ and $x = a$ and subjected to a distributed load q, can be obtained as a result of superimposing the deflection surfaces expressed by Eqs. (15.47) and (15.57).

The maximum deflection due to the bending moments at the clamped edges can be obtained from Eq. (15.57), putting $x = a/2$, $y = 0$:

$$w_{\max} = \frac{qa^3b}{D\pi^4}\sum_{m=1,3,\ldots}^{\infty}(-1)^{(m-1)/2}\frac{1}{m^4}\frac{\sinh 2u_m - 2u_m}{\sinh 2u_m + 2u_m}\frac{\sinh u_m}{\cosh^2 u_m}$$

For a square plate, for instance, when $a = b$ and $u_m = m\pi/a$, retaining just one term in this series, we have

$$w_{\max} = \frac{qa^4}{D\pi^4} \times 0.20918 = 0.002147\,\frac{qa^4}{D}$$

This result is in good agreement with the results obtained in Section 15.2.5 where M. Lévy's method was applied directly.

If a plate is clamped along all the four edges, then we should apply edge-bending moments to all the edges and superimpose the solutions obtained for the moments and for the distributed load. It is often more convenient, however, to apply an approximate variational method, such as, for instance, the Rayleigh–Ritz or Bubnov–Galerkin method. These are described in the next subsection.

15.2.8 Application of the Rayleigh–Ritz and Bubnov–Galerkin Methods

We proceed from Eq. (5.1) for the strain energy per unit volume of an elastic body. In accordance with the hypothesis of straight normals, the strains γ_{xz} and γ_{yz} are zero, as well as the stress σ_z. Since the strains γ_{xz} and γ_{yz} are zero, then the stresses τ_{xy} and τ_{yz} are zero as well. With $\sigma_z = \tau_{xz} = \tau_{yz} = 0$, the formula (5.1) yields

$$W = \tfrac{1}{2}E(\sigma_x^2 + \sigma_y^2 - 2\nu\sigma_x\sigma_y) + 2(1+\nu)\tau_{xy}^2$$

Using the formulas

$$\left.\begin{aligned}
\sigma_x &= -\frac{Ez}{1-\nu^2}\left(\frac{\partial^2 w}{\partial x^2} + \nu\frac{\partial^2 w}{\partial y^2}\right)\\
\sigma_y &= -\frac{Ez}{1-\nu^2}\left(\frac{\partial^2 w}{\partial y^2} + \nu\frac{\partial^2 w}{\partial x^2}\right)\\
\tau_{xy} &= -\frac{Ez}{1+\nu}\frac{\partial^2 w}{\partial x\partial y}
\end{aligned}\right\}$$

from Section 15.2.1, we rewrite the formula for the strain energy as

$$W = \frac{Ez^2}{2(1-\nu^2)^2}\left\{(1+\nu^2)\left[\left(\frac{\partial w}{\partial x^2}\right)^2 + \left(\frac{\partial^2 w}{\partial y^2}\right)^2\right] + 4\nu\frac{\partial^2 w}{\partial x^2}\frac{\partial^2 w}{\partial y^2}\right\} + \frac{2Ez^2}{1+\nu}\left(\frac{\partial^2 w}{\partial x\partial y}\right)^2$$

Integrating this expression over the height of the plate, we obtain the following formula for the strain energy per unit plate surface:

$$dU = \int_{-h/2}^{h/2} W\,dz = \frac{1}{2}D\left\{\left(\frac{\partial^2 w}{\partial x^2} + \frac{\partial^2 w}{\partial y^2}\right)^2 - 2(1-\nu)\left[\frac{\partial^2 w}{\partial x^2}\frac{\partial^2 w}{\partial y^2} - \left(\frac{\partial^2 w}{\partial x \partial y}\right)^2\right]\right\}$$

The strain energy of the plate is, therefore,

$$U = \frac{1}{2}D \iint_A \left\{\left(\frac{\partial^2 w}{\partial x^2} + \frac{\partial^2 w}{\partial y^2}\right)^2 - 2(1-\nu)\left[\frac{\partial^2 w}{\partial x^2}\frac{\partial^2 w}{\partial y^2} - \left(\frac{\partial^2 w}{\partial x \partial y}\right)^2\right]\right\} dx\,dy \tag{15.58}$$

where A is the surface of the plate. The work of the distributed lateral load acting on the plate is

$$W = \iint_A q(x, y) w\,dx\,dy \tag{15.59}$$

In accordance with the procedure of the *Rayleigh–Ritz method*, we seek the deflection surface in the form of a series,

$$w = \sum_{k=1}^{\infty} a_k \Phi_k(x, y) \tag{15.60}$$

where the functions $\Phi_k(x, y)$ should be chosen in such a way that the kinematic boundary conditions are fulfilled and a_k are coefficients to be determined.

Forming an *energy functional*

$$\Pi = U - W$$

substituting in this formula the expansion (15.60), and using the conditions (13.23), we obtain the following system of algebraic equations for the parameters a_k:

$$\sum_{k=1}^{\infty} \delta_{kj} a_k = \Delta_j \qquad j = 1, 2, \ldots \tag{15.61}$$

where

$$\delta_{kj} - \iint\limits_A D\left[\frac{\partial^2\Phi_k}{\partial x^2}\frac{\partial^2\Phi_j}{\partial x^2} + \frac{\partial^2\Phi_k}{\partial y^2}\frac{\partial^2\Phi_j}{\partial y^2} + \nu\left(\frac{\partial^2\Phi_k}{\partial x^2}\frac{\partial^2\Phi_j}{\partial y^2} + \frac{\partial^2\Phi_k}{\partial y^2}\frac{\partial^2\Phi_j}{\partial x^2}\right)\right.$$

$$\left. + 2(1-\nu)\frac{\partial^2\Phi_k}{\partial x\partial y}\frac{\partial^2\Phi_j}{\partial x\partial y}\right] dx\, dy \tag{15.62}$$

and

$$\Delta_j = \iint\limits_A q(x, y)\Phi_j(x, y)\, dx\, dy \tag{15.63}$$

Equations (15.61) coincide with Eqs. (13.28), obtained for the case of a beam, and Eqs. (15.62) and (15.63) are simply two-dimensional analogs of the formulas (13.29) and (13.30). The system of equations (15.61) enables us to determine all the parameters a_k.

In the first approximation, we can use just one equation,

$$\delta_{11}a_1 = \Delta_1$$

so that

$$a_1 = \frac{\Delta_1}{\delta_{11}}$$

In the second approximation, two equations,

$$\left.\begin{aligned} \delta_{11}a_1 + \delta_{21}a_2 &= \Delta_1 \\ \delta_{21}a_1 + \delta_{22}a_2 &= \Delta_2 \end{aligned}\right\}$$

are used, and so on.

If *Bubnov–Galerkin's method* is applied, then we should proceed from Eq. (15.14), whereas the deflection surface is still sought in the form (15.60). In accordance with Bubnov–Galerkin's procedure, we substitute the expansion (15.60) in Eq. (15.14), multiply the obtained expression by the coordinate function $\Phi_j(x, y)$, integrate over the plate's surface A, and demand that the obtained integrals are equal to zero. This results in Eqs. (15.61), where the coefficients δ_{kj} are expressed as

$$\delta_{kj} = \delta_{jk} = \iint_A D(\nabla^4\Phi_k)\Phi_j \, dx \, dy \tag{15.64}$$

and the load terms Δ_j are given by Eq. (15.63). We can easily show that the expressions (15.64) are equivalent to Eq. (15.62), so that the Rayleigh–Ritz and Bubnov–Galerkin methods result in identical basic equations (15.61) for the unknown parameters a_k.

As an example, examine a rectangular plate clamped along all the edges and subjected to a uniformly distributed load q (Fig. 15.8). Let the coordinate functions be

$$\Phi_1 = \left(x^2 - \frac{a^2}{4}\right)^2\left(y^2 - \frac{b^2}{4}\right)^2$$

$$\Phi_2 = \left(x^2 - \frac{a^2}{4}\right)^2\left(y^2 - \frac{b^2}{4}\right)^3$$

$$\Phi_3 = \left(x^2 - \frac{a^2}{4}\right)^3\left(y^2 - \frac{b^2}{4}\right)^3$$

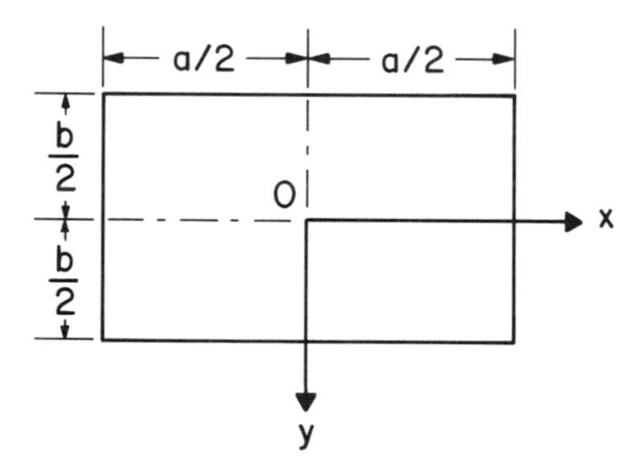

FIGURE 15.8 Rectangular plate.

and so on. Then the deflection function can be presented as

$$w = a_1\left(x^2 - \frac{a^2}{4}\right)^2\left(y^2 - \frac{b^2}{4}\right)^2 + a_2\left(x^2 - \frac{a^2}{4}\right)^2\left(y^2 - \frac{b^2}{4}\right)^3$$
$$+ a_3\left(x^2 - \frac{a^2}{4}\right)^3\left(y^2 - \frac{b^2}{4}\right)^2 + a_4\left(x^2 - \frac{a^2}{4}\right)^3\left(y^2 - \frac{b^2}{4}\right)^3 + \cdots$$

This function satisfies the boundary conditions

$$w = 0 \qquad \frac{\partial w}{\partial x} = 0 \qquad \text{at } x = -\frac{a}{2} \qquad \text{and } x = \frac{a}{2}$$

$$w = 0 \qquad \frac{\partial w}{\partial y} = 0 \qquad \text{at } y = -\frac{b}{2} \qquad \text{and } y = \frac{b}{2}$$

Let us limit ourselves to the first approximation only. Using Bubnov–Galerkin's method, wc have

$$\delta_{11}a_1 = \Delta_1$$

where

$$\delta_{11} = D\int_{-a/2}^{a/2}\int_{-b/2}^{b/2}(\nabla^4\Phi_1)\Phi_1\,dx\,dy \qquad \Delta_1 = \int_{-a/2}^{a/2}\int_{-b/2}^{b/2} q\Phi_1\,dx\,dy$$

Since

$$\Phi_1 = \left(x^2 - \frac{a^2}{4}\right)^2\left(y^2 - \frac{b^2}{4}\right)^2$$

then

$$\frac{\partial^4\Phi_1}{\partial x^4} = 24\left(y^2 - \frac{b^2}{4}\right)^2 \qquad \frac{\partial^2\Phi_1}{\partial y^2} = 24\left(x^2 - \frac{a^2}{4}\right)^2$$
$$\frac{\partial^2\Phi_1}{\partial x^2\partial y^2} = 16\left(3x^2 - \frac{a^2}{4}\right)\left(3y^2 - \frac{b^2}{4}\right)$$

and

$$\delta_{11} = D \int_{-a/2}^{a/2} \int_{-b/2}^{b/2} \left(\frac{\partial^4 \Phi_1}{\partial x^4} \Phi_1 + 2 \frac{\partial^4 \Phi_1}{\partial x^2 \partial y^2} \Phi_1 + \frac{\partial^4 \Phi_1}{\partial y^4} \Phi_1 \right) dx\, dy$$

$$= 4D \int_0^{a/2} \int_0^{b/2} \left[24\left(x^2 - \frac{a^2}{4}\right)^2 \left(y^2 - \frac{b^2}{4}\right)^4 \right.$$

$$+ 32\left(3x^2 - \frac{a^2}{4}\right)\left(3y^2 - \frac{b^2}{4}\right)\left(x^2 - \frac{a^2}{4}\right)^2\left(y^2 - \frac{b^2}{4}\right)^2$$

$$\left. + 24\left(x^2 - \frac{a^2}{4}\right)^4\left(y^2 - \frac{b^2}{4}\right)^2 \right] dx\, dy$$

$$= \frac{4 \times 128 \times 64}{9 \times 7 \times 5 \times 5} \left(\frac{a}{2}\right)^5 \left(\frac{b}{2}\right)^5 \left[\left(\frac{a}{2}\right)^4 + \frac{4}{7}\left(\frac{a}{2}\right)^2\left(\frac{b}{2}\right)^2 + \left(\frac{b}{2}\right)^4 \right]$$

$$\Delta_1 = 4q \int_0^{a/2} \int_0^{b/2} \left(x^2 - \frac{a^2}{4}\right)^2 \left(y^2 - \frac{b^2}{4}\right)^2 dx\, dy = 4q \frac{64}{225} \left(\frac{a}{2}\right)^5 \left(\frac{b}{2}\right)^5$$

Then the parameter a_1 is

$$a_1 = \frac{\Delta_1}{\delta_{11}} = \frac{7q}{128\left[\left(\frac{a}{2}\right)^4 + \frac{4}{7}\left(\frac{a}{2}\right)^2\left(\frac{b}{2}\right)^2 + \left(\frac{b}{2}\right)^4 \right] D}$$

and the deflection surface is given by the equation

$$w = a_1 \Phi_1(x, y) = \frac{7q}{8(a^4 + \frac{4}{7}a^2b^2 + b^4)D} \left(x^2 - \frac{a^2}{4}\right)^2 \left(y^2 - \frac{b^2}{4}\right)^2$$

The maximum deflection is in the center of the plate ($x = y = 0$) and is

$$w_{\max} = \frac{7qa^4b^4}{2048(a^4 + \frac{4}{7}a^2b^2 + b^4)D}$$

For a square plate ($a = b$), we have

$$w_{\max} = 0.001331 \frac{qa^4}{D}$$

Now, let us examine the same problem with different coordinate functions, namely,

$$\Phi_1 = \cos^2 \frac{3\pi x}{a} \cos^2 \frac{3\pi y}{b} \qquad \Phi_2 = \cos^2 \frac{3\pi x}{a} \cos^2 \frac{3\pi y}{b}$$

$$\Phi_3 = \cos^2 \frac{5\pi x}{a} \cos^2 \frac{5\pi y}{b}$$

and so on. Then the deflection function w is as follows:

$$w = a_1 \cos^2 \frac{\pi x}{a} \cos^2 \frac{\pi y}{b} + a_2 \cos^2 \frac{3\pi x}{a} \cos^2 \frac{3\pi y}{b} + a_3 \cos^2 \frac{5\pi x}{a} \cos^2 \frac{5\pi y}{b} + \cdots$$

This function satisfies all the boundary conditions of the problem. Limiting ourselves to the first approximation only, we have

$$\Phi_1 = \cos^2 \frac{\pi x}{a} \cos^2 \frac{\pi y}{b}$$

so that

$$\delta_{11} = \frac{\pi^4 D}{64a^3b^3}(3a^4 + 2a^2b^2 + 3b^4) \qquad \Delta_1 = \frac{qab}{4}$$

Then we have

$$a_1 = \frac{\Delta_1}{\delta_{11}} = \frac{16a^4b^4q}{D\pi^4(3a^4 + 2a^2b^2 + 3b^4)}$$

and the deflection function is

$$w = \frac{16a^4b^4q}{D\pi^4(3a^4 + 2a^2b^2 + 3b^4)} \cos^2 \frac{\pi x}{a} \cos^2 \frac{\pi y}{b}$$

The maximum deflection is in the center of the plate ($x = y = 0$) and, in the case of a square plate ($a = b$), is

$$w_{\max} = 0.001283 \frac{qa^4}{D}$$

which differs by about 3.5 percent from the result obtained using polynomials.

15.2.9 Large Deflections of Plates: von Kármán's Equations

In the preceding analyses, we assumed that the bending of a plate is due exclusively to lateral loading. In many cases, however, there are forces acting in the middle surface of the plate. These forces are usually referred to as *membrane forces*. They can affect essentially the deflections and the stresses in the plate. If, for instance, a printed circuit board is supported by an immovable contour (which is, as a rule, the case) and is subjected to a large lateral load, then this contour applies reactive forces to the plate. These forces not only influence the deflections and stresses but also make the problem of finding the elastic surface nonlinear. We encountered a similar situation in Section 12.9, where nonlinear bending of beams under the combined action of lateral and axial loading was examined. The situation is significantly more complicated for plates because of the two-dimensional nature of the problem.

To obtain the governing equations in the case in which the membrane forces are present, we examine the equilibrium of a small element cut from the plate by two pairs of planes parallel to the coordinate planes xz and yz (Fig. 15.9). The difference between the case in question and the case considered previously in the first subsection under Section 15.2 is that now we have forces acting in the middle surface of the plate. As we can see from Fig. 15.9, there are three types of these forces: T_x, T_y, and $T_{xy} = T_{yx}$. Taking the projections of these forces on the axes x and y, we have the following equilibrium equations:

$$\frac{\partial T_x}{\partial x} + \frac{\partial T_{xy}}{\partial y} = 0 \qquad \frac{\partial T_{xy}}{\partial x} + \frac{\partial T_y}{\partial y} = 0 \tag{15.65}$$

As far as the projection of the membrane forces on the z axis is concerned, we should take into account the fact that, because of bending, the angles that the directions of the forces T_x and $T_x + (\partial T_x/\partial x)\,dx$ form with the coordinate axes are different. Indeed, if the angle that the force T_x forms with the x axis is simply $\partial w/\partial x$, the angle that the force $T_x + (\partial T_x/\partial x)\,dx$ forms with the x axis is $(\partial w/\partial x) + \partial/\partial x\,(\partial w/\partial x)\,dx$. Then the projection of the force T_x on the z axis is

$$T_x \cos\left(\frac{\pi}{2} - \frac{\partial w}{\partial x}\right) dy = T_x \sin\frac{\partial w}{\partial x}\,dy \cong T_x \frac{\partial w}{\partial x}\,dy$$

and the projection of the force $T_x + (\partial T_x/\partial x)\,dx$ on the z axis is

$$\left(T_x + \frac{\partial T_x}{\partial x}\,dx\right)\left(\frac{\partial w}{\partial x} + \frac{\partial^2 w}{\partial x^2}\,dx\right) dy \cong \left(T_x \frac{\partial w}{\partial x} + T_x \frac{\partial^2 w}{\partial x^2}\,dx + \frac{\partial T_x}{\partial x}\frac{\partial w}{\partial x}\,dx\right) dy$$

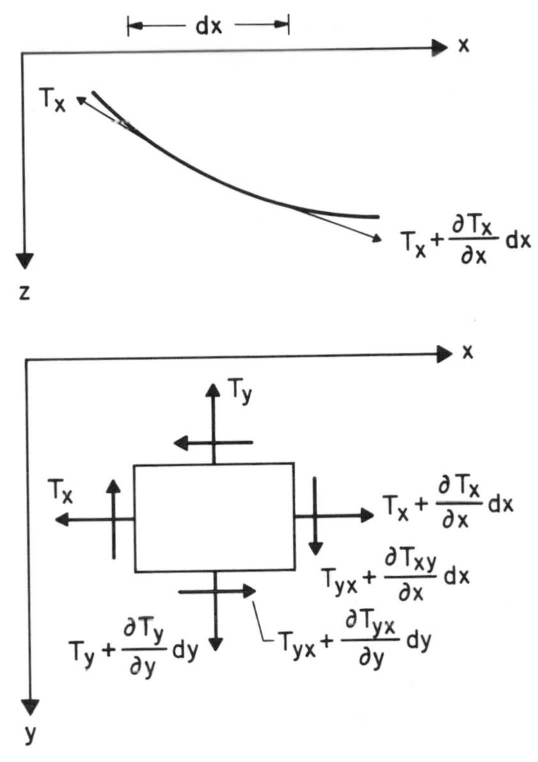

FIGURE 15.9 Variation of the in-plane ("membrane") forces over an element of a flat plate.

Therefore, the total projection of the forces T_x on the z axis is

$$\left(T_x \frac{\partial^2 w}{\partial x^2} + \frac{\partial T_x}{\partial x}\frac{\partial w}{\partial x}\right) dx\, dy$$

Similarly, for the projection of the forces T_y, we have

$$\left(T_y \frac{\partial^2 w}{\partial y^2} + \frac{\partial T_y}{\partial y}\frac{\partial w}{\partial y}\right) dx\, dy$$

In order to obtain the projection of the forces T_{xy} on the z axis, we should consider the fact that, by analogy with the above, the angle that the elements' sides form with the axis x changes from $\partial w/\partial y$ for the force T_{xy} to $(\partial w/\partial y) + \partial/\partial x\,(\partial w/\partial y)\, dx$ for the force $T_{xy} + (\partial T_{xy}/\partial x)\, dx$, so that the total projection of the forces T_{xy} on the z axis is

$$-T_{xy}\frac{\partial w}{\partial y}dy+\left(T_{xy}+\frac{\partial T_{xy}}{\partial x}dx\right)\left(\frac{\partial w}{\partial y}+\frac{\partial^2 w}{\partial x\partial y}dx\right)dy$$

$$\cong\left(T_{xy}\frac{\partial^2 w}{\partial x\partial y}+\frac{\partial T_{xy}}{\partial x}\frac{\partial w}{\partial y}\right)dx\,dy$$

An identical equation can be obtained for the projection of the force T_{xy} on the z axis. Then the final expression for the projection of all the membrane forces on the z axis is as follows:

$$\left[T_x\frac{\partial^2 w}{\partial x^2}+\frac{\partial T_x}{\partial x}\frac{\partial w}{\partial x}+T_y\frac{\partial T_y}{\partial y}\frac{\partial w}{\partial y}+T_{xy}\frac{\partial^2 w}{\partial x\partial y}+\frac{\partial T_{xy}}{\partial x}\frac{\partial w}{\partial y}\right.$$
$$\left.+T_{xy}\frac{\partial^2 w}{\partial x\partial y}+\frac{\partial T_{xy}}{\partial y}\frac{\partial w}{\partial x}\right]dx\,dy=\left[T_x\frac{\partial^2 w}{\partial x^2}+T_y\frac{\partial^2 w}{\partial y^2}+2T_{xy}\frac{\partial^2 w}{\partial x\partial y}\right.$$
$$\left.+\left(\frac{\partial T_x}{\partial x}+\frac{\partial T_{xy}}{\partial y}\right)\frac{\partial w}{\partial x}+\left(\frac{\partial T_{xy}}{\partial y}+\frac{\partial T_{xy}}{\partial x}\right)\frac{\partial w}{\partial y}\right]dx\,dy$$

The expressions in the parentheses, in accordance with Eqs. (15.65), are zero and, therefore, the membrane forces give the following projection on the z axis:

$$T_x\frac{\partial^2 w}{\partial x^2}+T_y\frac{\partial^2 w}{\partial y^2}+2T_{xy}\frac{\partial^2 w}{\partial x\partial y}$$

Adding this projection to the lateral load q, we obtain the governing equation (15.14) for the plate's deflections in the form

$$\nabla^4 w=\frac{\partial^4 w}{\partial x^4}+2\frac{\partial^4 w}{\partial x^2\partial y^2}+\frac{\partial^4 w}{\partial y^4}=\frac{1}{D}\left(q+T_x\frac{\partial^2 w}{\partial x^2}+T_y\frac{\partial^2 w}{\partial x^2}+2T_{xy}\frac{\partial^2 w}{\partial x\partial y}\right)\tag{15.66}$$

When the forces T_x, T_y, and T_{xy} are known, and especially if they are constant, obtaining a solution to this equation does not cause significant difficulties. If these forces are affected by the stretching of the middle surface of the plate as a result of large deflections, however, an additional equation is needed to relate the membrane forces to the deflections.

To derive this equation, examine a linear element AB at the midplane of a plate (Fig. 15.10a). As a result of the bending of the plate, this element is

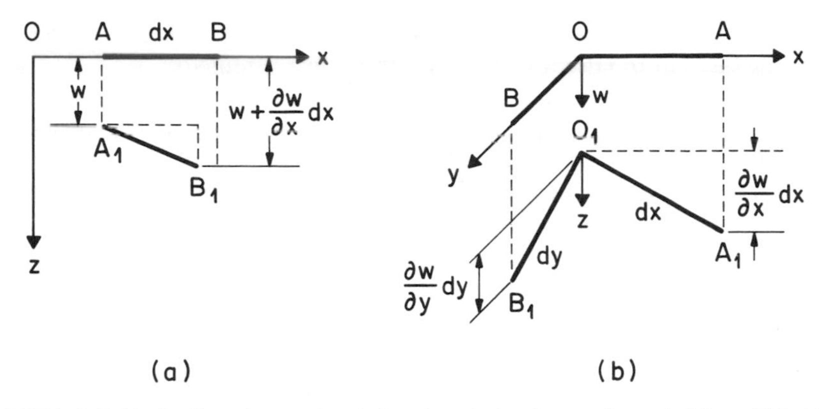

FIGURE 15.10 Evaluation of normal and shearing strains for an element at the midplane of a plate.

displaced to become A_1B_1. If no midplane stressing occurs, then the length A_1B_1 would be equal to the initial length AB, and its projection on the x axis would be expressed as

$$\sqrt{dx^2 - \left(\frac{\partial w}{\partial x}\, dx\right)^2} = dx\sqrt{1 - \left(\frac{\partial w}{\partial x}\right)^2} \cong \left[1 - \frac{1}{2}\left(\frac{\partial w}{\partial x}\right)^2\right] dx$$

where the binomial expansion of the square root is used. If, however, the element AB is stretched in such a way that its projection on the x axis remains unchanged, the projection of the element's elongation on this axis is

$$du = dx - \left[1 - \frac{1}{2}\left(\frac{\partial w}{\partial x}\right)^2\right] dx = \frac{1}{2}\left(\frac{\partial w}{\partial x}\right)^2 dx$$

and the resulting strain is

$$\varepsilon_x = \frac{du}{dx} = \frac{1}{2}\left(\frac{\partial w}{\partial x}\right)^2 \tag{15.67}$$

Similarly, the strain in the y direction is

$$\varepsilon_y = \frac{1}{2}\left(\frac{\partial w}{\partial y}\right)^2 \tag{15.68}$$

Note that a similar relationship between the longitudinal strain and the deflection was obtained previously when the bending of beams under the combined action of lateral and axial forces was examined.

In order to determine the shearing strain due to plate bending, consider two linear elements OA and OB oriented in the x and y directions, respectively (Fig. 15.10b). Let the new positions of these elements due to bending of the plate be O_1A_1 and O_1B_1. The direction cosines of these new positions are

$$l_1 = \frac{\sqrt{dx^2 + \left(\dfrac{\partial w}{\partial x}dx\right)^2}}{dx} \cong 1 - \frac{1}{2}\left(\frac{\partial w}{\partial x}\right)^2 \qquad m_1 = 0 \qquad n_1 = \frac{\dfrac{\partial w}{\partial x}dx}{dx} = \frac{\partial w}{\partial x}$$

and

$$l_2 = 0 \qquad m_2 \cong 1 - \frac{1}{2}\left(\frac{\partial w}{\partial y}\right)^2 \qquad n_2 = \frac{\partial w}{\partial y}$$

The shearing strain γ_{xy} can be evaluated as

$$\begin{aligned}\gamma &= \frac{\pi}{2} - \angle A_1O_1B = \sin\left(\frac{\pi}{2} - \angle A_1O_1B\right)\\ &= \cos \angle A_1O_1B = l_1l_2 + m_1m_2 + n_1n_2\\ &= n_1n_2 = \frac{\partial w}{\partial x}\frac{\partial w}{\partial y}\end{aligned} \tag{15.69}$$

Adding the strains (15.67)–(15.69) to the corresponding strains due to the displacements u and v in the midplane, we have the following formulas for the total strains in the midplane:

$$\left.\begin{aligned}\varepsilon_x &= \frac{\partial u}{\partial x} + \frac{1}{2}\left(\frac{\partial w}{\partial x}\right)^2\\ \varepsilon_y &= \frac{\partial v}{\partial y} + \frac{1}{2}\left(\frac{\partial w}{\partial y}\right)^2\\ \gamma_{xy} &= \frac{\partial u}{\partial y} + \frac{\partial v}{\partial x} + \frac{\partial w}{\partial x}\frac{\partial w}{\partial y}\end{aligned}\right\} \tag{15.70}$$

Differentiating the first of these equations twice with respect to y, the second equation with respect to x, and the third equation with respect to x

and y, and summing up the obtained results, we obtain the following compatibility condition:

$$\frac{\partial^2 \varepsilon_x}{\partial y^2} + \frac{\partial^2 \varepsilon_y}{\partial x^2} = \left(\frac{\partial^2 w}{\partial x \partial y}\right)^2 - \frac{\partial^2 w}{\partial x^2}\frac{\partial^2 w}{\partial y^2} \tag{15.71}$$

On the other hand, the strains in a two-dimensional state of stress are expressed as

$$\left.\begin{aligned} \varepsilon_x &= \frac{1}{E}(\sigma_x - \nu\sigma_y) = \frac{1}{Eh}(T_x - \nu T_y) \\ \varepsilon_y &= \frac{1}{E}(\sigma_y - \nu\sigma_x) = \frac{1}{Eh}(T_y - \nu T_x) \\ \gamma_{xy} &= \frac{\tau_{xy}}{G} = \frac{T_{xy}}{Gh} = \frac{2(1+\nu)}{Eh}T_{xy} \end{aligned}\right\} \tag{15.72}$$

Introducing these relationships in Eq. (15.70), we have

$$\begin{aligned} &\frac{\partial^2 T_x}{\partial y^2} - \nu\frac{\partial^2 T_y}{\partial y^2} + \frac{\partial^2 T_y}{\partial x^2} - \nu\frac{\partial^2 T_x}{\partial x^2} - 2(1+\nu)\frac{\partial^2 T_{xy}}{\partial x \partial y} \\ &\quad = Eh\left[\left(\frac{\partial^2 w}{\partial x \partial y}\right)^2 - \frac{\partial^2 w}{\partial x^2}\frac{\partial^2 w}{\partial y^2}\right] \end{aligned} \tag{15.73}$$

Introduce a stress function ϕ as follows:

$$\sigma_x = \frac{T_x}{h} = \frac{\partial^2 \phi}{\partial y^2} \qquad \sigma_y = \frac{T_y}{h} = \frac{\partial^2 \phi}{\partial x^2} \qquad \tau_{xy} = \frac{T_{xy}}{h} = -\frac{\partial^2 \phi}{\partial x \partial y} \tag{15.74}$$

Using these notations, we can present the equation of equilibrium (15.66) and the compatibility equation (15.71) in the form of a system of the following two nonlinear differential equations in partial derivatives:

$$\left.\begin{aligned} \nabla^4 w &= \frac{\partial^4 w}{\partial x^4} + 2\frac{\partial^4 w}{\partial x^2 \partial y^2} + \frac{\partial^4 w}{\partial y^4} \\ &= \frac{h}{D}\left(\frac{q}{h} + \frac{\partial^2 \phi}{\partial y^2}\frac{\partial^2 w}{\partial x^2} - 2\frac{\partial^2 \phi}{\partial x \partial y}\frac{\partial^2 w}{\partial x \partial y} + \frac{\partial^2 \phi}{\partial x^2}\frac{\partial^2 w}{\partial y^2}\right) \\ \nabla^4 \phi &= \frac{\partial^4 \phi}{\partial x^4} + 2\frac{\partial^4 \phi}{\partial x^2 \partial y^2} + \frac{\partial^4 \phi}{\partial y^4} = E\left[\left(\frac{\partial^2 w}{\partial x \partial y}\right)^2 - \frac{\partial^2 w}{\partial x^2}\frac{\partial^2 w}{\partial y^2}\right] \end{aligned}\right\} \tag{15.75}$$

These are known as *von Kármán's equations* of large deflections of plates. When the deflections w are small, the stress function ϕ is next to zero, and the first equation results in the Lagrange equation (15.14) of linear deflection of plates.

Equations (15.75) can be written in a more compact form if the following operator is introduced:

$$L(w, \phi) = \frac{\partial^2 \phi}{\partial y^2}\frac{\partial^2 w}{\partial x^2} - 2\frac{\partial^2 \phi}{\partial x \partial y}\frac{\partial^2 w}{\partial x \partial y} + \frac{\partial^2 \phi}{\partial x^2}\frac{\partial^2 w}{\partial y^2}$$

Obviously,

$$L(w, w) = 2\frac{\partial^2 w}{\partial x^2}\frac{\partial^2 w}{\partial y^2} - 2\left(\frac{\partial^2 w}{\partial x \partial y}\right)^2$$

Then we have

$$\left.\begin{aligned} D\nabla^4 w &= hL(w, \phi) + q \\ \frac{1}{E}\nabla^4 \phi &= \frac{1}{2}L(w, w) \end{aligned}\right\} \tag{15.76}$$

15.2.10 Large Deflections of Plates Under Uniformly Distributed Lateral Load

The integration of von Kármán's equations is generally associated with enormous difficulties. An approximate solution for a rectangular plate simply supported on a nondeformable contour and subjected to a uniform lateral loading can be obtained, assuming

$$w(x, y) = f\bar{w}(x, y) \qquad \phi(x, y) = f^2\bar{\phi}(x, y) \tag{15.77}$$

where f is the maximum displacement of the plate, and $\bar{w}(x, y)$ and $\bar{\phi}(x, y)$ are thus far unknown functions. After introducing these formulas into the second equation in Eqs. (15.76), we conclude that the functions $\bar{w}(x, y)$ and $\bar{\phi}(x, y)$ are related as follows:

$$\nabla^4\bar{\phi}(x, y) = -\frac{E}{2}L(\bar{w}, \bar{w}) \tag{15.78}$$

To solve the first equation in Eqs. (15.76), we use the Bubnov–Galerkin method. In accordance with the procedure for this method, we substitute

Eqs. (15.77) into this equation, multiply the obtained expression by the coordinate function $\bar{w}(x, y)$, and integrate over the plate's surface A. Then we obtain

$$f + \mu_0 f^3 = A \tag{15.79}$$

where

$$\mu_0 = \frac{h}{D} \frac{\iint_A \bar{w} L(\bar{w}, \phi)\, dA}{\iint_A \bar{w} \nabla^4 \bar{w}\, dA} \qquad A = \frac{q}{D} \frac{\iint_A \bar{w}\, dA}{\iint_A \bar{w} \nabla^4 \bar{w}\, dA} \tag{15.80}$$

Assuming (Fig. 15.8)

$$\bar{w} = \cos \frac{\pi x}{a} \cos \frac{\pi y}{b} \tag{15.81}$$

we obtain Eq. (15.78) as

$$\nabla^4 \phi(x, y) = -\frac{\pi^4 E}{2a^2 b^2} \left(\cos \frac{2\pi x}{a} + \cos \frac{2\pi y}{b} \right) \tag{15.82}$$

We seek the solution to this equation in the form

$$\bar{\phi} = D_1 x^2 + D_2 y^2 + D_3 \cos \frac{2\pi x}{a} + D_4 \cos \frac{2\pi y}{b} \tag{15.83}$$

After introducing this equation in Eq. (15.82), we find

$$D_3 = -\frac{E}{32} \frac{a^2}{b^2} \qquad D_4 = -\frac{E}{32} \frac{b^2}{a^2}$$

The constants D_1 and D_2 can be determined from the following formulas for the displacements of the arbitrary points of the middle surface of the plate:

$$\left.\begin{aligned} u &= \int_0^x \left[\frac{1}{E} \left(\frac{\partial^2 \phi}{\partial y^2} - \nu \frac{\partial^2 \phi}{\partial x^2} \right) - \frac{1}{2} \left(\frac{\partial w}{\partial x} \right)^2 \right] dx \\ v &= \int_0^y \left[\frac{1}{E} \left(\frac{\partial^2 \phi}{\partial x^2} - \nu \frac{\partial^2 \phi}{\partial y^2} \right) - \frac{1}{2} \left(\frac{\partial w}{\partial y} \right)^2 \right] dy \end{aligned}\right\} \tag{15.84}$$

These formulas can be obtained from Eqs. (15.70), (15.72), and (15.74). For a nondeformable contour, the following conditions must be fulfilled:

$$u\left(\pm\frac{a}{2}, y\right) = 0 \qquad v\left(x, \pm\frac{b}{2}\right) = 0$$

Then the formulas (15.84) yield

$$\left.\begin{aligned}\int_0^{a/2}\left[\frac{1}{E}\left(\frac{\partial^2\phi}{\partial y^2} - v\frac{\partial^2\phi}{\partial x^2}\right) - \frac{1}{2}\left(\frac{\partial w}{\partial x}\right)^2\right]dx = 0\\ \int_0^{b/2}\left[\frac{1}{E}\left(\frac{\partial^2\phi}{\partial x^2} - v\frac{\partial^2\phi}{\partial y^2}\right) - \frac{1}{2}\left(\frac{\partial w}{\partial y}\right)^2\right]dy = 0\end{aligned}\right\} \tag{15.85}$$

Substituting Eqs. (15.81) and (15.83) in the conditions (15.85), we find

$$D_1 = \frac{\pi^2 E(a^2 + vb^2)}{16(1 - v^2)a^2b^2} \qquad D_2 = \frac{\pi^2 E(b^2 + va^2)}{16(1 - v^2)a^2b^2}$$

Then Eq. (15.83) results in the following formula for the dimensionless stress function:

$$\begin{aligned}\bar{\phi}(x, y) = \frac{E}{32a^2b^2}\left\{\frac{2\phi^2}{1 - v^2}[(a^2 + vb^2)x^2 + (b^2 + va^2)y^2]\right.\\ \left. - \left(a^2\cos\frac{2\pi x}{a} + b^2\cos\frac{2\pi y}{b}\right)\right\}\end{aligned} \tag{15.86}$$

Now, introducing Eqs. (15.81) and (15.86) into Eqs. (15.80), we obtain

$$\mu_0 = \frac{3}{4h^2}\frac{(3 - v^2)(a^4 + b^4) + 4va^2b^2}{(a^2 + b^2)^2} \qquad A = \frac{16a^4b^4q}{\pi^6 D(a^2 + b^2)^2} \tag{15.87}$$

For a square plate ($a = b$), these formulas yield

$$\mu_0 = \frac{3}{8h^2}(1 + v)(3 - v) \qquad A = \frac{4a^4q}{\pi^6 D}a$$

As we can see from Eq. (15.79), when the deflection f is small compared to the thickness of the plate, the nonlinear term is also small, and the deflection in the center of the plate is

$$f = A = \frac{4qa^4}{\pi^6 D} = 0.004161 \frac{qa^4}{D}$$

This result is in fairly good agreement with the exact solution obtained in Section 15.2.4 for a square simply supported plate.

As follows from Eq. (15.79), a linear solution gives sufficiently accurate results if the value $\mu_0 f^2$ is substantially smaller than unity. Using the first formula in Eq. (15.87), we find that, in terms of the deflection-to-thickness ratio, this requirement is equivalent to the requirement that this ratio should be substantially smaller than

$$\left(\frac{f}{h}\right)_c = \frac{2(a^2 + b^2)}{\sqrt{3[(3 - \nu^2)(a^4 + b^4) + 4\nu a^2 b^2]}}$$

In the case of an elongated plate ($b \gg a$), this ratio, assuming $\nu = 0.3$, is

$$\left(\frac{f}{h}\right)_c = \frac{2}{\sqrt{3(3 - \nu^2)}} = 0.677$$

In the case of a square plate ($a = b$), with $\nu = 0.3$, we obtain

$$\left(\frac{f}{h}\right)_c = \sqrt{\frac{8}{3(1 + \nu)(3 - \nu)}} = 0.872$$

Hence, the nonlinear effects on the maximum deflection are somewhat greater for elongated plates. This is due to the fact that such plates experience larger deflections for the given loading. If, for instance, the calculated maximum deflection in an elongated plate is such that the deflection-to-thickness ratio is about 0.2, then the application of a linear approach results in the overestimation of the actual maximum deflection by

$$\left(\frac{f}{h}\right)^2 \div \left(\frac{f}{h}\right)_c^2 = \left(\frac{0.2}{0.677}\right)^2 = 0.0873 = 8.73\%$$

This might be acceptable in engineering practice.

The situation might be different, however, when the maximum stresses are of primary interest. Although a linear approach, considering only the bending stresses (which are proportional to the maximum deflection), always overestimates these stresses, such an approach does not consider at all the membrane stresses (which are proportional to the maximum deflection squared). Obviously, when the deflections, as well as the membrane stresses, are small, a linear approach is acceptable. For large deflections, however, it can lead to underestimation of the actual stresses.

In order to assess the limitations of the linear approach in the problem in question, i.e., in the case of a rectangular plate simply supported on a nondeformable contour and subjected to a uniformly distributed load, we evaluate the total stresses using the formulas:

$$\sigma_x = \sigma_x^0 + \frac{12M_x}{h^3} z \qquad \sigma_y = \sigma_y^0 + \frac{12M_y}{h^3} z \qquad \tau_{xy} = \tau_{xy}^0 + \frac{12M_{xy}}{h^3} z$$

In these formulas, σ_x^0, σ_y^0, and τ_{xy}^0 are membrane stresses acting in the middle surface of the plate, and M_x, M_y, and M_{xy} are the bending and the twisting moments due to the plate's bending. The membrane stresses are given by the formulas (15.74). Using the second formula in Eqs. (15.77), the formulas (15.74) can be presented as

$$\sigma_x^0 = f^2 \frac{\partial^2 \bar{\phi}}{\partial y^2} \qquad \sigma_y^0 = f^2 \frac{\partial^2 \bar{\phi}}{\partial x^2} \qquad \tau_{xy}^0 = -f^2 \frac{\partial^2 \bar{\phi}}{\partial x \partial y} \tag{15.88}$$

The maximum bending stresses occur at $z = \pm(h/2)$ and are as follows:

$$\sigma_x^0 = \frac{6M_x}{h^2} \qquad \sigma_y^0 = \frac{6M_y}{h^2} \qquad \tau_{xy}^0 = \frac{6M_{xy}}{h^2}$$

Clearly, the presence of the membrane stresses makes the distribution of the total stresses asymmetric with respect to the midplane of the plate. Therefore, the maximum total stresses in the problem in question are tensile and occur at the lower surface of the plate. Using the formulas (15.8) for the bending moments, the formula (15.13) for the twisting moment, and considering the first formula in Eqs. (15.77), we have the following expressions for the maximum bending stresses:

$$\left.\begin{aligned} \sigma_x^b &= -\frac{6D}{h^2} f\left(\frac{\partial^2 \bar{w}}{\partial y^2} + v\frac{\partial^2 \bar{w}}{\partial y^2}\right) \\ \sigma_y^b &= -\frac{6D}{h^2} f\left(\frac{\partial^2 \bar{w}}{\partial y^2} + v\frac{\partial^2 \bar{w}}{\partial x^2}\right) \\ \tau_{xy}^b &= 6\frac{D}{h^2}(1-v) f \frac{\partial^2 \bar{w}}{\partial x \partial y} \end{aligned}\right\} \tag{15.89}$$

Using the expressions (15.81) and (15.86) for the functions $\bar{w}$ and $\bar{\phi}$, and combining the membrane stresses (15.88) with the bending stresses (15.89), we obtain the following formulas for the maximum total normal stresses at the center of the plate ($x = y = 0$):

$$\left.\begin{aligned} (\sigma_x)_{\max} &= \frac{6\pi^2 D}{h^2 a^2} f\left(1 + \nu\gamma^2 + \frac{2 - \nu^2 + \nu\gamma^2}{4}\frac{f}{h}\right) \\ (\sigma_y)_{\max} &= \frac{6\pi^2 D}{h^2 a^2} f\left[\nu + \gamma^2 + \frac{(2 - \nu^2)\gamma^2 + \nu}{4}\frac{f}{h}\right] \end{aligned}\right\} \tag{15.90}$$

where $\gamma = a/b$ is the plate's aspect ratio. The maximum shearing stresses occur at the corners of the plate and are

$$(\tau_{xy})_{\max} = \frac{6\pi^2 D}{h^2 a^2} f(1 - \nu)\gamma \tag{15.91}$$

These stresses in the case in question are due to bending only and are proportional to the nonlinear deflections.

In the case of an elongated plate ($b \to \infty, \gamma = 0$), the formulas (15.90) and (15.91) yield

$$\left.\begin{aligned} (\sigma_x)_{\max} &= \frac{6\pi^2 D}{h^2 a^2} f\left(1 + \frac{2 - \nu^2}{4}\frac{f}{h}\right) \\ (\sigma_y)_{\max} &= \frac{6\pi^2 D}{h^2 a^2} f\nu\left(1 + \frac{f}{4h}\right) \\ (\tau_{xy})_{\max} &= 0 \end{aligned}\right\} \tag{15.92}$$

In the case of a square plate ($a = b, \gamma = 1$), the formulas (15.90) and (15.91) result in the equations

$$\left.\begin{aligned} (\sigma_x)_{\max} = (\sigma_y)_{\max} &= \frac{6\pi^2 D}{h^2 a^2}(1 + \nu) f\left(1 + \frac{2 - \nu}{4}\frac{f}{h}\right) \\ (\tau_{xy})_{\max} &= -\frac{6\pi^2 D}{h^2 a^2}(1 - \nu) f \end{aligned}\right\} \tag{15.93}$$

Comparing the first formulas in Eqs. (15.92) and (15.93), we can see that, for the given maximum deflection, the role of the membrane stresses is somewhat greater in the case of a square plate.

In order to assess the "break-even" point, beyond which the application of the linear theory leads to the underestimation of the maximum stresses, we proceed from the first formula in Eqs. (15.90). When a linear approach is used, the maximum stress $(\sigma_x)_{\max}$ uses the formula

$$(\sigma_x)_{\max} = \frac{6\pi^2 D}{h^2 a^2}(1 + \nu\gamma^2)A$$

Then the sought "break-even" point is characterized by the equation

$$\frac{f}{h}\left(1 + \nu\gamma^2 + \frac{2 - \nu^2 + \nu\gamma^2}{4}\frac{f}{h}\right) = (1 + \nu\gamma^2)\frac{A}{h}$$

Using Eq. (15.79), we obtain the following equation for the dimensionless "break-even" nonlinear deflection f/h:

$$\frac{f}{h} = \frac{(2 - \nu^2 + \nu\gamma^2)(1 + \gamma^2)^2}{3(1 + \nu\gamma^2)[(3 - \nu^2)(1 + \gamma^4) + 4\nu\gamma^2]}$$

This equation leads to the formula

$$\frac{f}{h} = \frac{2 - \nu^2}{3(3 - \nu^2)}$$

in the case of an elongated plate ($\gamma = 0$) and to the formula

$$\frac{f}{h} = \frac{2 - \nu^2}{3(1 + \nu)(3 - \nu)}$$

in the case of a square plate ($\gamma = 1$). If, for instance, $\nu = 0.3$, then the "break-even" f/h value is 0.219 in the case of an elongated plate and 0.323 in the case of a square plate. The corresponding deflection-to-thickness ratios calculated on the basis of a linear theory are $A/h = 0.242$ and $A/h = 0.367$, respectively. Thus, if the maximum deflections calculated on the basis of a linear theory are smaller than about one fifth of the plate's thickness, then this theory will lead to conservative results not only for the maximum deflections but for the maximum stresses as well. Note that, in the case of a plate with a finite aspect ratio, and especially in the case of a square plate, even larger deflections, up to one third of the thickness, may be allowed without making a nonlinear approach necessary. It should be emphasized, however, that the foregoing analysis and conclusions are based on an assumption that the material behaves within the elastic limits for any stress level. Of course, if, at a certain stress level, the material exceeds its yield stress, then the problem becomes not only geometrically but also physically nonlinear.

15.2.11 Classification of Thin Plates from the Viewpoint of Their Analysis

In Section 15.1, we pointed out that plates are usually classified, from the viewpoint of their thickness, into thin and thick plates and that plates that are so thin that they do not resist bending at all and experience tensile stresses only are called membranes. In this section, we will classify thin plates from the viewpoint of the method of their analyses. We proceed from von Kármán's equations, the most general equations for large deflections of plates.

If the plates are so stiff that the membrane stresses have a negligible effect on bending, then we can neglect the terms depending on the stress function ϕ. The calculation of such plates can be based on Lagrange's equation

$$\nabla^4 w = \frac{\partial^4 w}{\partial x^4} + 2\frac{\partial^4 w}{\partial x^2 \partial y^2} + \frac{\partial^4 w}{\partial y^4} = \frac{q(x, y)}{D} \tag{15.94}$$

These are often referred to as *absolutely stiff* plates.

If the lateral loading, as well as the deflections w, is relatively small, then we may not consider the effect of bending on the membrane stresses, and we put the right part of the second equation in Eqs. (15.75) equal to zero. If, at the same time, the plate is so thin that the effect of the membrane stresses on the deflection function is essential, then the first equation in Eqs. (15.75) cannot be simplified, and the system of the governing equations of the bending of a plate is

$$\left.\begin{aligned} \nabla^4 w &= \frac{\partial^4 w}{\partial x^4} + 2\frac{\partial^4 w}{\partial x^2 \partial y^2} + \frac{\partial^4 w}{\partial y^4} \\ &= \frac{h}{D}\left(\frac{q}{h} + \frac{\partial^2 \phi}{\partial y^2}\frac{\partial^2 w}{\partial y^2} - 2\frac{\partial^2 \phi}{\partial x \partial y}\frac{\partial^2 w}{\partial x \partial y} + \frac{\partial^2 \phi}{\partial x^2}\frac{\partial^2 w}{\partial y^2}\right) \\ \nabla^4 \phi &= 0 \end{aligned}\right\} \tag{15.95}$$

These equations can be integrated quite easily, since the second equation for the stress function can be solved first, and then the obtained solution can be substituted in the equation for the deflections. Plates whose behavior can be described by Eqs. (15.95) are called *flexible plates of small deflection.* The theory of such plates was developed by B. Saint-Venant.

In the most general case, when the deflections are so large that coupling effects between the deflections and the membrane stresses are important in both ways and cannot be ignored, then the calculation should be based on the general von Kármán's equations (15.95). Such plates are called *flexible plates of large deflection*, or *plates of finite rigidity.*

When the thickness of a plate is so small that its flexural rigidity can be put equal to zero, then von Kármán's equations yield

$$\left.\begin{aligned} &\frac{\partial^2\phi}{\partial y^2}\frac{\partial^2 w}{\partial x^2} - 2\frac{\partial^2\phi}{\partial x\partial y}\frac{\partial^2 w}{\partial x\partial y} + \frac{\partial^2\phi}{\partial x^2}\frac{\partial^2 w}{\partial y^2} = -\frac{q(x, y)}{h} \\ &\nabla^4\phi = E\left[\left(\frac{\partial^2 w}{\partial x\partial y}\right)^2 - \frac{\partial^2 w}{\partial x^2}\frac{\partial^2 w}{\partial y^2}\right] \end{aligned}\right\} \tag{15.96}$$

Plates described by these equations are called *absolutely flexible plates*, or *membranes.*

If an initial large tension is created in a membrane, then the membrane stresses—the stresses in the middle surface—are independent of the deflections, and the analysis can be based on the following simplified system of equations:

$$\left.\begin{aligned} &\frac{\partial^2\phi}{\partial y^2}\frac{\partial^2 w}{\partial x^2} - 2\frac{\partial^2\phi}{\partial x\partial y}\frac{\partial^2 w}{\partial x\partial y} + \frac{\partial^2\phi}{\partial x^2}\frac{\partial^2 w}{\partial y^2} = -\frac{q(x, y)}{h} \\ &\nabla^4\phi = 0 \end{aligned}\right\} \tag{15.97}$$

In these equations, the stress function can also be determined independently of the deflections, as in the case of plates of small deflections. In a special case in which $T_{xy} = 0$ and $T_x = T_y = T = \text{const}$, the preceding system of equations reduces to the following simple equation for the deflection function w:

$$\frac{\partial^2 w}{\partial x^2} + \frac{\partial^2 w}{\partial y^2} = -\frac{q(x, y)}{T} \tag{15.98}$$

This equation was first obtained and analyzed by L. Euler.

As an example, let us show how Eqs. (15.95) of the plates of small deflection can be applied to the problem examined in Section 15.2.10. If Eqs. (15.95) are used, then the dimensionless stress function $\bar{\phi}$ can be sought in the form

$$\bar{\phi} = D_1 x^2 + D_2 y^2$$

where the constants D_1 and D_2 can be determined from the conditions (15.85) of the nondeformability of the plate's contour. Note that, whereas the theory of flexible plates of large deflection "automatically" averages the values of these constants over the plate's edges, the theory of plates of small deflection gives different values of these constants for different points on the plate's edges. Averaging the first equation in Eqs. (15.85) along the y axis, and the second equation along the x axis, we find the expression for the function $\bar{\phi}$ as follows:

$$\bar{\phi}(x, y) = \frac{3\pi^4 D}{4h^3 a^2} [(\nu + \gamma^2)x^2 + (1 + \nu\gamma^2)y^2]$$

where $\gamma = a/b$ is the plate's aspect ratio. Then we have

$$\mu_0 = \frac{3}{2h^2} \frac{1 + 2\nu\gamma^2 + \gamma^4}{(1 + \gamma^2)^2}$$

For a square plate ($\gamma = 1$),

$$\mu_0 = \frac{3}{4h^2}(1 + \nu)$$

The stresses obtained on the basis of the theory of plates of small deflections are as follows:

$$(\sigma_x)_{\max} = \frac{6\pi^2 D}{h^2 a^2} f(1 + \nu\gamma^2)\left(1 + \frac{f}{4h}\right) \qquad x = y = 0$$

$$(\sigma_y)_{\max} = \frac{6\pi^2 D}{h^2 a^2} f(\nu + \gamma^2)\left(1 + \frac{f}{4h}\right) \qquad x = y = 0$$

$$(\tau_{xy})_{\max} = -\frac{6\pi^2 D}{h^2 a^2} f\gamma(1 - \nu) \qquad x = \pm\frac{a}{2} \quad y = \pm\frac{b}{2}$$

The results obtained on the basis of this theory lie between the results predicted by the theory of absolutely stiff plates and the theory of flexible plates of large deflections.

15.3 BENDING OF CIRCULAR PLATES

15.3.1 Basic Equations

Structural elements that can be treated as circular plates are quite often encountered in microelectronic and fiber-optic systems. Silicon wafers and pads of plated through-holes are just two examples of such elements. In this section, we present the basic equations and relationships for circular plates.

The von Kármán equations of large deflections of circular plates can be presented in the form similar to Eqs. (15.77) for rectangular plates, as follows:

$$\left.\begin{aligned} D\nabla^4 w &= hL(w, \phi) + q \\ \frac{1}{E}\nabla^4\phi &= -\frac{1}{2}L(w, w) \end{aligned}\right\} \qquad (15.99)$$

where the operators ∇^4 and L, expressed in polar coordinates, are

$$\left.\begin{aligned} \nabla^4 &= \left(\frac{\partial^2}{\partial r^2} + \frac{1}{r}\frac{\partial}{\partial r} + \frac{1}{r^2}\frac{\partial^2}{\partial \theta^2}\right)^2 \\ L &= \frac{\partial^2 w}{\partial r^2}\left(\frac{1}{r}\frac{\partial \phi}{\partial r} + \frac{1}{r^2}\frac{\partial^2 \phi}{\partial \theta^2}\right) - 2\frac{\partial}{\partial r}\left(\frac{1}{r}\frac{\partial \phi}{\partial \theta}\right)\frac{\partial}{\partial r}\left(\frac{1}{r}\frac{\partial w}{\partial \theta}\right) \\ &\quad + \left(\frac{1}{r}\frac{\partial w}{\partial r} + \frac{1}{r^2}\frac{\partial^2 w}{\partial \theta^2}\right)\frac{\partial^2 \phi}{\partial r^2} \end{aligned}\right\} \qquad (15.100)$$

These operators are analogous to operators (15.14) and (15.76) in rectangular coordinates. In Section 7.1, we dealt with the biharmonic operator ∇^4 and examined the basic relationships of the two-dimensional problem of the elasticity theory in polar coordinates.

The bending moments M_r and M_θ in the radial and circumferential directions and the twisting moment $M_{r\theta}$ are expressed as follows (Fig. 15.11):

$$\left.\begin{aligned} M_r &= -D\left[\frac{\partial^2 w}{\partial r^2} + \nu\left(\frac{1}{r}\frac{\partial w}{\partial r} + \frac{1}{r^2}\frac{\partial^2 w}{\partial \theta^2}\right)\right] \\ M_\theta &= -D\left[\frac{1}{r}\frac{\partial w}{\partial r} + \frac{1}{r^2}\frac{\partial^2 w}{\partial \theta^2} + \nu\frac{\partial^2 w}{\partial r^2}\right] \\ M_{r\theta} &= M_{\theta r} = -D(1-\nu)\left(\frac{1}{r}\frac{\partial w}{\partial \theta}\right) \end{aligned}\right\} \qquad (15.101)$$

These formulas are analogous to Eqs. (15.8) and (15.13) in rectangular coordinates.

The shearing forces acting in the cross sections perpendicular to the plate's radius are

$$N_r = -D\frac{\partial}{\partial r}(\nabla^2 w) \qquad N_\theta = -D\frac{1}{r}\frac{\partial}{\partial \theta}(\nabla^2 w) \qquad (15.102)$$

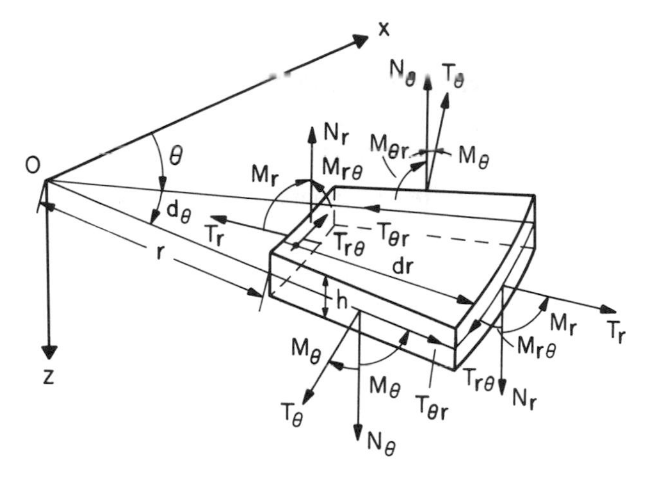

FIGURE 15.11 Variations in moments, forces, and stresses over an element of a flat plate in polar coordinates.

where

$$\nabla^2 = \frac{\partial^2}{\partial r^2} + \frac{1}{r}\frac{\partial}{\partial r} + \frac{1}{r^2}\frac{\partial^2}{\partial \theta^2}$$

is the Laplacean operator in polar coordinates. The formulas (15.102) are analogous to the formulas (15.16).

The membrane stresses acting in the middle surface of the plate are related to the stress function ϕ by the equations

$$\sigma_r^0 = \frac{1}{r}\frac{\partial \phi}{\partial r} + \frac{1}{r^2}\frac{\partial^2 \phi}{\partial \theta^2} \qquad \sigma_\theta^0 = \frac{\partial^2 \phi}{\partial r^2} \qquad \tau_{r\theta}^0 = -\frac{\partial}{\partial r}\left(\frac{1}{r}\frac{\partial \phi}{\partial \theta}\right) \tag{15.103}$$

These equations were introduced in Section 7.1 in connection with the two-dimensional problem of the elasticity theory in polar coordinates. The bending stresses are given by the formulas

$$\sigma_r = \frac{12M_r}{h^3}z \qquad \sigma_\theta = \frac{12M_\theta}{h^3}z \qquad \tau_{r\theta} = \frac{12M_{r\theta}}{h^3}z \tag{15.104}$$

The maximum bending stresses occur at the extreme points of the cross sections and are as follows:

$$\sigma_r = \frac{6M_r}{h^2} \qquad \sigma_\theta = \frac{6M_\theta}{h^2} \qquad \tau_{r\theta} = \frac{6M_{r\theta}}{h^2} \tag{15.105}$$

The displacements u and v in the radial and the circumferential directions are expressed as

$$u = u^0 - z\frac{\partial w}{\partial r} \qquad v = v^0 - z\frac{1}{r}\frac{\partial w}{\partial \theta} \tag{15.106}$$

where u^0 and v^0 are displacements of the points located on the middle surface of the plate, and

$$\alpha_r = \frac{\partial w}{\partial r} \qquad \alpha_\theta = \frac{1}{r}\frac{\partial w}{\partial \theta} \tag{15.107}$$

are the angles of rotation of the normal in the corresponding directions. The curvatures in the radial and the circumferential directions are

$$\frac{1}{\rho_r} = -\frac{\partial^2 w}{\partial r^2} \qquad \frac{1}{\rho_\theta} = -\frac{1}{r}\frac{\partial w}{\partial r} - \frac{1}{r^2}\frac{\partial^2 w}{\partial \theta^2} \tag{15.108}$$

These formulas are analogous to Eqs. (15.11).

The normal strains ε_r, ε_θ and the shearing strain $\gamma_{r\theta}$ are expressed as

$$\left.\begin{aligned} \varepsilon_r &= \varepsilon_r^0 - z\frac{\partial^2 w}{\partial r^2} \qquad \varepsilon_\theta = \varepsilon_\theta^0 - z\left(\frac{1}{r}\frac{\partial w}{\partial r} + \frac{1}{r^2}\frac{\partial^2 w}{\partial \theta^2}\right) \\ \gamma_{r\theta} &= \gamma_{r\theta}^0 - 2z\frac{\partial}{\partial r}\left(\frac{1}{r}\frac{\partial w}{\partial \theta}\right) \end{aligned}\right\} \tag{15.109}$$

where

$$\left.\begin{aligned} \varepsilon_r^0 &= \frac{\partial u^0}{\partial r} + \frac{1}{2}\left(\frac{\partial w}{\partial r}\right)^2 \\ \varepsilon_\theta^0 &= \frac{1}{r}\frac{\partial v^0}{\partial \theta} + \frac{1}{2}\left(\frac{1}{r}\frac{\partial w}{\partial \theta}\right)^2 \\ \gamma_{r\theta}^0 &= \frac{1}{r}\frac{\partial u^0}{\partial \theta} + \frac{\partial v^0}{\partial r} + \frac{1}{r}\frac{\partial w}{\partial r}\frac{\partial w}{\partial \theta} \end{aligned}\right\} \tag{15.110}$$

are the normal and the shearing strains in the middle surface of the plate.

In an important special case of an axisymmetric loading, all the relationships become independent of the angle θ. In this case, Eqs. (15.99) are as follows:

$$\left.\begin{aligned} D\nabla^4 w &= \frac{h}{r}\frac{d}{dr}\left(\frac{d\phi}{dr}\frac{dw}{dr}\right) + q \\ \frac{1}{E}\nabla^4\phi &= -\frac{1}{r}\frac{dw}{dr}\frac{d^2w}{dr^2} \end{aligned}\right\} \qquad (15.111)$$

and the biharmonic operator ∇^4 is

$$\nabla^4 = \left(\frac{d^2}{dr^2} + \frac{1}{r}\frac{d}{dr}\right)^2 = \frac{1}{r}\frac{d}{dr}\left\{r\frac{d}{dr}\left[\frac{1}{r}\frac{d}{dr}\left(r\frac{d}{dr}\right)\right]\right\} \qquad (15.112)$$

The shearing moment in the case of axisymmetric bending is zero, and the bending moments are expressed as

$$\left.\begin{aligned} M_r &= -D\left(\frac{d^2w}{dr^2} + \frac{v}{r}\frac{dw}{dr}\right) \\ M_\theta &= -D\left(\frac{1}{r}\frac{dw}{dr} + v\frac{d^2w}{dr^2}\right) \end{aligned}\right\} \qquad (15.113)$$

The membrane stresses are

$$\sigma_r^0 = \frac{1}{r}\frac{d\phi}{dr} \qquad \sigma_\theta^0 = \frac{d^2\phi}{dr^2} \qquad \tau_{r\theta}^0 = 0 \qquad (15.114)$$

and the maximum bending stresses can be calculated by the formulas

$$\sigma_r = \frac{6M_r}{h^2} \qquad \sigma_\theta = \frac{6M_\theta}{h^2} \qquad (15.115)$$

The shearing stress is zero for the axisymmetric bending.

If the deflections are small, then the membrane stresses can be neglected, and Eqs. (15.111) reduce to the following simple equation:

$$D\nabla^4 w = q \qquad (15.116)$$

which is analogous to Eq. (15.14) for rectangular plates.

15.3.2 Large Axisymmetric Deflections under Uniform Pressure

Let a circular plate be loaded by a uniformly distributed lateral load q. Presenting the deflection function w and the stress function ϕ in the form

$$w = f\bar{w}(r) \qquad \phi = f^2\bar{\phi}(r) \tag{15.117}$$

and introducing these equations in the second equation in Eqs. (15.111), we have

$$\frac{1}{E}\nabla^4\bar{\phi} = -\frac{1}{r}\, d\,\frac{\bar{w}}{dr}\frac{d^2\bar{w}}{dr^2} \tag{15.118}$$

The dimensionless deflection function $\bar{w}$ can be sought as

$$\bar{w} = 1 + C_1 r^2 + C_2 r^4 \tag{15.119}$$

After substituting this equation in Eq. (15.18), we obtain the following equation for the function $\bar{\phi}$:

$$\nabla^4\bar{\phi} = -4E(C_1^2 + 8C_1C_2r^2 + 12C_2^2r^4) \tag{15.120}$$

The solution to this equation can be sought in the form

$$\bar{\phi} = B_1r^2 + B_2r^4 + B_3r^6 + B_4r^8 \tag{15.121}$$

Introducing this solution in Eq. (15.120), we find

$$B_2 = -\frac{E}{16}C_1^2 \qquad B_3 = -\frac{E}{18}C_1C_2 \qquad B_4 = -\frac{E}{48}C_2^2 \tag{15.122}$$

For a simply supported plate, the functions $\bar{w}$ and $\bar{\phi}$ must satisfy the following boundary conditions at the contour of the plate:

$$\bar{w} = 0 \qquad \frac{d^2\bar{w}}{dr^2} + \frac{\nu}{r}\frac{d\bar{w}}{dr} = 0 \tag{15.123}$$

$$\frac{d^2\bar{\phi}}{dr^2} - \frac{\nu}{r}\frac{d\bar{\phi}}{dr} = 0 \tag{15.124}$$

The preceding conditions for the dimensionless deflection function $\bar{w}$ indicate that this function must be zero at the contour of the plate, as is the bending moment M_r. These conditions result in the following formulas for the constants C_1 and C_2:

$$C_1 = -\frac{2}{r_0^2}\frac{3+\nu}{5+\nu} \qquad C_2 - \frac{1}{r_0^4}\frac{1+\nu}{5+\nu} \tag{15.125}$$

where r_0 is the radius of the plate. The condition (15.124) for the stress function $\bar{\phi}$ reflects the fact that the contour of the plate is nondeformable. Indeed, the condition

$$u|_{r=r_0} = 0$$

and Eq. (7.3) for the circumferential normal strain

$$\varepsilon_\theta^0 = \frac{u}{r} = \frac{1}{E}(\sigma_\theta^0 - \nu\sigma_r^0) = \frac{1}{E}\left(\frac{d^2\phi}{dr^2} - \frac{\nu}{r}\frac{d\phi}{dr}\right)$$
$$= \frac{1}{E}f^2\left(\frac{d^2\bar{\phi}}{dr^2} - \frac{\nu}{r}\frac{d\bar{\phi}}{dr}\right)$$

result in the condition (15.124).

Introducing Eq. (15.121) into the condition (15.124), we find

$$B_1 = \frac{Er_0^2}{24(1-\nu)}[3(3-\nu)C_1^2 + 4(5-\nu)C_1C_2r_0^2 + 2(7-\nu)C_2^2r_0^4]$$

Then the function $\bar{\phi}$ is as follows:

$$\bar{\phi}(r) = \frac{E}{144(1-\nu)}\{6r_0^2[3(3-\nu)C_1^2 + 4(5-\nu)C_1C_2r_0^2 + 2(7-\nu)C_2^2r_0^4]r^2$$
$$-(1-\nu)(9C_1^2r^4 + 8C_1C_2r^6 + 3C_2^2r^8)\} \tag{15.126}$$

With the formulas (15.117), the first equation in Eqs. (15.111) yields

$$Df\nabla^4\bar{w} = \frac{h}{r}f^3\frac{d}{dr}\left(\frac{d\bar{\phi}}{dr}\frac{d\bar{w}}{dr}\right) + q \tag{15.127}$$

Using the procedure of the Bubnov–Galerkin method, we introduce the expressions (15.119) and (15.126) into (15.127), multiply the obtained equation by $\bar{w}$, and integrate over the surface of the plate. Then we have

$$f + \mu f^3 = A \tag{15.128}$$

where

$$\mu = 0.0251 \frac{(1-\nu)(31 + 11\nu + \nu^2)}{(6+\nu)h^2} \tag{15.129}$$

$$A = \frac{3(1-\nu)(5+\nu)r_0^4}{16Eh^3} q = \frac{(5+\nu)r_0^4}{64(1+\nu)D} q \tag{15.130}$$

Obviously, the formula (15.130) gives the maximum deflection in the center of an absolutely rigid plate. The second term in Eq. (15.128) reflects the nonlinear effect of the membrane stresses. This effect can be neglected if the value μf^2 is significantly smaller than unity. If, for instance, the maximum deflection is equal to 0.2 of the plate thickness, then, assuming that $\nu = 0.3$, we obtain $\mu f^2 = 0.00384$; i.e., the error due to neglecting membrane stresses is as little as about 0.4 percent. The membrane stresses in the problem in question, calculated using formulas (15.114), are

$$\sigma_r^0 = \frac{1}{r}\frac{d\phi}{dr} = \frac{Ef^2}{12(1-\nu)}\{r_0^2[3(3-\nu)C_1^2 + 4(5-\nu)C_1C_2r_0^2 + 2(7-\nu)C_2^2r_0^4]$$
$$-(1-\nu)(3C_1^2 + 4C_1C_2r^2 + 2C_2^2r^4)r^2\}$$

$$\sigma_\theta^0 = \frac{d^2\phi}{dr^2} = \frac{Ef^2}{12(1-\nu)}\{r_0^2[3(3-\nu)C_1^2 + 4(5-\nu)C_1C_2r_0^2 + 2(7-\nu)C_2^2r_0^4]$$
$$-(1-\nu)(9C_1^2 + 20C_1C_2r^2 + 14C_2^2r^4)r^2\}$$

and the bending moments acting in the radial and circumferential cross sections are

$$M_r = M_\theta = \frac{4D}{r_0^2} f \frac{(1+\nu)(3+\nu)}{5+\nu}\left(1 - \frac{r^2}{r_0^2}\right)$$

The maximum stresses at the center of the plate due to the combined action of the membrane and the bending stresses are

$$(\sigma_r)_{\max} = (\sigma_\theta)_{\max} = \frac{2Eh(3+\nu)}{(1-\nu)(5+\nu)r_0^2} f\left[1 + \frac{f}{h}\frac{109 - \nu - 17\nu^2 - 3\nu^3}{12(3+\nu)(5+\nu)}\right] \tag{15.131}$$

Clearly, in the absence of the membrane stresses, the total stresses are due to bending only and are as follows:

$$(\sigma_r)_{\max} = (\sigma_\theta)_{\max} = \frac{2Eh(3+\nu)}{(1-\nu)(5+\nu)r_0^2} f$$

In a linear case (absolutely rigid plate), when the maximum deflection is given by Eq. (15.130), this formula yields

$$(\sigma_r)_{\max} = (\sigma_\theta)_{\max} = \frac{3(3+\nu)r_0^2}{8h^2} q \tag{15.132}$$

Note that the second term in the brackets in formula (15.131) is equal to 0.102 for $\nu = 0.3$ and $f/h = 0.2$. Hence, as shown earlier for the rectangular plates, the consideration of the effect of the membrane stresses is substantially more important for the stresses than for the deflections.

In the case of a clamped plate, the deflection function should satisfy the conditions

$$w = 0 \qquad \frac{dw}{dr} = 0 \qquad \text{for } r = r_0$$

and the stress function the condition (15.124). Then the deflection function and the stress function arc as follows:

$$w = f\left(1 - \frac{r^2}{r_0^2}\right)^2$$

$$\phi = \frac{Ef^2}{144}\left(12\frac{5-3\nu}{1-\nu}\frac{r^2}{r_0^2} - 36\frac{r^4}{r_0^4} + 16\frac{r^6}{r_0^6} - 3\frac{r^8}{r_0^8}\right)$$

The equation for the maximum deflection is

$$f + \mu f^3 = A$$

where

$$\mu = \frac{(1+\nu)(23-9\nu)}{56} \tag{15.133}$$

and

$$A = \frac{3(1-\nu^2)}{16}\frac{q}{E}\left(\frac{r_0}{h}\right)^4 = \frac{qr_0^4}{64Dh} \tag{15.134}$$

The membrane stresses are expressed by the formulas

$$\left.\begin{aligned} \sigma_r^0 &= \frac{Ef^2}{6r_0^2}\left(\frac{5-3\nu}{1-\nu} - 6\frac{r^2}{r_0^2} + 4\frac{r^4}{r_0^4} - \frac{r^6}{r_0^6}\right) \\ \sigma_\theta^0 &= \frac{Ef^2}{6r_0^2}\left(\frac{5-3\nu}{1-\nu} - 18\frac{r^2}{r_0^2} + 20\frac{r^4}{r_0^4} - 7\frac{r^6}{r_0^6}\right) \end{aligned}\right\} \tag{15.135}$$

In the center of the plate ($r = 0$),

$$\sigma_r^0 = \sigma_\theta^0 = \frac{Ef^2}{6r_0^2}\frac{5-3\nu}{1-\nu} \tag{15.136}$$

On the contour ($r = r_0$),

$$\sigma_r^0 = \frac{Ef^2}{3(1-\nu)r_0^2} \qquad \sigma_\theta^0 = \nu\frac{Ef^2}{3(1-\nu)r_0^2} \tag{15.137}$$

The bending moments are expressed as follows:

$$\left.\begin{aligned} M_r &= 4D\frac{f}{r_0^2}\left[(1+\nu) - (3+\nu)\frac{r^2}{r_0^2}\right] \\ M_\theta &= 4D\frac{f}{r_0^2}\left[(1+\nu) - (1+3\nu)\frac{r^2}{r_0^2}\right] \end{aligned}\right\} \tag{15.138}$$

In the center of the plate ($r = 0$), the bending moments and the corresponding stresses are

$$M_r = M_\theta = 4D\frac{f}{r_0^2}(1+\nu) \qquad \sigma_r^b = \sigma_\theta^b = \frac{2Ehf}{(1-\nu)r_0^2} \tag{15.139}$$

On the contour ($r = r_0$),

$$\left.\begin{aligned} M_r &= -8D\frac{f}{r_0^2} \qquad & M_\theta &= -8\nu D\frac{f}{r_0^2} \\ \sigma_r &= -\frac{4Ehf}{(1-\nu^2)r_0^2} \qquad & \sigma_\theta &= -\frac{4\nu Ehf}{(1-\nu^2)r_0^2} \end{aligned}\right\} \tag{15.140}$$

In the second volume, we will show how these solutions can be used to evaluate the elastic constants of materials from the measured forces and deflections.

15.3.3 Axisymmetric Bending of Absolutely Rigid Plates

In many cases, the deflections of plates are sufficiently small, so that nonlinear analyses of the kind examined previously are not necessary. In this section, we will deal with absolutely rigid plates, i.e., with plates experiencing small linear deflections. The case of a uniformly distributed lateral loading was examined previously as a special case of large deflections of flexible plates. Here we examine some additional problems associated with axisymmetric bending of absolutely rigid circular plates. The solutions to these problems can be of help in the analysis of some structural elements in microelectronic and fiber-optic systems.

Examine a plate with a hole in the center (Fig. 15.12). Let this plate be subjected to the bending moments M_1 and M_2 uniformly distributed along the inner and the outer boundaries, respectively.

We proceed from the first equation in Eqs. (15.102). In the case in question, the force N_r is equal to zero, and the Laplacean operator is a function of the radius only, so that the differential equation of bending is

$$\frac{d}{dr}\left(\frac{d^2w}{dr^2} + \frac{1}{r}\frac{dw}{dr}\right) = 0$$

or

$$\frac{d^2w}{dr^2} + \frac{1}{r}\frac{dw}{dr} = C_1$$

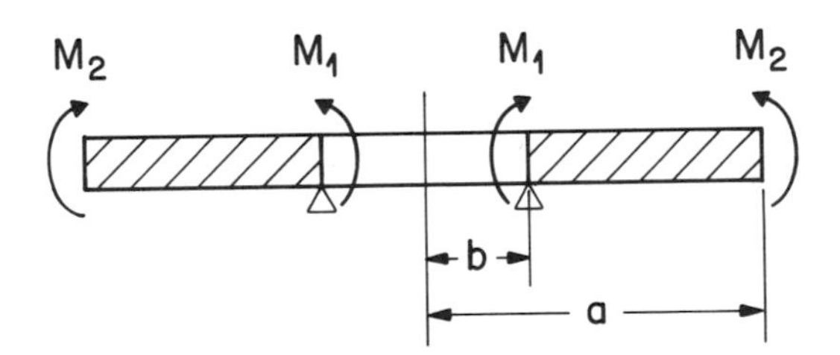

FIGURE 15.12 Circular ring subjected to the bending moments distributed over the boundaries.

where C_1 is the constant of integration. This equation can be rewritten as follows:

$$\frac{d}{dr}\left(r\frac{dw}{dr}\right) = C_1 r$$

Integrating this equation, we have

$$\frac{dw}{dr} = \frac{1}{2}C_1 r + \frac{C_2}{r}$$

Finally, we obtain the following equation for the deflection function:

$$w = \frac{1}{4}C_1 r^2 + C_2 \ln\frac{r}{a} + C_3 \tag{15.141}$$

where the constants C_1 and C_2 can be determined from the boundary conditions for the moments, and the constant C_3 should be determined from the boundary conditions for the deflection function w.

The constants C_1 and C_2 are determined from the boundary conditions

$$M_r = M_2 \text{ for } r = a \qquad M_r = -M_1 \text{ for } r = b \tag{15.142}$$

Using the first formula in Eqs. (15.113) for the radial bending moment, and introducing Eq. (15.141) into this formula, we have

$$M_r = D\left[\frac{C_1}{2} - \frac{C_2}{r^2} + \nu\left(\frac{C_1}{2} + \frac{C_2}{r^2}\right)\right]$$

The boundary conditions (15.142) result in the equations

$$\left.\begin{aligned} \frac{1+\nu}{2}C_1 - \frac{1-\nu}{b^2}C_2 &= \frac{M_1}{D} \\ \frac{1+\nu}{2}C_1 - \frac{1-\nu}{a^2}C_2 &= \frac{M_2}{D} \end{aligned}\right\}$$

from which

$$C_1 = \frac{2(a^2M_1 - b^2M_2)}{(1+\nu)D(a^2 - b^2)} \qquad C_2 = \frac{a^2b^2(M_2 - M_1)}{(1-\nu)D(a^2 - b^2)} \tag{15.143}$$

If the plate is simply supported along the inner contour, then the formula (15.141) yields

$$C_3 = \frac{1}{4} C_1 b^2 - C_2 \ln \frac{b}{a}$$

In a special case, when $M_2 = 0$, we have

$$\left. \begin{aligned} C_1 &= \frac{2a^2 M_1}{(1+\nu)D(a^2-b^2)} \qquad C_2 = -\frac{a^2 b^2 M_1}{(1-\nu)D(a^2-b^2)} \\ C_3 &= \frac{a^2 b^2 M_1}{D(a^2-b^2)} \left[\frac{1}{2(1+\nu)} + \frac{1}{1-\nu} \ln \frac{b}{a} \right] \end{aligned} \right\} \tag{15.144}$$

These results can be used to model the mechanical behavior of pads in plated through-hole structures.

If the plate is simply supported along the outer contour, then the formula (15.141) yields

$$C_3 = -\frac{1}{4} C_1 a^2 = -\frac{a^2(a^2 M_2 - b^2 M_1)}{2(1+\nu)D(a^2-b^2)}$$

If, in addition, $M_2 = 0$, then

$$\left. \begin{aligned} C_1 &= \frac{2b^2 M_1}{(1+\nu)D(a^2-b^2)} \qquad C_2 = \frac{a^2 b^2 M_1}{(1-\nu)D(a^2-b^2)} \\ C_3 &= \frac{a^2 b^2 M_1}{2(1+\nu)D(a^2-b^2)} \end{aligned} \right\} \tag{15.145}$$

With the determined values of the constants of integration, the deflection function w and all the other characteristics of bending can easily be evaluated.

Now, examine a situation shown in Fig. 15.13, where a circular plate with a hole is loaded by shearing forces uniformly distributed along the inner contour. The solution to this problem can also be used for the evaluation of stresses in a pad of a plated-through hole. In this case, the forces N_0 are applied to the pad from the barrel and are due to the thermal contraction mismatch of the plated-through hole and printed-wiring-board materials.

The shearing force per unit length of the circumference of the radius r is

$$N_r = \frac{N_0 b}{r} = \frac{P}{2\pi r}$$

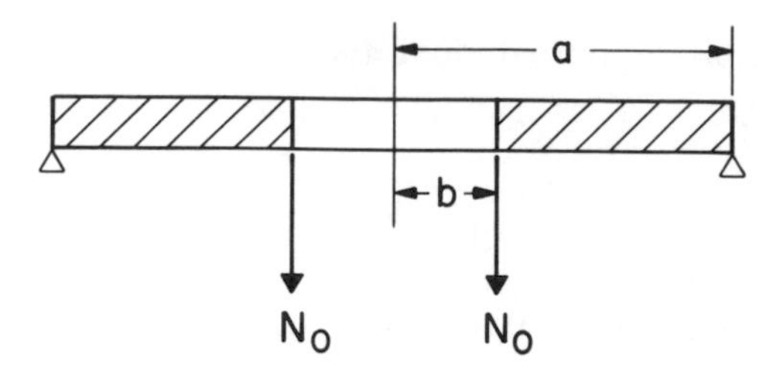

FIGURE 15.13 Circular ring subjected to the lateral forces distributed over the inner boundary.

where $P = 2\pi b\, N_0$ is the total load applied to the inner boundary. After substituting this formula into the first equation in Eqs. (15.102) and omitting, in the expression for the Laplacean operator, the term containing a derivative with respect to θ, we have the following differential equation of bending:

$$\frac{d}{dr}\left(\frac{d^2w}{dr^2} + \frac{1}{r}\frac{dw}{dr}\right) = -\frac{N_r}{D} = -\frac{P}{2\pi Dr}$$

The first integration yields

$$\frac{d^2w}{dr^2} + \frac{1}{r}\frac{dw}{dr} = -\frac{P}{2\pi D}\ln\frac{r}{a} + C_1$$

The next integration results in the formula

$$\frac{dw}{dr} = -\frac{Pr}{8\pi D}\left(2\ln\frac{r}{a} - 1\right) - \frac{1}{2}C_1 r - \frac{C_2}{r}$$

so that

$$w = -\frac{Pr^2}{8\pi D}\left(\ln\frac{r}{a} - 1\right) - \frac{1}{4}C_1 r^2 - C_2\ln\frac{r}{a} + C_3 \qquad (15.146)$$

The constants of integration can be determined from the boundary conditions on the contours of the plate. These conditions are

$$w = 0 \qquad M_r = -D\left(\frac{d^2w}{dr^2} + \frac{\nu}{r}\frac{dw}{dr}\right) = 0 \qquad \text{for } r = a$$

and

$$M_r = -D\left(\frac{d^2w}{dr^2} + \frac{\nu}{r}\frac{dw}{dr}\right) = 0 \qquad \text{for } r = b$$

These conditions reflect the fact that the plate is simply supported along the outer contour and that there are no bending moments acting at the inner contour. After substituting the obtained equation for the deflection function into the boundary conditions, we obtain three algebraic equations for the unknowns C_1, C_2, and C_3. Solving these equations, we find

$$\left.\begin{aligned} C_1 &= -\frac{P}{2\pi D}\left(\frac{1-\nu}{1+\nu} - \frac{2b^2}{a^2-b^2}\ln\frac{b}{a}\right) \\ C_2 &= \frac{(1+\nu)P}{4\pi(1-\nu)D}\frac{a^2b^2}{a^2-b^2}\ln\frac{b}{a} \\ C_3 &= -\frac{Pa^2}{8\pi D}\left(1 + \frac{1}{2}\frac{1-\nu}{1+\nu} - \frac{b^2}{a^2-b^2}\ln\frac{b}{a}\right) \end{aligned}\right\}$$

Thus, the deflection function is determined.

As a final example, examine a plate under a concentrated load applied at its center (Fig. 15.14). The deflection surface of the plate is sought in the form

$$w = C_1 r^2 \ln r + C_2 r^2 + C_3 \tag{15.147}$$

In the case of a simply supported plate, the constants C_1, C_2, and C_3 should be determined from the boundary conditions

$$w = 0 \qquad M_r = 0 \qquad \text{for } r = a$$

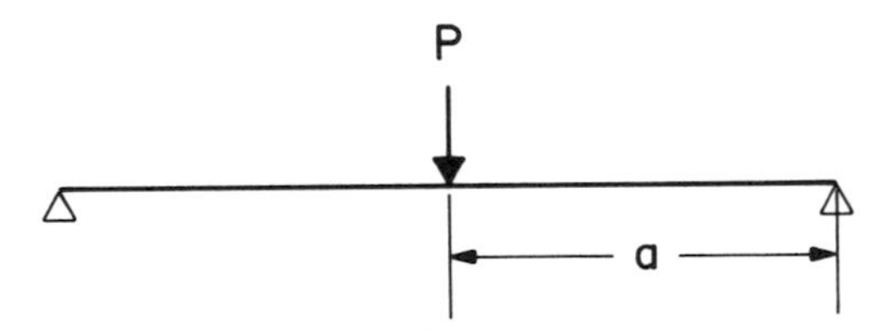

FIGURE 15.14 Circular plate under concentrated load applied at its center.

and the equation

$$N_r = -\frac{P}{2\pi r}$$

These relationships result in the formula

$$w = \frac{P}{16\pi D}\left[2r^2 \ln\frac{r}{a} + \frac{3+v}{1+v}(a^2 - r^2)\right] \tag{15.148}$$

The maximum deflection in the center ($r = 0$) is

$$w_{\max} = \frac{Pa^2}{16\pi D}\frac{3+v}{1+v} \tag{15.149}$$

The radial and the circumferential stresses are

$$\left.\begin{aligned} \sigma_r &= \frac{3P}{2\pi h^2}(1+v)\ln\frac{a}{r} \\ \sigma_\theta &= \frac{3P}{2\pi h^2}\left[(1+v)\ln\frac{a}{r} + 1 - v\right] \end{aligned}\right\} \tag{15.150}$$

In the case of a clamped edge, the deflection function can still be sought in the form (15.147); however, the boundary conditions are

$$w = 0 \qquad \frac{\partial w}{\partial r} = 0 \qquad \text{for } r = a$$

After determining the constants of integration and substituting the obtained results in Eq. (15.147), we obtain

$$w = \frac{P}{16\pi D}\left(2r^2 \ln\frac{r}{a} + a^2 - r^2\right) \tag{15.151}$$

The maximum deflection is

$$w_{\max} = \frac{Pa^2}{16\pi D} \tag{15.152}$$

and the stresses are as follows:

$$\left.\begin{aligned}\sigma_r &= \frac{3P}{2\pi h^2}\left[(1+\nu)\ln\frac{a}{r} - 1\right] \\ \sigma_\theta &= \frac{3P}{2\pi h^2}\left[(1+\nu)\ln\frac{a}{r} - \nu\right]\end{aligned}\right\} \tag{15.153}$$

As we can see from Eqs. (15.150) and (15.153), the stresses grow without limit as the radius r approaches zero. Analysis by a more elaborate method indicates, however, that the actual stress caused by a force P on a very small area of radius $r_c < 1.7h$ can be obtained by replacing the original r_c by an equivalent radius r_e, given by the approximate formula

$$r_e = \sqrt{1.6r_c^2 + h^2} - 0.675h$$

16

Buckling

16.1 BUCKLING OF BARS

16.1.1 Euler's Formulas

Elastic buckling is a phenomenon that defines a failure mode in which the system "collapses" without necessarily having any portion of the structure exceeding the yield strength of the material. When the load is released, the system returns to its unstressed condition and is undamaged.

The elastic stability of single-span bars (columns) was first studied by Euler in the middle of the eighteenth century. Examine a beam subjected to compressive loads T (Fig. 16.1a). The approach used by Euler is as follows. Suppose that there is a small lateral disturbance caused by an initial curvature of the beam, as shown in Fig. 16.1b. Then the forces T result in a bending moment

$$M = Tw$$

and the differential equation of the elastic curve is

$$EIw'' = M$$

Hence,

$$EIw'' + Tw = 0$$

or

$$EIw^{\text{IV}} + Tw'' = 0$$

This equation can be rewritten as

$$w^{\text{IV}} + k^2w'' = 0 \tag{16.1}$$

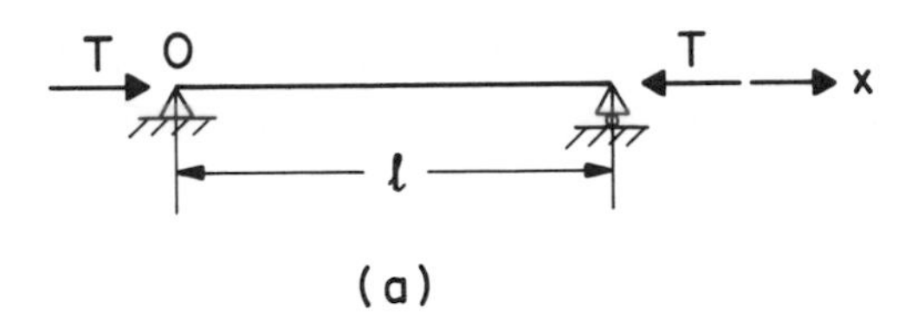

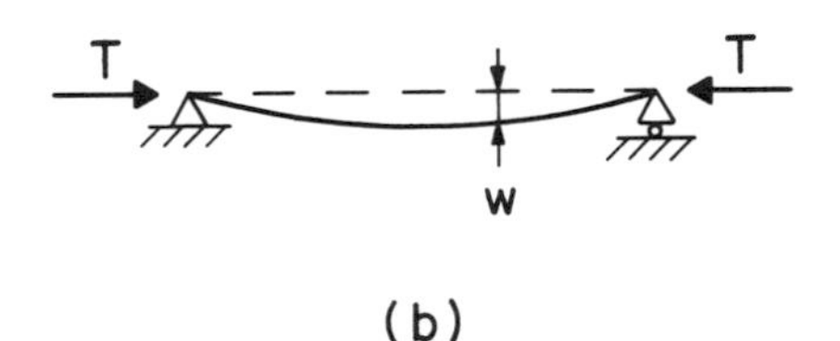

FIGURE 16.1 Beam under a pair of compressive loads: (a) before buckling; (b) after buckling.

where the eigenvalue of the problem k is

$$k = \sqrt{\frac{T}{EI}} \tag{16.2}$$

Equation (16.1) has the following solution:

$$w = C_0 + C_1 kx + C_2 \cos kx + C_3 \sin kx \tag{16.3}$$

where the constants of integration can be determined from the boundary conditions.

In the case of a simply supported bar (Fig. 16.1a), the constants C_0 and C_2 should be put equal to zero, and the constants C_1 and C_3 can be found from the boundary conditions

$$w = 0 \qquad w'' = 0$$

at the ends of the bar. These conditions are fulfilled automatically at the end $x = 0$. At the end $x = l$, they yield

$$\left.\begin{aligned} C_1 kl + C_3 \sin kl = 0 \\ k^2 C_3 \sin kl = 0 \end{aligned}\right\} .$$

The constant C_3 in the second equation cannot be equal to zero since, in this case, the constant C_1 in the first equation will also be zero, and the formula (16.3) would not describe a curved line at all. Therefore, we should admit that

$$\sin kl = 0$$

The smallest nonzero root of this equation is

$$kl = \pi$$

Using the notation (16.4), we have

$$T_e = \frac{\pi^2 EI}{l^2} \tag{16.4}$$

This is *Euler's formula* for a simply supported bar. The obtained result can be interpreted as follows. If the actual compressive force acting on the bar is smaller than T_e, then the bar deflected by a small lateral force will return to its initially straight position after this disturbance is removed. However, if the actual compressive force equals or exceeds the critical value T_e, then the bar will remain in the deflected position after the removal of the lateral disturbance.

In the case of a bar clamped at both ends, the conditions

$$w = 0 \qquad w' = 0 \qquad \text{for } x = 0 \text{ and } x = l$$

result in the equations

$$\left.\begin{aligned} C_0(1 - \cos kl) + C_1(kl - \sin kl) = 0 \\ C_0 \sin kl + C_1(1 - \cos kl) = 0 \end{aligned}\right\}$$

The constants C_1 and C_3 have nonzero values, if the determinant of this system of homogeneous algebraic equations is zero, i.e., if

$$\begin{vmatrix} 1 - \cos kl & kl - \sin kl \\ \sin kl & 1 - \cos kl \end{vmatrix} = 0$$

or

$$\left(\tan\frac{kl}{2} - \frac{kl}{2}\right)\sin\frac{kl}{2} = 0$$

The equation

$$\sin\frac{kl}{2} = 0$$

results in smaller roots than the equation

$$\tan\frac{kl}{2} - \frac{kl}{2} = 0$$

and, therefore,

$$\frac{kl}{2} = \pi a$$

This leads to the following formula for the critical force:

$$T_c = \frac{4\pi^2 EI}{l^2} \tag{16.5}$$

This force is four times larger than the critical force for a simply supported bar.

In a similar way, we can obtain the formulas for the critical (Euler's) force for other boundary conditions. This force can be presented by the following general formula:

$$T_e = \frac{\pi^2 EI}{(\mu l)^2} \tag{16.6}$$

where $\mu = 1$ for a simply supported bar, $\mu = 0.5$ for a clamped bar, $\mu = 2$ for a cantilever bar, and $\mu = 0.7$ for a clamped-pinned bar.

An electronics engineer encounters the phenomenon of buckling, for instance, in the structures of the flexible ("Euler") test probes.

16.1.2 Bars on Elastic Foundations

For a bar lying on an elastic foundation (Fig. 16.2), the governing differential equation is

$$EIw^{\mathrm{IV}} + Tw'' + Kw = 0 \tag{16.7}$$

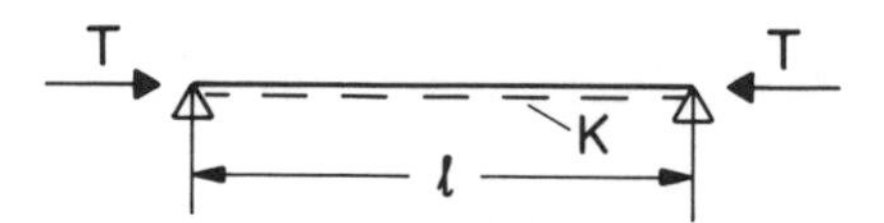

FIGURE 16.2 Beam on an elastic foundation under compressive loads.

This equation can be obtained from Eq. (16.1) by supplementing it with the term Kw, reflecting the reaction of the elastic foundation. In a simply supported bar, the boundary conditions are

$$w(0) = w(l) = 0 \qquad w''(0) = w''(l) = 0$$

In order to satisfy these conditions, we present the elastic curve of the buckled bar in the form

$$w = A \sin \frac{n\pi x}{l} \tag{16.8}$$

Introducing this formula in Eq. (16.7), we obtain

$$EI\left(\frac{n\pi}{l}\right)^4 - T\left(\frac{n\pi}{l}\right)^2 + K = 0$$

so that

$$T = \frac{\pi^2 EI}{l^2}\left(n^2 + \frac{\gamma}{n^2}\right) \tag{16.9}$$

where

$$\gamma = \frac{Kl^4}{\pi^4 EI} \tag{16.10}$$

The critical force is equal to the minimum value of the force T (see Fig. 16.3). Therefore, the integer n should be determined in such a way that the

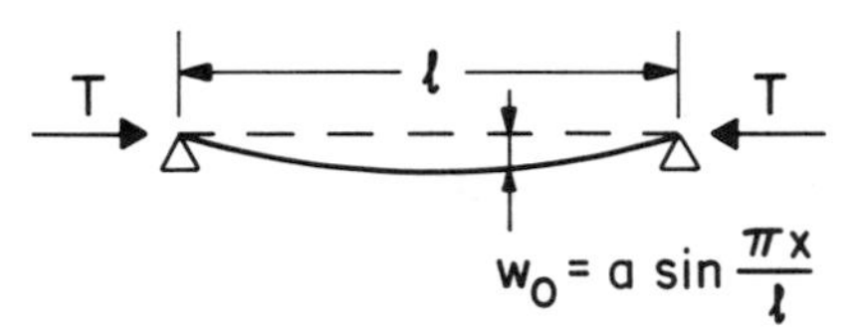

FIGURE 16.3 Initial deflection of a beam subjected to compressive loads.

expression (16.9) becomes minimum. The number n can be found from the obvious unequality

$$\left[(n-1)^2 + \frac{\gamma}{(n-1)^2}\right] > n^2 + \frac{\gamma}{n^2} < \left[(n+1)^2 + \frac{\gamma}{(n+1)^2}\right]$$

which can be simplified as follows:

$$n^2(n-1)^2 < \gamma < n^2(n+1)^2 \tag{16.11}$$

From this relationship, we conclude that $0 \leq \gamma \leq 4$ for $n = 1$, $4 \leq \gamma \leq 36$ for $n = 2$, $36 \leq \gamma \leq 144$ for $n = 3$, and so on. Hence, for the γ values within the foregoing limits, buckling will occur with one, two, three, and more "half-waves."

When the spring constant K of the foundation is large, the number of half-waves in a buckled bar is also large, and the condition (16.11) yields $\gamma = n^4$. Introducing $n^2 = \sqrt{\gamma}$ into the formula (16.9), we have

$$T = 2\sqrt{\gamma}\,\frac{\pi^2 EI}{l^2} = 2\sqrt{KEI} \tag{16.12}$$

This formula indicates that, in the case of a relatively stiff elastic foundation, the critical force becomes independent of the length of the bar. We encounter such a situation in the problem of low-temperature microbending of optical fibers.

16.1.3 The Effect of Initial Curvature

Let a compressed bar have an initial curvature given by the equation (see Fig. 16.3):

$$w_0 = f_0 \sin\frac{\pi x}{l}$$

and additional deflections

$$w_1 = f_1 \sin\frac{\pi x}{l}$$

due to compressive forces T. Then the bending moment in the cross section x is

$$M = T(w_0 + w_1)$$

The deflections w_1 due to the applied forces T can be obtained from the equation

$$EIw_1'' = -T(w_0 + w_1)$$

or

$$w_1'' + k^2 w_1 = -k^2 f_0 \sin\frac{\pi x}{l} \tag{16.13}$$

where

$$k = \sqrt{\frac{T}{EI}}$$

Equation (16.13) has the following solution:

$$w_1 = A\sin kx + B\cos kx + \frac{1}{(\pi^2/k^2 l^2) - 1} f_0 \sin\frac{\pi x}{l}$$

The boundary condition $w_1 = 0$ for $x = 0$ and $x = l$ will be satisfied if we put $A = B = 0$. Then we have

$$w_1 = \frac{f_0}{(\pi^2/k^2 l^2) - 1}\sin\frac{\pi x}{l} = \frac{f_0}{(\pi^2 EI/Tl^2) - 1}\sin\frac{\pi x}{l} = \frac{f_0}{(T_e/T) - 1}\sin\frac{\pi x}{l} \tag{16.14}$$

where T_e is the critical compressive force and T the actual compressive force. The total deflection is

$$w = w_0 + w_1 = \frac{f_0}{1 - (T/T_e)}\sin\frac{\pi x}{l} \tag{16.15}$$

As evident from this formula, an initially curved bar buckles under the same critical load as an initially straight bar. The initial curvature, however, affects the prebuckling behavior of the bar: whereas the deflections of an initially curved bar increase gradually until the compressive force becomes very close to its critical value, the initially straight bar remains straight up to the very moment of buckling. When the compressive force in an initially curved bar approaches its critical value, the deflections increase without any limit.

The buckling behavior of initially curved bars will be examined in greater detail in the second volume in connection with low-temperature microbending of optical fibers.

16.1.4 Large Deflections of Buckled Bars (the Elastica)

In Section 12.10, we examined a problem of large deflections of a cantilever beam subjected to bending. In this section, we address large deflections of buckled bars. The solution to this problem can be of help, for instance, in the evaluation of the curvature of an optical fiber of a fiber-optic-guided missile at the peel point.

In previous analyses of the buckling behavior of bars, we did not try to evaluate the actual deflection of the buckled bar. This knowledge was not necessary to determine the critical force. The deflections could have any value, provided they were small. No other conclusion could be reached on the basis of linear differential equations used in our discussions. What made these equations linear is the assumption of small deflections, which enabled us to use an approximate expression d^2w/dx^2 for the curvature of the bar. In order to be able to predict the elastic curve of a buckled bar, shown in Fig. 16.4, we use the following exact differential equation of the deflection curve:

$$EI\frac{d\theta}{ds} = -Ty \tag{16.16}$$

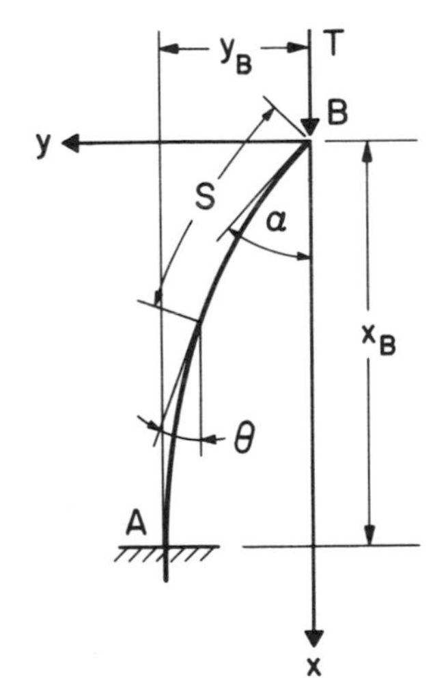

α	0°	20°	40°	60°	80°	100°	120°
T/T_e	1	1.015	1.063	1.152	1.293	1.518	0.884
x_B/l	0	0.220	0.422	0.593	0.719	0.792	0.803
y_B/l	1	0.970	0.881	0.741	0.560	0.349	0.123

FIGURE 16.4 Large deflections of a cantilever beam under a compressive end force.

where $d\theta/ds$ is the curvature of the bar.
Using the relation

$$\frac{dy}{ds} = \sin\theta$$

we obtain Eq. (16.16) in the form

$$\frac{d^2\theta}{ds^2} + k^2 \sin\theta = 0 \tag{16.17}$$

where

$$k = \sqrt{\frac{T}{EI}}$$

Equation (16.17) can be rewritten as

$$\frac{d}{ds}\left[\frac{1}{2}\left(\frac{d\theta}{ds}\right)^2 - k^2\cos\theta\right] = 0$$

and, therefore,

$$\frac{1}{2}\left(\frac{d\theta}{ds}\right)^2 - k^2\cos\theta = C \tag{16.18}$$

The constant of integration C can be determined from the conditions at the upper end of the bar, where

$$\frac{d\theta}{ds} = 0 \qquad \text{for } \theta = \alpha$$

This condition reflects the fact that there is no bending moment at $\theta = \alpha$. This gives

$$C = -k^2 \cos\alpha$$

and (16.18) yields

$$\frac{d\theta}{ds} = -k\sqrt{2(\cos\theta - \cos\alpha)}$$

The negative sign is used to indicate that the curvature $d\theta/ds$ is always negative. Thus, the radius of curvature is

$$\frac{ds}{d\theta} = -\frac{1}{k\sqrt{2(\cos\theta - \cos\alpha)}}$$

The total length of the curved bar can then be evaluated as

$$l = \int_\alpha^0 ds = \frac{1}{k}\int_0^\alpha \frac{d\theta}{\sqrt{2(\cos\theta - \cos\alpha)}} = \frac{1}{2k}\int_0^\alpha \frac{d\theta}{\sqrt{\sin^2(\alpha/2) - \sin^2(\theta/2)}}$$

Introducing a new variable ζ in such a way that

$$\sin\frac{\theta}{2} = \sin\frac{\alpha}{2}\sin\zeta$$

we can simplify the obtained integral as follows:

$$l = \frac{1}{k}\int_0^{\pi/2} \frac{d\phi}{\sqrt{1 - p^2\sin^2\phi}} = \frac{1}{k}K(p) \tag{16.19}$$

where the following notation is used:

$$p = \sin\frac{\alpha}{2}$$

$$K(p) = \int_0^{\pi/2} \frac{d\phi}{\sqrt{1 - p^2\sin^2\phi}}$$

The function $K(p)$ is known as a *complete elliptic integral of the first kind.* The values of this function are tabulated, so that the value of p (and, obviously, the value of the angle α of rotation of the top cross section) can be determined readily for any given load T. In the case of small deflections, the angle α is also small, as is the p value. Then the second term under the rest in the expression for the elliptic integral becomes small compared to unity, so that $K(p)$ becomes $\pi/2$. Then the formula (16.19) yields

$$l = \frac{\pi}{2k} = \frac{\pi}{2}\sqrt{\frac{EI}{T}}$$

Solving this formula for T, we have

$$T_e = \frac{\pi^2 EI}{4l^2}$$

which is the formula for the critical force of a cantilever bar.

The calculated ratios T/T_e for various angles α are given in the table in Fig. 16.4. The ratios x_B/l and y_B/l, which are also given in this table, can be evaluated as follows.

Since

$$dy = \sin\theta\, ds = -\frac{\sin\theta\, d\theta}{k\sqrt{2(\cos\theta - \cos\alpha)}}$$

then the total deflection of the top of the bar in the horizontal direction is

$$y_B = \frac{1}{2k}\int_0^{\alpha} \frac{\sin\theta\, d\theta}{\sqrt{\sin(2\alpha/2) - \sin^2(\theta/2)}} = \frac{2p}{k}\int_0^{\pi/2} \sin\phi\, d\phi = \frac{2p}{k}$$

The deflection of the top of the bar can be calculated by first selecting a value of α, then evaluating p and k, and finally finding y_B from the preceding equation. The coordinate x_B can be determined in a similar way, and the result is

$$x_B = \frac{2}{k}\int_0^{\pi/2} \sqrt{1 - p^2 \sin^2\phi}\, d\phi - l = \frac{2}{k} E(p) - l$$

where

$$E(p) = \int_0^{\pi/2} \sqrt{1 - p^2 \sin^2\phi}\, d\phi$$

is the *complete elliptic integral of the second kind.* Its values are also tabulated, and the obtained results can be found in many mathematical handbooks.

16.1.5 Buckling of Curved Beams

The results of the analysis of buckling of curved beams can be of help, for instance, when it is necessary to evaluate the mechanical behavior of wire bonds during an epoxy molding encapsulation process.

Let a curved beam AB of a constant curvature a be subjected to a

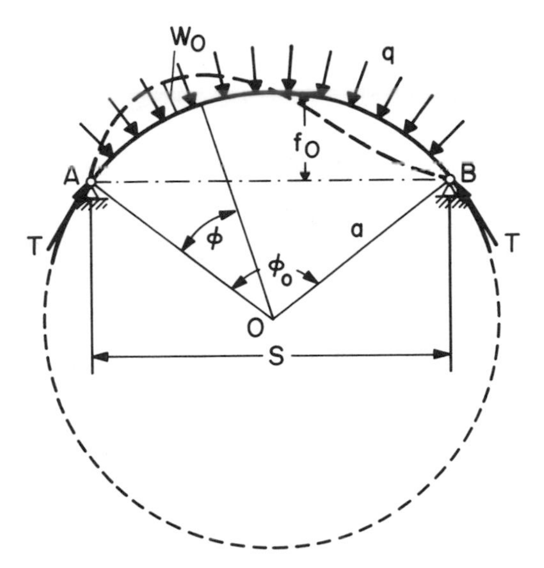

FIGURE 16.5 Curved beam subjected to lateral uniform pressure.

distributed load q, applied as shown in Fig. 16.5. This load results in reactive forces T, which can cause buckling of the beam if the load q is sufficiently great. The shape of a buckled beam is shown as a broken line.

The compressive forces T can be evaluated on the basis of the following equilibrium condition of the beam:

$$2T \sin \frac{\phi_0}{2} = 2a \int_0^{\phi_0/2} q \cos \phi \, d\phi = 2aq \sin \frac{\phi_0}{2}$$

so that

$$T = qa \tag{16.20}$$

The bending moment in a buckled beam can be evaluated as

$$M = -EI\left(\frac{1}{\rho} - \frac{1}{a}\right) = -EI\kappa \tag{16.21}$$

where EI is the flexural rigidity of the beam, ρ is the radius of curvature after pressure q is applied, and κ is the total curvature.

In order to express the radius ρ through the lateral deflections w of the beam, let us examine a situation shown in Fig. 16.6, where A_0B_0 is the initial

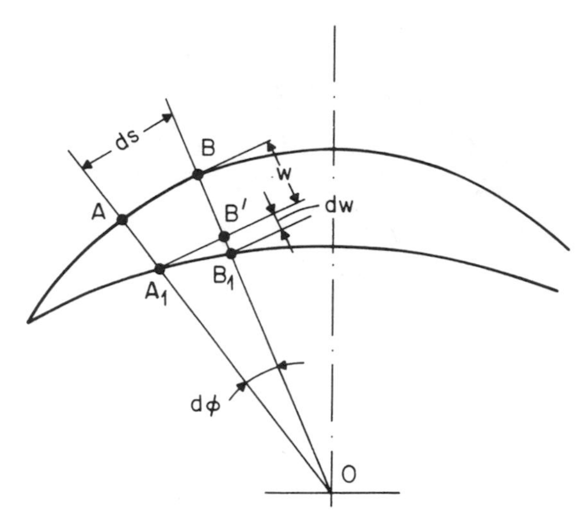

FIGURE 16.6 Initial (A_0B_0) and final (A_1B_1) position of an element of a curved beam subjected to buckling.

position of an element of the beam, and A_1B_1 is its final position after buckling. This final position is due to the uniform pressure q (which simply shortens the element A_0B_0, bringing it to the length A_1B') and to the rotation of the element A_1B' around point A_1. Accordingly, the total change in curvature consists of two components that correspond to the foregoing displacements.

The application of the uniform pressure, when no rotation is involved, results in the following change in curvature:

$$\kappa_1 = \frac{1}{a - w} - \frac{1}{a} = \frac{w}{a(a - w)} \cong \frac{w}{a^2} \tag{16.22}$$

The change in curvature due to the rotation of an element A_1B' around point A_1 is caused by the nonuniform distribution of the deflections w along the buckled beam. In accordance with the general definition, the curvature is expressed as the derivative

$$\kappa = \frac{d\theta}{ds} \tag{16.23}$$

where $d\theta$ is the angle of rotation of the cross section at point B_1 with respect to the cross section at point B'. Assuming that the distance A_1B' is equal to

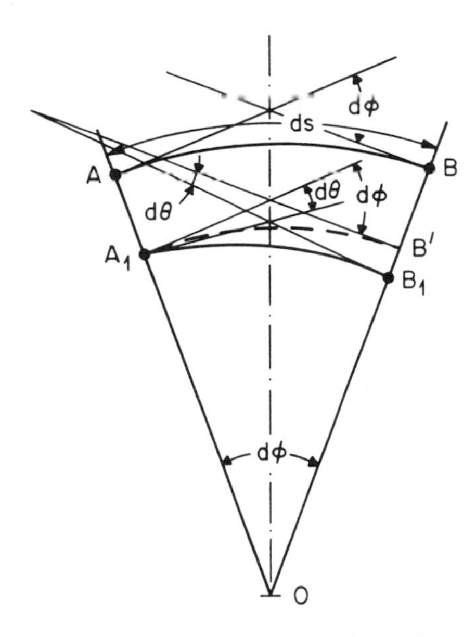

FIGURE 16.7 Evaluation of the critical lateral pressure.

the distance A_1B_1, we conclude (see Fig. 16.7) that when point A is displaced into position A_1, the tangent rotates by an angle

$$\frac{\partial w}{\partial s} = \frac{1}{a}\frac{\partial w}{\partial \phi}$$

The angle of rotation of the tangent at point B, whose distance from point A is $ds = a\,d\phi$, is

$$\frac{1}{a}\left(\frac{\partial w}{\partial \phi} + \frac{\partial^2 w}{\partial \phi^2}\,d\phi\right)$$

Then the angle between the tangents at points A_1 and B_1 is expressed by the formula

$$d\theta = \frac{1}{a}\frac{\partial^2 w}{\partial \phi^2}\,d\phi$$

In accordance with Eq. (16.23), the change in curvature due to the element rotation is

$$\kappa_2 = \frac{1}{a^2}\frac{\partial^2 w}{\partial \phi^2}$$

and the total change in curvature, with consideration of Eq. (16.22), is

$$\kappa = \kappa_1 + \kappa_2 = \frac{1}{a^2}\left(w + \frac{\partial^2 w}{\partial \phi^2}\right)$$

Using this expression, we present the bending moment given by Eq. (16.21) in the form

$$M = -\frac{EI}{a^2}\left(w + \frac{\partial^2 w}{\partial \phi^2}\right)$$

Since, on the other hand, the bending moment can be approximately presented as

$$M \cong Tw$$

we have the following equation of bending:

$$\frac{\partial^2 w}{\partial \phi^2} + k^2 w = 0 \tag{16.24}$$

where

$$k = \sqrt{1 + \frac{qa^3}{EI}} \tag{16.25}$$

The solution to Eq. (16.24) can be sought for a simply supported beam in the form

$$w = w_0 \sin\frac{2n\pi\phi}{\phi_0} \qquad n = 1, 2, 3, \ldots \tag{16.26}$$

where w_0 is the maximum deflection of a buckled beam, and the angle ϕ is counted from point A. After substituting this solution in Eq. (16.24), we find

$$k = \frac{2n\pi}{\phi_0} \qquad n = 1, 2, 3, \ldots$$

Then the formula (16.25) results in the following expression for the critical load q:

$$q_c = \frac{EI}{a^3}\left(\frac{4n^2\pi^2}{\phi_0^2} - 1\right)$$

The minimum critical load takes place for $n = 1$ and is

$$q_e = \frac{EI}{a^3}\left(\frac{4\pi^2}{\phi_0^2} - 1\right) \tag{16.27}$$

This formula can also be used for curved beams whose shape deviates from circular. In this case, the radius a and the angle ϕ_0 can be expressed through the beam's span s and the initial maximum deflection f_0 as follows:

$$a = \frac{f_0}{2}\left[1 + \left(\frac{s}{2f_0}\right)^2\right] \qquad \phi_0 = 2 \arcsin \frac{s}{2a} \tag{16.28}$$

When the deflection f_0 is small compared to the span s, we have

$$a = \frac{s^2}{8f_0} \qquad \phi_0 = \frac{8f_0}{s}$$

and the formula (16.27) yields

$$q_e = \frac{32\pi^2 f_0 EI}{s^4}$$

The compressive force, according to Eq. (16.20), is

$$T_e = q_e a = \frac{4\pi^2 EI}{s^4}$$

This load is four times greater than the critical load for a straight simply supported beam subjected to axial compression. This is due to the fact that the lowest critical force in a curved beam occurs when the deflection curve has two half-waves, i.e., the mode of the buckled initially curved beam is such that the deflection at its midpoint is zero. This requires a larger critical load than in the case of an initially straight beam.

16.2 BUCKLING OF PLATES

16.2.1 Buckling of a Rectangular Plate Uniformly Compressed in One Direction

The first major contribution in the theory of elastic stability of thin plates came in 1891 from G. H. Bryan in England. Bryan solved the problem of a rectangular plate, simply supported along the contour and compressed in one direction. He showed that such a plate buckles into a number of doubly sinusoidal hills and cups, with straight nodal lines dividing the main rectangle into a number of subrectangles in the manner of double Fourier series. Examine the Bryan solution.

We proceed from the equation (15.66) of bending of rectangular plates subjected to forces acting in their middle surface. In the absence of the lateral load, this equation yields

$$D\nabla^4 w + T_x \frac{\partial^2 w}{\partial x^2} + T_y \frac{\partial^2 w}{\partial y^2} - 2T_{xy} \frac{\partial^2 w}{\partial x \partial y} = 0 \tag{16.29}$$

where the signs of the membrane forces T_x and T_y in Eq. (15.66) were changed, since we deal here with compressive forces only. When a plate is uniformly compressed in one direction (Fig. 16.8), then $T_y = 0$, $T_{xy} = 0$, and Eq. (16.29) is

$$D\nabla^4 w + T \frac{\partial^2 w}{\partial x^2} = 0 \tag{16.30}$$

where the forces $T = T_x$ are uniformly distributed along the sides $x = 0$ and $x = a$.

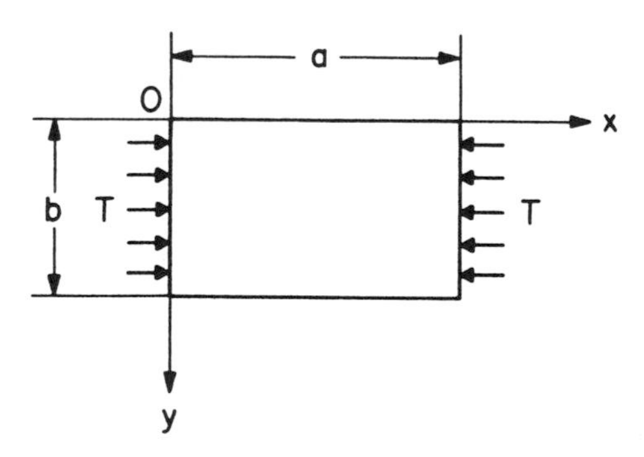

FIGURE 16.8 Rectangular plate uniformly compressed in one direction.

We seek the nonzero solution of Eq. (16.30) in the form of a double Fourier series:

$$w = \sum_{m=1}^{\infty} \sum_{n=1}^{\infty} A_{mn} \sin \alpha_m x \sin \beta_n y \tag{16.31}$$

where

$$\alpha_m = \frac{m\pi}{a} \qquad \beta_n = \frac{n\pi}{b}$$

Substituting Eq. (16.31) in Eq. (16.30) and demanding that the coefficients A_{mn} are nonzero, we conclude that the forces T must be expressed as follows:

$$T = \frac{\pi^2 D(m^2\gamma^2 + n^2)^2}{b^2 n^2 \gamma^2} \tag{16.32}$$

where $\gamma = a/b$ is the aspect ratio of the plate.

In order to determine the critical load T, we should choose the whole numbers m and n in the expression (16.32) in such a way that the load T is minimum. As far as the number of the half-waves m is concerned, the minimum of the expression (16.32) will always be at $m = 1$; i.e., the plate will always buckle with just one half-wave in the direction perpendicular to the direction of compression. Then we have

$$T = \frac{\pi^2 D}{b^2}\left(\frac{\gamma}{n} + \frac{n}{\gamma}\right)^2 \tag{16.33}$$

If a certain number n of half-waves results in the minimum of this expression, i.e., in the minimum (and actual) critical force, then the T values calculated for $n - 1$ and $n + 1$ half-waves will be larger than the force (16.33), so that

$$\left(\frac{\gamma}{n-1} + \frac{n-1}{\gamma}\right) > \left(\frac{\gamma}{n} + \frac{n}{\gamma}\right) < \left(\frac{\gamma}{n+1} + \frac{n+1}{\gamma}\right)$$

Note that we encountered a similar situation in Section 16.1.2, where the elastic stability of bars on elastic foundations was examined. The preceding inequalities can be reduced to

$$\sqrt{n(n-1)} < \gamma < \sqrt{n(n+1)}$$

These inequalities enable us to find the intervals of the ratios γ for different numbers of half-waves n of buckling. If $n = 1$, then $0 < \gamma < \sqrt{2}$; if $n = \alpha$, then $\sqrt{2} < \gamma < \sqrt{6}$; if $n = 3$, then $\sqrt{6} < \gamma < \sqrt{12}$; and so on. Within each interval of the change in the ratio γ, when the number n of half-waves is the same, the critical load T depends only on the aspect ratio γ. This relationship is illustrated by a diagram in Fig. 16.9, where the function

$$K_n(\gamma) = \frac{\gamma}{n} + \frac{n}{\gamma} \tag{16.34}$$

is plotted versus the γ value. As we can see from this plot, the K_n value is close to its minimum value $K_n = 2$ if the ratio γ is greater than unity. The maximum deviation from the minimum value takes place at $\gamma = \sqrt{2}$ and $n = 1$, when

$$K_n(\gamma) = K_n(\sqrt{2}) = \frac{1}{\sqrt{2}} + \sqrt{2} = 2.13$$

Therefore, being conservative, we may assume that $K_n(\gamma) \cong 2$ when the aspect ratio γ is greater than unity. Then the formula (16.33) yields

$$T_e = \frac{4\pi^2 D}{b^2} \tag{16.35}$$

and the corresponding critical stress is

$$\sigma_e = \frac{4\pi^2 D}{b^2 h} \tag{16.36}$$

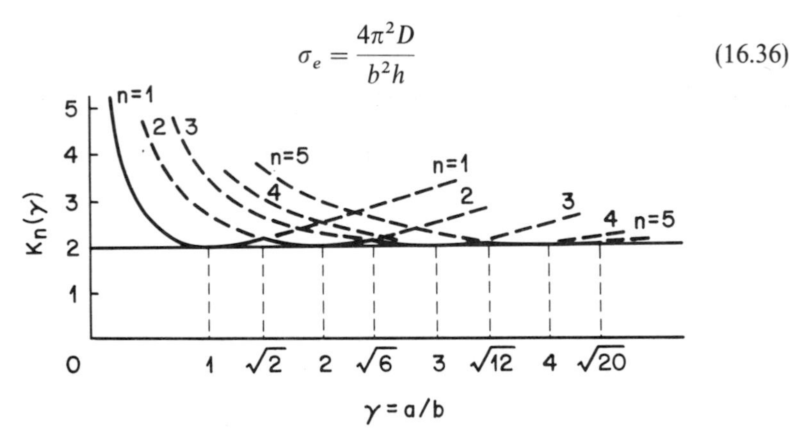

FIGURE 16.9 Function determining the number n of half-waves during buckling of a rectangular plate with an aspect ratio γ.

When $\gamma < 1$, then $n = 1$, and the formula (16.33) yields

$$T_e = \frac{\pi^2 D}{a^2}(1 + \gamma^2)^2 \tag{16.37}$$

so that

$$\sigma_e = \frac{\pi^2 D}{a^2 h}(1 + \gamma^2)^2$$

For an elongated plate compressed along its short sides ($b \gg a, \gamma \ll 1$), we have

$$T_e = \frac{\pi^2 D}{a^2} \qquad \sigma_e = \frac{\pi^2 D}{a^2 h} \tag{16.38}$$

i.e., the initial load is evaluated as for a simply supported strip beam of a span a.

Thus, for rectangular simply supported plates compressed along their long sides ($\gamma > 1$), we can use the formulas (16.35) and (16.36) to calculate the critical load. For plates compressed along their short sides, the formula (16.37) should be applied.

16.2.2 Buckling of a Rectangular Plate Simply Supported Along Opposite Edges

As an example of the application of the M. Lévy solution, we examine buckling of a rectangular plate simply supported along the sides parallel to the direction of loading (Fig. 16.10).

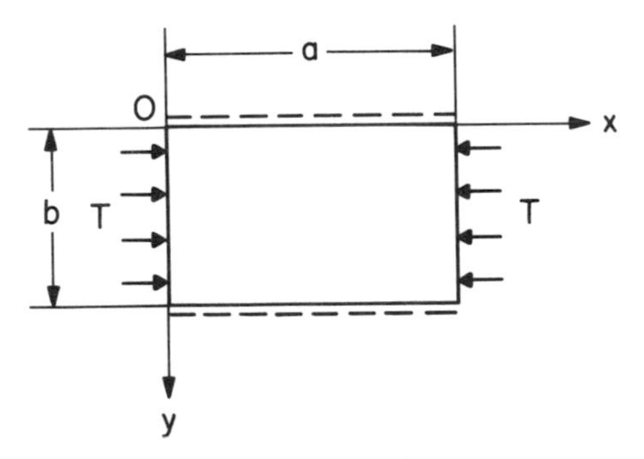

FIGURE 16.10 Rectangular plate simply supported along the edges parallel to the direction of compression.

Let the edges $x = 0$ and $x = a$ be simply supported, and the edges $y = 0$ and $y = b$ have arbitrary boundary conditions. We seek the solution to Eq. (16.30) in the form

$$w = \sum_{m=1}^{\infty} f_m(y) \sin \alpha_m x \tag{16.39}$$

where

$$\alpha_m = \frac{m\pi}{a}$$

After substituting Eq. (16.39) in Eq. (16.30), we have

$$f_m^{\mathrm{IV}}(y) - 2\alpha_m^2 f''_m(y) + \alpha_m^2\left(\alpha_m^2 - \frac{T}{D}\right) f_m(y) = 0 \tag{16.40}$$

If the plate were absolutely free on the edges $y = 0$ and $y = b$, then the critical force would be

$$T_e = \alpha_m^2 D = \left(\frac{m\pi}{a}\right)^2 D$$

In all the other cases of support along these edges, the critical force is larger than this value. Therefore, the characteristic equation corresponding to Eq. (16.40) will have two real and two imaginary roots, and the general solution of this equation is

$$f_m(y) = A_m \cosh \gamma_m y + B_m \sinh \gamma_m y + C_m \cos \delta_m y + D_m \sin \delta_m y \tag{16.41}$$

where

$$\gamma_m = \sqrt{\alpha_m^2 + \alpha_m \sqrt{\frac{T}{D}}} \qquad \delta_m = \sqrt{-\alpha_m^2 + \alpha_m \sqrt{\frac{T}{D}}} \tag{16.42}$$

and A_m, B_m, C_m, and D_m are constants of integration.

If, for instance, the edge $y = 0$ is simply supported and the edge $y = b$ is free, then the boundary conditions are as follows:

$$w = 0 \qquad \frac{\partial^2 w}{\partial y^2} = 0 \qquad \text{for } y = 0$$

$$\frac{\partial^2 w}{\partial y^2} + \nu \frac{\partial^2 w}{\partial x^2} = 0 \qquad \frac{\partial^3 w}{\partial y^3} + (2 - \nu)\frac{\partial^3 w}{\partial x^2 \partial y} = 0 \qquad \text{for } y = b$$

These conditions result in the following conditions for the functions $f_m(y)$:

$$f_m(0) = 0 \qquad f''_m(0) = 0 \tag{16.43}$$

$$f''_m(b) - \nu\alpha_m^2 f_m(b) = 0 \qquad f'''_m(b) - (2 - \nu)\alpha_m^2 f'_m(b) = 0 \tag{16.44}$$

Introducing the expression (16.41) into the conditions (16.43), we find that $A_m = C_m = 0$, so that

$$f_m(y) = B_m \sinh \gamma_m y + D_m \sin \delta_m y \tag{16.45}$$

After substituting this equation in the conditions (16.44), we obtain the following system of algebraic equations for the unknowns B_m and D_m:

$$\left.\begin{aligned} &B_m(\gamma_m^2 - \nu\alpha_m^2)\sinh \gamma_m b - D_m(\delta_m^2 + \nu\alpha_m^2)\sin \delta_m b = 0 \\ &B\gamma_m[\gamma_m^2 - (2 - \nu)\alpha_m^2]\cosh \gamma_m b - D_m\delta_m[\delta_m^2 + (2 - \nu)\alpha_m^2]\cos \delta_m b = 0 \end{aligned}\right\}$$

The constants B_m and D_m have nonzero values if the determinant of this system of equations is zero:

$$\delta_m(\gamma_m^2 - \nu\alpha_m^2)\tanh \gamma_m b = \gamma_m(\delta_m^2 + \nu\alpha_m^2)\tan \delta_m b = 0 \tag{16.46}$$

The smallest root T of this equation, which enters it through the parameters δ_m and γ_m, determines the critical value of the compressive load. This root takes place always for $n = 1$, regardless of the aspect ratio γ. Equation (16.46) can be solved numerically, and the results can be presented in the form

$$\sigma_e = \frac{T_e}{h} = k\frac{\pi^2 D}{b^2 h}$$

where the factor k is given in the following table as a function of the ratio γ.

$\gamma = \frac{a}{b}$	1	1.5	2.0	3.0
k	2.36	2.30	2.10	1.72

The solution (16.39) enables us to consider other possible cases of boundary conditions at the edges $y = 0$ and $y = b$.

16.2.3 A Rectangular Plate Compressed in Two Opposite Directions

Let a simply supported plate be subjected to compressive stresses σ_x and σ_y uniformly distributed along the corresponding edges (Fig. 16.11). This problem was also solved by Bryan.

Equation (16.29) can be presented as

$$D\nabla^4 w + \sigma_x h \frac{\partial^2 w}{\partial x^2} + \sigma_y h \frac{\partial^2 w}{\partial y^2} = 0 \tag{16.47}$$

We seek the solution to this equation in the form

$$w = A_{mn} \sin \alpha_m x \sin \beta_n y \tag{16.48}$$

where

$$\alpha_m = \frac{m\pi}{a} \qquad \beta_m = \frac{n\pi}{b}$$

Introducing Eq. (16.48) into Eq. (16.47), we have

$$D\left[\left(\frac{m}{a}\right)^2 + \left(\frac{n}{b}\right)^2\right]^2 = \frac{h}{\pi^2}\left[\sigma_x\left(\frac{m}{a}\right)^2 + \sigma_y\left(\frac{n}{b}\right)^2\right] \tag{16.49}$$

Using notation

$$\gamma = \frac{a}{b} \qquad \phi_0 = \frac{\sigma_y}{\sigma_x}$$

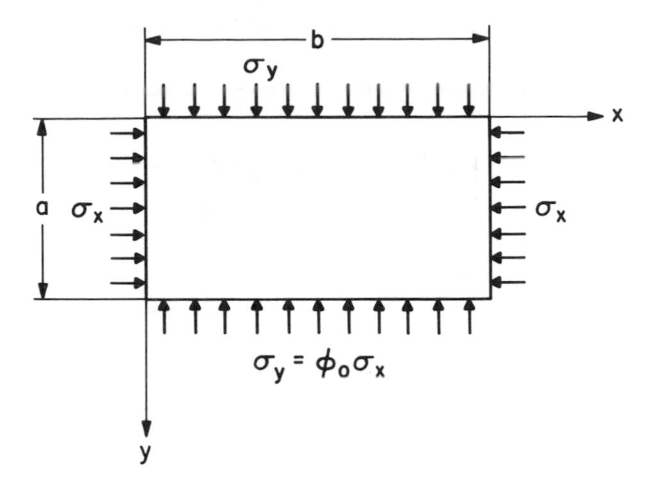

FIGURE 16.11 Rectangular plate compressed in two opposite directions.

we present the critical stresses in the forms

$$\sigma_{x,e} = K_x \frac{\pi^2 D}{b^2 h} \qquad \sigma_{y,e} = K_y \frac{\pi^2 D}{b^2 h} = \phi_0 \sigma_{x,e}$$

Then Eq. (16.49) can be written as

$$K_x = \frac{(m^2/\gamma^2 + n^2)^2}{\dfrac{m^2}{\gamma^2} + \phi_0 n^2} \tag{16.50}$$

The minimum K_x values are plotted in Fig. 16.12 versus γ values for different ϕ_0.

In the case of a square plate, $m = n = 1$, and the dimensionless critical stresses are

$$K_x = \frac{4}{1 + \phi_0} \qquad K_y = \frac{4\phi_0}{1 + \phi_0} \tag{16.51}$$

Note that, in this case,

$$K_x + K_y = 4$$

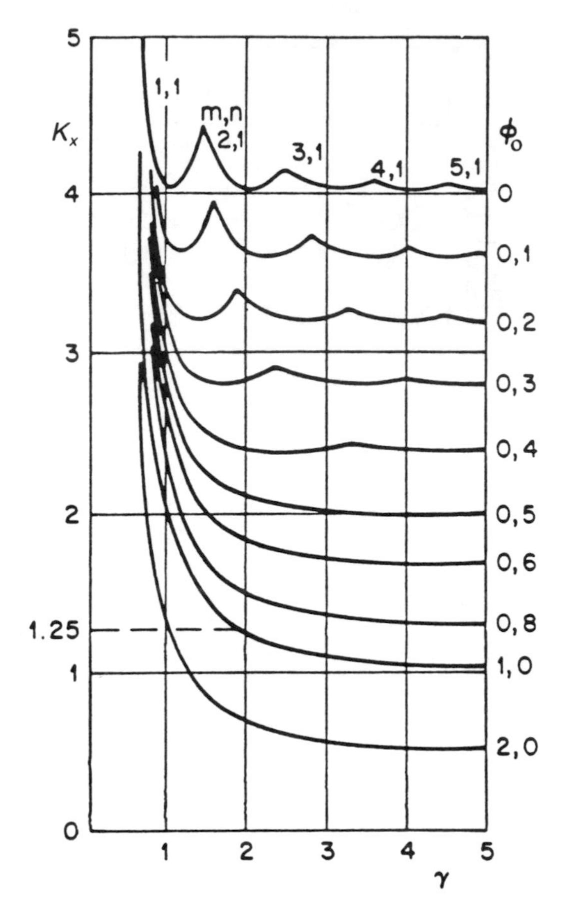

FIGURE 16.12 Factor K_x (dimensionless critical stress) in the buckling of a rectangular plate compressed in two opposite directions.

16.2.4 Buckling of Circular Plates

Let a clamped circular plate of radius r_0 be subjected to radial pressure q uniformly distributed along the plate's contour (Fig. 16.13). The points at the plate's contour are free to move in the plane of the plate.

In the case of an axisymmetric bending, Eqs. (15.102) reduce to the following single equation:

$$N_r = -D \frac{\partial}{\partial r} (\nabla^2 w) \tag{16.52}$$

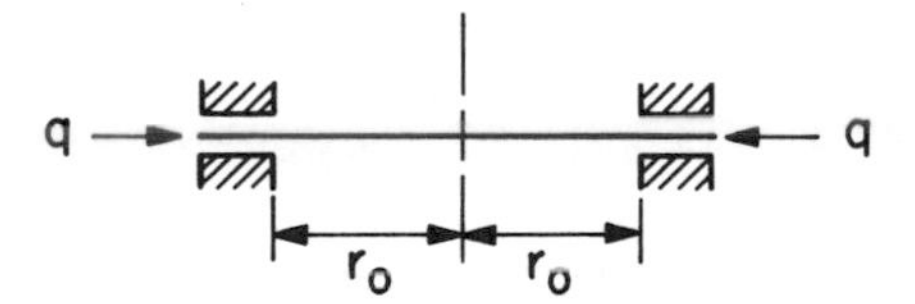

FIGURE 16.13 Uniformly compressed clamped circular plate.

where the Laplacean operator ∇^2 is expressed as

$$\nabla^2 = \frac{\partial^2}{\partial r^2} + \frac{1}{r}\frac{\partial}{\partial r} \tag{16.53}$$

Equation (16.52) can be rewritten in terms of the angle θ between the axis of the deflection surface, which is a surface of revolution, and any normal to the plate. This angle is related to the deflection w as

$$\theta = -\frac{\partial w}{\partial r}$$

so that

$$\nabla^2 w = -\frac{\partial \theta}{\partial r} - \frac{\theta}{r}$$

Then Eq. (16.52) yields

$$N_r = D\frac{\partial}{\partial r}\left(\frac{\partial \theta}{\partial r} + \frac{\theta}{r}\right) = D\left(\frac{\partial^2 \theta}{\partial r^2} + \frac{1}{r}\frac{\partial \theta}{\partial r} - \frac{1}{r^2}\theta\right) \tag{16.54}$$

Introducing a new variable,

$$u = r\sqrt{\frac{N_r}{D}} \tag{16.55}$$

we have

$$\frac{\partial \theta}{\partial r} = \frac{d\theta}{dr} = \frac{d\theta}{du}\frac{du}{dr} = \frac{d\theta}{du}\sqrt{\frac{N_r}{D}}$$

or

$$\frac{\partial^2\theta}{\partial r^2} = \frac{d}{dr}\left(\frac{d\theta}{dr}\right) = \frac{d}{du}\left(\frac{d\theta}{dr}\right)\frac{du}{dr} = \frac{d^2\theta}{du^2}\frac{N_r}{D}$$

so that Eq. (16.54) becomes

$$u^2\frac{d^2\theta}{du^2} + r\frac{d\theta}{du} + (u^2 - 1)\theta = 0 \tag{16.56}$$

Note that the partial derivatives were replaced in the preceding relationships by the ordinary derivatives since the angle θ is a function of only one variable, the radius of the plate.

Equation (16.56) is known as the *Bessel equation*, and its solution can be presented as

$$\theta = C_1 J_1(u) + C_2 Y_1(u)$$

where $J_1(u)$ and $Y_1(u)$ are *Bessel functions* of first order of the first and second kinds, respectively. The function $Y_1(u)$ tends to infinity as the radius r and the variable u approach zero. In order to obtain finite values of the angle θ, we should put $C_2 = 0$, so that

$$\theta = C_1 J_1(u)$$

Since the plate is clamped at the contour, the condition $\theta(r_0) = 0$ must be fulfilled. This results in the equation

$$J_1(u_0) = 0 \tag{16.57}$$

where

$$u_0 = r_0\sqrt{\frac{N_r}{D}}$$

The smallest root of Eq. (16.57) is 3.832, which leads to the following formula for the critical N_r value:

$$N_{r,e} = \frac{(3.832)^2 D}{r_0^2} = \frac{14.68D}{r_0^2} \tag{16.58}$$

Note that the critical compressive force for a strip of unit width with clamped ends and having the length equal to the diameter $2r_0$ of the plate is $\pi^2 D/r_0^2$, which is about 50 percent lower than in the case of a circular plate.

When the plate is simply supported on the contour, the boundary condition for the angle θ is

$$\frac{d\theta}{dr} + \nu\frac{\theta}{r} = 0 \qquad \text{for } r = r_0$$

The function $J_1(u)$ obeys the following rule of differentiation:

$$\frac{dJ_1}{du} = J_0 - \frac{J_1}{u}$$

where J_0 is the Bessel function of zero order. Then the preceding boundary condition yields

$$u_0 J_0(u_0) - (1 - \nu)J_1(u_0) = 0 \tag{16.59}$$

Using tables of the Bessel functions, we find the smallest root of the transcendental equation (16.59) to be 2.050. Then the critical value of the compressive force, assuming that $\nu = 0.3$, is

$$N_{r,e} = \frac{(2.050)^2 D}{r_0^2} = \frac{4.20D}{r_0^2} \tag{16.60}$$

This value is about three-and-a-half times smaller than in the case of a plate with a clamped edge.

The solutions just obtained are exact solutions. Now, let us show how the Bubnov–Galerkin method can be used to obtain an approximate solution for the critical force. The deflection function of the buckled plate clamped at the edge can be approximately presented as

$$w = w_0\left(1 - \frac{r^2}{r_0^2}\right)^2$$

This equation satisfies the boundary conditions

$$w = 0 \qquad \frac{dw}{dr} = -\theta = 0 \qquad \text{for } r = r_0$$

The angle θ is expressed as follows:

$$\theta = -\frac{dw}{dr} = \frac{4w_0}{r_0}\left(\frac{r}{r_0} - \frac{r^3}{r_0^3}\right) \tag{16.61}$$

Introduce the following notation:

$$L = D\left(\frac{d^2\theta}{dr^2} + \frac{1}{r}\frac{\partial\theta}{\partial r} - \frac{1}{r^2}\theta\right) - N_r \tag{16.62}$$

Then the procedure underlying the Bubnov–Galerkin method can be mathematically presented as follows:

$$\int_A L\theta r\, dr = 0 \tag{16.63}$$

where the integration is carried out over the plate's area A. After substituting Eq. (16.62), with consideration of Eq. (16.61), in Eq. (16.63), and executing the integrations, we find

$$N_{r,e} = \frac{16D}{r_0^2} \tag{16.64}$$

which is about 9 percent higher than the exact result (16.58).

17

Numerical Methods

17.1 FINITE DIFFERENCES

In many cases, the differential equations obtained cannot be solved analytically, even by using approximate methods. In this case, the numerical solutions can be found by employing a numerical differentiation scheme in which the differentials are replaced by *finite differences.* In the case of an ordinary differential equation, the finite difference is taken along one independent variable only whereas, for the partial differential equations, finite difference is taken along each variable. In general, a *mesh* is formed, and the differential equations are then expressed as *difference equations* corresponding to various mesh points. Various schemes are proposed to obtain higher accuracy by minimizing the error introduced by the approximation. In microelectronics, the finite difference, sometimes referred to as the *mesh method,* is in especially wide use for obtaining solutions to various heat-transfer problems.

The *numerical derivatives* of a function $f(x)$ can be calculated by referring to its Taylor expansion about the points x_n and x_{n+1} and subtracting the second expansion from the first one. If the values of the function $f(x)$ are spaced at equal intervals of length h, then the first derivative of this function calculated for $x = x_n$ is

$$f'_n = \left(\frac{df}{dx}\right)_{x=x_n} = \frac{f_{n+1} - f_n}{h} - \frac{h^2}{2!}\frac{d^2f}{dx^2}$$

where

$$f_n = f(a + nh)$$

is the value of the function $f(x)$ for $x = x_n$ (Fig. 17.1). The *first, second,*

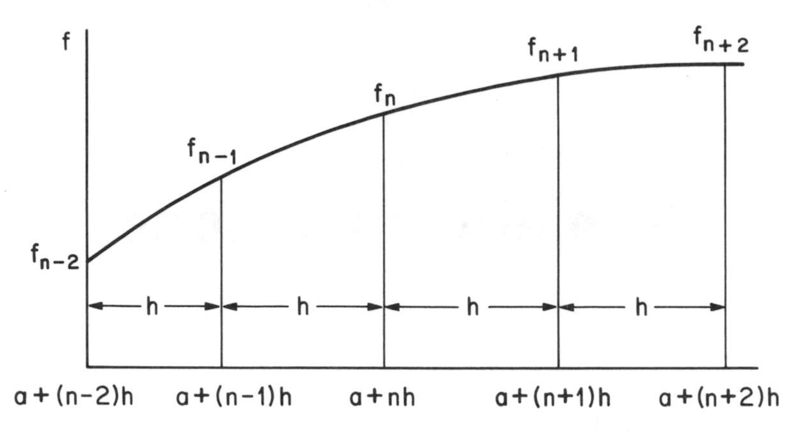

FIGURE 17.1 Uniformly distributed points in a one-dimensional solution domain.

..., n*th forward differences* of the function f_n are denoted as

$$\left.\begin{aligned}
\Delta f_n &= f_{n+1} - f_n \\
\Delta^2 f_n &= \Delta\Delta f_n = (f_{n+2} - f_{n+1}) - (f_{n+1} - f_n) = f_{n+2} - 2f_{n+1} + f_n \\
\Delta^3 f_n &= f_{n+3} - 3f_{n+2} + 3f_{n+1} - f_n \\
\Delta^4 f_n &= f_{n+4} - 4f_{n+3} + 6f_{n+2} - 4f_{n+1} + f_n \\
\Delta^5 f_n &= f_{n+5} - 5f_{n+4} + 10f_{n+3} - 10f_{n+2} + 5f_{n+1} - f_n
\end{aligned}\right\} \tag{17.1}$$

Then the function $f(x)$ can be approximately replaced within an interval $a < x < a + nh$ by the following *interpolation polynomial*:

$$f(x) = f(a) + \frac{x-a}{h}\Delta f_0 + \frac{(x-a)(x-a-h)}{2!\,h^2}\Delta^2 f_0 + \cdots + \frac{(x-a)(x-a-h)\cdots[x-z-(n-1)h]}{n!\,h^n}\Delta^n f_0 \tag{17.2}$$

where the forward differences are evaluated for $x_0 = a$, $x_1 = a + h, \ldots, x_n = a + nh$. This formula is known also as *Newton's formula for forward interpolation.*

Similarly, we can define

$$\delta f_{n+1} = \frac{f_{n+1} - f_{n-1}}{2}$$

$$\delta^2 f_{n+1} = \frac{f_{n+2} - f_n}{2} - \frac{f_{n+1} - f_{n-1}}{2} = \frac{f_{n+2} - f_{n+1} - f_n + f_{n-1}}{2}$$

as *central differences*, and

$$\Delta f_{n-1} = f_n - f_{n-1}$$

$$\Delta^2 f_{n-1} = (f_n - f_{n-1}) - (f_{n-1} - f_{n-2}) = f_n - 2f_{n-1} + f_{n-2}$$

as *backward differences*. Clearly, the central differences can be obtained by subtracting the two Taylor expansions about points x_{n-1} and x_{n+1}, and the backward differences can be obtained by subtracting the two Taylor expansions about x_n and x_{n-1}.

The purpose of the finite-difference method, starting from an initial value of the function f, is given by a differential equation, to calculate its values at the nodal points $n = 1, 2, \ldots$ on the basis of the relationship defined by this equation.

In the simplest *first-order methods*, the following *point-slope formula* can be applied:

$$f_{n+1} \cong f_n + hf'_n$$

or

$$f_{n+1} \cong f_{n-1} + 2hf'_n$$

If, for instance, the differential equation

$$w''(x) = \frac{P}{EI}x$$

of bending of a cantilever beam subjected to a concentrated force P applied to its free end (Fig. 12.12) is solved numerically, then, using the following first-order equation for the rotation angle

$$w'(x) = \frac{P}{2EI}x^2$$

we have, in accordance with the point-slope formula,

$$\Delta w_n = w_{n+1} - w_n \qquad \Delta x = h$$

The differential equation for the deflection function $w(x)$ enables us to calculate w_{n+1} from the previous w_n as

$$w_{n+1} = w_n + Hx_n^2$$

where $H = Ph/2EI$. In FORTRAN language,

```
I = 1
DO 20I = 1,N
W(I + 1) = W(I) + (X(I)**2)*H
20 CONTINUE
```

If the *trapezoidal formula* is used, the following formula can be used to calculate the f_{n+1} value from the f_n value:

$$f_{n+1} \cong f_n + \frac{h}{2}(f'_{n+1} + f'_n)$$

Formulas that enable us to calculate a certain value f_n from the previous values $f_{n-1}, f_{n-2}, \ldots$, are called *recurrence formulas.*

The widespread *Runge–Kutta method* is an example of a *second-order method,* in which the following recurrence formula is used:

$$f_{n+1} = f_n + \frac{k_1 + k_2}{2}$$

where the value k_1 is evaluated from the given differential equation as a function of x_n and f_n, and the value k_2 is calculated as a function of $x_n + h$ and $f_n + k_1$. There are also *higher-order methods* in which the order of the error term is improved.

The methods discussed here have been extended to systems of ordinary differential equations in which there are multiple equations corresponding to multiple dependent variables and to partial differential equations. In the latter case, the geometry of the space of independent variables is partitioned as a mesh to apply various finite-difference methods.

In engineering practice, relatively simple interpolation polynomials are used. These can be obtained from Eq. (17.2) by differentiation. If, for instance, only three terms are retained in the final expressions for the derivatives, then the following formulas can be obtained for the "right" derivatives at the given nodal point n:

$$
\left.\begin{aligned}
f'_n &= \frac{1}{2h}(-3f_n + 4f_{n+1} - f_{n+2}) \\
f''_n &= \frac{1}{h^2}(2f_n - 5f_{n+1} + 4f_{n+2} - f_{n+3}) \\
f'''_n &= \frac{1}{2h^3}(-5f_n + 18f_{n+1} - 24f_{n+2} + 14f_{n+3} - 3f_{n+4}) \\
f^{\mathrm{IV}}_n &= \frac{1}{h^4}(3f_n - 14f_{n+1} + 26f_{n+2} - 24f_{n+3} + 11f_{n+4} - 2f_{n+5})
\end{aligned}\right\} \qquad (17.3)
$$

Similar formulas can be obtained for the "left" derivatives:

$$
\left.\begin{aligned}
f'_n &= \frac{1}{2h}(3f_n - 4f_{n-1} + f_{n+2}) \\
f''_n &= \frac{1}{h^2}(2f_n - 5f_{n-1} + 4f_{n-2} - f_{n-3}) \\
f'''_n &= \frac{1}{2h^3}(5f_n - 18f_{n-1} + 24f_{n-2} - 14f_{n-3} + 3f_{n-4}) \\
f^{\mathrm{IV}}_n &= \frac{1}{h^4}(3f_n - 14f_{n-1} + 26f_{n-2} - 24f_{n-3} + 11f_{n-4} - 2f_{n-5})
\end{aligned}\right\} \qquad (17.4)
$$

The formulas (17.3) and (17.4) are used to replace the boundary conditions given for the derivatives of the deflection function by the finite-difference equations.

The "central" derivatives can be determined as

$$
\left.\begin{aligned}
f'_n &= \frac{1}{2h}(f_{n+1} - f_{n-1}) \\
f''_n &= \frac{1}{h^2}(f_{n+1} - 2f_n + f_{n-1}) \\
f'''_n &= \frac{1}{2h^3}(f_{n+2} - 2f_{n+1} + 2f_{n-1} - f_{n-2}) \\
f^{\mathrm{IV}}_n &= \frac{1}{h^2}(f_{n+2} - 4f_{n+1} + 6f_n - 4f_{n-1} + f_{n-2})
\end{aligned}\right\} \qquad (17.5)
$$

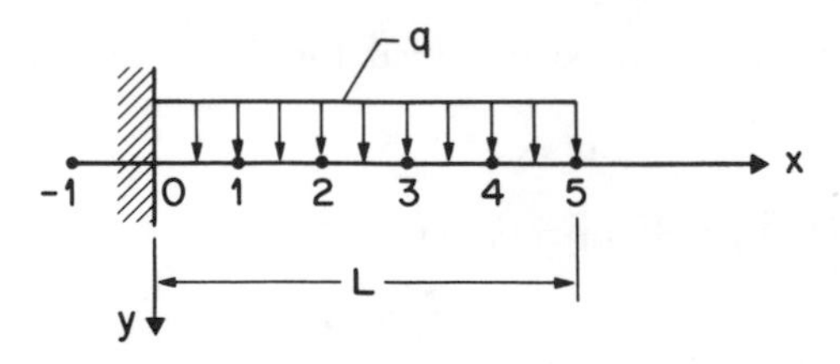

FIGURE 17.2 Uniformly distributed points in a one-dimensional solution domain for a cantilever beam under a uniform lateral load.

These formulas can be used to replace the differential equations and the boundary conditions by the equations in finite differences.

As an example, examine the beam shown in Fig. 17.2. Its bending can be described by the equation

$$EIw^{\mathrm{IV}}(x) = q \tag{17.6}$$

with the following boundary conditions:

$$w(0) = 0 \qquad w'(0) = 0 \qquad w''(L) = 0 \qquad w'''(L) = 0$$

Let us break down the beam into five equal portions. Using the formulas (17.4) and (17.5), we rewrite the preceding boundary conditions as

$$w_0 = 0 \qquad -w_{-1} + w_1 = 0 \qquad \text{at } x = 0$$

$$w_4 - 2w_5 + w_6 = 0 \qquad 3w_1 - 14w_2 + 24w_3 - 18w_4 + 5w_5 = 0 \qquad \text{at } x = L \tag{17.7}$$

The equilibrium equation (17.6) written for the nth nodal point results in the following equation in finite differences:

$$w_{n-2} - 4w_{n-1} + 6w_n - 4w_{n+1} + w_{n+2} = \frac{qh^4}{EI} \tag{17.8}$$

The four conditions (17.7) and Eqs. (17.8), written for the four nodal points $n = 1, 2, 3, 4$, form a system of eight linear algebraic equations for eight unknown nodal displacements w_n $(n = -1, 0, 1, \ldots, 6)$.

17.2 COLLOCATION METHOD

This method, unlike the finite-difference method, enables us to obtain an approximate solution to the given boundary-value problem in the form of functions rather than as an array of numbers. We illustrate this method by using an example of a beam shown in Fig. 17.3.

The mechanical behavior of the beam is described by Eq. (17.6), with the following boundary conditions:

$$w(0) = 0 \qquad w'(0) = 0 \qquad w(L) = 0 \qquad w''(L) = 0 \tag{17.9}$$

We seek the deflection function in the form of the power series

$$w(x) = \sum_{n=0}^{N} a_n \left(\frac{x}{L}\right)^n \tag{17.10}$$

For arbitrary a_n values, this expression does not satisfy either the differential equation (17.6) or the boundary conditions (17.9). In accordance with the collocation method, the coefficients a_n should be chosen in such a way that the governing differential equation is fulfilled only in the finite amount of *collocation points*. In addition, all the boundary conditions must be fulfilled.

If, for instance, three collocation points are chosen, then we have at our disposal, together with the four boundary conditions, seven independent equations. Then, retaining seven terms in the series (17.10), so that

$$w(x) = a_0 + a_1 \frac{x}{L} + a_2\left(\frac{x}{L}\right)^2 + a_3\left(\frac{x}{L}\right)^3 + a_4\left(\frac{x}{L}\right)^4 + a_5\left(\frac{x}{L}\right)^5 + a_6\left(\frac{x}{L}\right)^6$$

and introducing this polynomial into the boundary conditions (17.9), we have

$$\left.\begin{aligned} &a_0 = 0 \qquad a_1 = 0 \\ &a_0 + a_1 + a_2 + a_3 + a_4 + a_5 + a_6 = 0 \\ &2a_2 + 6a_3 + 12a_4 + 20a_5 + 30a_6 \ = 0 \end{aligned}\right\}$$

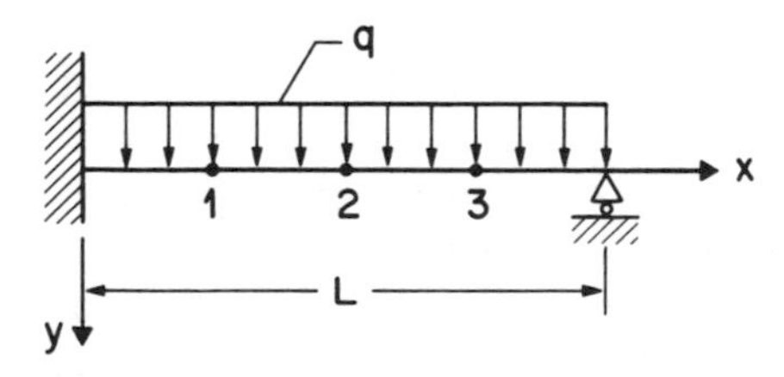

FIGURE 17.3 Uniformly distributed points in a solution domain for a clamped/simply supported beam under a uniform lateral load.

Substituting the preceding polynomial in Eq. (17.6) and requiring that this equation is fulfilled at points $x_1 = L/4$, $x_2 = L/2$, $x_3 = 3L/4$, we have

$$\left.\begin{aligned} 24a_4 + 30a_5 + 22.5a_6 &= \frac{qL^4}{EI} \\ 24a_4 + 60a_5 + 90a_6 &= \frac{qL^4}{EI} \\ 24a_4 + 90a_5 + 202.5a_6 &= \frac{qL^4}{EI} \end{aligned}\right\}$$

Then, solving the obtained equations for the coefficients a_n, we find

$$a_0 = a_1 = 0 \qquad a_2 = \frac{qL^4}{16EI} \qquad a_3 = -\frac{5qL^4}{48EI} \qquad a_4 = \frac{qL^4}{24EI} \qquad a_5 = a_6 = 0$$

Introducing these values into the series (17.10), we obtain the following formula for the deflection function:

$$w(x) = \frac{qL^4}{16}\left(\frac{x}{L}\right)^2\left[1 - \frac{5}{3}\frac{x}{L} + \frac{2}{3}\left(\frac{x}{L}\right)^2\right] \tag{17.11}$$

This formula provides an exact solution to the problem. It satisfies both the governing equation (17.6) and the boundary conditions (17.9).

17.3 FINITE ELEMENTS

17.3.1 Finite Elements

The *finite-element method* (FEM) is the most powerful and the most widespread numerical method in structural analysis, including the areas of microelectronics and fiber optics. The wide and versatile application of this method is due, first of all, to the availability of numerous software packages in which the tedious and time-consuming procedures of forming and solving systems of algebraic equations, as well as presenting the finite-element meshes, is "automated." In addition, the method provides minimum requirements for the input data and presents the output data in a very convenient format. The FEM enables us to consider all the peculiarities of the geometry of the structure and loading conditions. It can take into account, when necessary, the time and temperature dependence of material properties, boundary conditions, and applied forces.

In the FEM, the continuum medium, which can be a plate, a frame, or a beam, is replaced by a finite number of elements of finite size (*finite elements*), which interact in a finite number of *nodal points*. This replacement is performed in such a way that the total strain energy of the structure remains unchanged. Therefore, the FEM is often treated as a possible numerical extension of the variational Rayleigh–Ritz method, which is based on the condition of minimum of the energy functional (see Section 13.4). Indeed, the basic equation (13.50) of the FEM, when the nodal displacements are chosen as the redundant unknowns, is similar to the canonical equations (13.28) of the Rayleigh–Ritz method. The only essential difference between the two methods is that, in the Rayleigh–Ritz method, a change in any of the coordinate functions affects the displacements throughout the structure whereas, in the FEM, a change in any of the nodal displacements results in a change of the displacement field only for these finite elements that adjoin this node. As in the Rayleigh–Ritz method, for some structures, the utilization of the energy balance condition in the FEM results in a discrete model that provides an exact solution to the given problem. Clearly, this is typical for those structures that already consist of separate elements connected at a discrete number of points, such as trusses, frames, or scantlings.

The breakdown of a structure into finite elements is the first step, and a very important one, requiring a good feeling for the mechanical behavior of the structure. The number of finite elements should be increased in those areas in which a rapid change in the stress and strain fields is expected. If, for instance, a two-dimensional truss is assembled as a collocation of straight rods connected by hinges, then each rod experiences only tension or compression. If these rods are chosen as finite elements, then the idealized discrete model will adequately present the actual structure. When the displacement method is used, i.e., when the displacements of the nodes that coincide with the hinge locations are chosen as the basic unknowns, then the general solutions of the FEM equations center on finding the nodal displacements. Other information, such as the forces and stresses acting in the finite elements, can be obtained quite easily, once the structural displacements are known. The equations for the evaluation of the nodal displacements include the equilibrium equations for the nodal points and the compatibility conditions for the nodal displacements. Clearly, the compatibility conditions are needed only in the case of statically indeterminate trusses. An example of a finite element of a truss with hinged nodes is shown in Fig. 17.4a. Such an element experiences only uniaxial tension or compression.

Let us take up several other examples of finite elements. If the rods of a truss are connected at their ends by knee elements, so that the truss becomes, in effect, a complex frame (see Section 14.2), then the interaction between

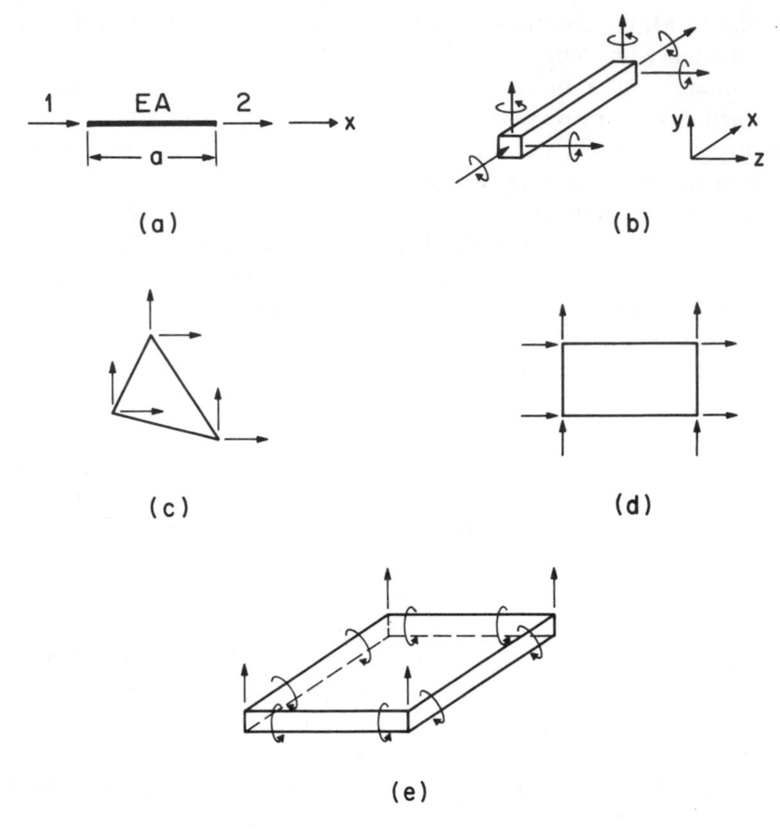

FIGURE 17.4 Typical finite elements: (a) element experiencing only uniaxial tension or compression; (b) element experiencing uniaxial tension or compression, bending in two mutually perpendicular planes, and torsion; (c) triangular element; (d) rectangular element; (e) plate element experiencing both in-plane and lateral forces.

the rod elements in the nodal points becomes different from the interaction in a truss, described in the preceding paragraph. Lateral forces and bending moments occur at the end cross sections of rods joined by knees. This results in a greater number of unknowns, thereby also requiring a greater number of equations for the evaluation of these unknowns. In such a situation, an element experiencing not only tension or compression but also bending should be introduced. In Fig. 17.4b an even more "powerful" finite element than is needed for the solution of this problem is shown. This element experiences tension or compression in the axial direction, bending in two mutually perpendicular planes, and tension. Whereas the element in Fig. 17.4a has just 2 *degrees of freedom* (tension or compression), the element in

Fig. 17.4b has 12 degrees of freedom (in each of the 2 end planes, there are an axial force, 2 lateral forces, 2 bending moments, and a torque).

Plate structures and two-dimensional theory-of-elasticity problems can be addressed by using finite elements, as shown in Figs. 17.4c and 17.4d. The number of degrees of freedom for the triangular element is 6, and for the rectangular element 8.

If, in addition to in-plane (membrane) forces, a plate is loaded by lateral (transverse) forces, then additional forces of interaction appear at the element boundaries. In this case, an element shown in Fig. 17.4c must be used. It has 12 degrees of freedom.

17.3.2 Stiffness Matrix

The idealization employed in the FEM reduces the actual structure to a set of finite elements that are connected at the nodal points only. Therefore, such an idealization anticipates that the state of stress in each element is determined through the magnitudes of the nodal displacements or the nodal forces. The relation between the nodal displacements and the nodal forces in a finite element is established by the *stiffness matrix* or the compliance matrix. This matrix contains a description of the elastic properties of the element. In dynamics problems, which are not examined in this book, the stiffness matrix contains also the inertia terms.

In Section 13.4, the stiffness matrix was obtained for a plane stress condition from the expression for the total strain energy. Let us show how this matrix can be obtained from the Lagrange theorem. Treating the nodal forces $\{F\} = \{F_1, F_2, \dots\}$ as generalized external forces and the nodal displacements $\{q\} = \{q_1, q_2, \dots\}$ as generalized displacements, we present the formula (13.5) as follows:

$$\sum_k F_k \delta q_k - \int_V (\sigma_x \delta\varepsilon_x + \sigma_y \delta\varepsilon_y + \cdots + \tau_{yz}\delta\gamma_{yz})\, dV = 0 \qquad (17.12)$$

This equation states that the sum of the works of all the external $\{F\}$ and the internal $\{\sigma\}$ forces moving through virtual displacements is zero. The second term in the preceding expression gives the increase in the potential energy, which is equal to the work of the internal forces but is opposite in sign. The integral should be taken throughout the volume of the finite element. In matrix form, the relationship (17.12) can be presented as

$$\{\delta q\}^T \{F\} = \int_V \{\delta\varepsilon\}^T \{\sigma\}\, dV \qquad (17.13)$$

where the symbol T denotes matrix transposition.

The Cauchy formulas (2.34) can be presented in the matrix form as follows:

$$\{\varepsilon\} = [\Psi]\{q\} \tag{17.14}$$

where $\{\varepsilon\}$ and $\{q\}$ are the column matrices of the strains and displacements, and $[\Psi]$ is the rectangular matrix whose elements depend on the coordinates of the given point. The Hooke law relationship (13.48) is

$$\{\sigma\} = [E]\{q\} \tag{17.15}$$

where $\{\sigma\}$ is the column matrix of the stresses, and $[E]$ is the rectangular matrix whose elements depend on the elastic constants of the material. These can be, generally speaking, also coordinate-dependent. After substituting Eqs. (17.14) and (17.15) in Eq. (17.13), we have

$$[K]\{q\} = \{F\} \tag{17.16}$$

where

$$[K] = \int_V [\Psi]^T[E]\, dV \tag{17.17}$$

is the stiffness matrix of the finite element. This is a square matrix. Its order is equal to the number of the degrees of freedom of the given finite element.

In practical applications, the stiffness matrix can be obtained in the simplest way from Klapeiron's theorem (13.11). In matrix form, Eq. (13.11) can be presented as follows:

$$U = \tfrac{1}{2}\{F\}\{q\}^T$$

After introducing the vector $\{F\}$ of nodal forces from Eq. (17.16) into this formula, we obtain

$$U = \tfrac{1}{2}\{q\}^T[K]\{q\}$$

or, in the expanded form,

$$U = \frac{1}{2}\sum_{i=1}^{m}\sum_{j=1}^{m} k_{ij}q_iq_j \tag{17.18}$$

where k_{ij} are the elements of the stiffness matrix $[K]$. Thus, in order to obtain the stiffness matrix for the given finite element, it is sufficient to have the expression for the potential energy of the finite element in the form.

As an example, let us obtain the stiffness matrix for an element experiencing uniaxial tension or compression (Fig. 17.4a). If the displacements q_1 and q_2 of the end cross sections of this element are known, the displacements of any cross section can be evaluated as

$$q(x) = q_1 H_1(x) + q_2 H_2(x) = \sum_{i=1}^{2} q_i H_i(x) \tag{17.19}$$

where

$$H_1(x) = 1 - \frac{x}{a} \qquad H_2(x) = \frac{x}{a} \tag{17.20}$$

are one-dimensional *Hermite functions.* Indeed, $q(0) = q_1$, and $q(a) = q_2$. The strain energy of the element in question is

$$U = \frac{1}{2} EA \int_0^a \varepsilon_x^2 \, dx = \frac{1}{2} EA \int_0^a [q'(x)]^2 \, dx \tag{17.21}$$

After introducing Eq. (17.19) into Eq. (17.20), we have

$$U = \frac{1}{2} EA \int_0^a \left[\sum_{i=1}^{2} q_i H_i'(x) \right]^2 dx = \frac{1}{2} \sum_{i=1}^{2} \sum_{j=1}^{2} k_{ij} q_i q_k$$

where

$$k_{ij} = EA \int_0^a H_i'(x) H_j'(x) \, dx$$

are the elements of the stiffness matrix. Using the formulas (17.20), we obtain $k_{11} = k_{22} = EA/a$, $k_{12} = k_{21} = -EA/a$, so that the relationship (17.16) is as follows:

$$\begin{bmatrix} \dfrac{EA}{a} & -\dfrac{EA}{a} \\ -\dfrac{EA}{a} & \dfrac{EA}{a} \end{bmatrix} \begin{Bmatrix} q_1 \\ q_2 \end{Bmatrix} = \begin{Bmatrix} F_1 \\ F_2 \end{Bmatrix} \tag{17.22}$$

where

$$[K] = \frac{EA}{a}\begin{bmatrix} 1 & -1 \\ -1 & 1 \end{bmatrix}$$

is the stiffness matrix of an element subjected to tension.

As another example, examine a beam element subjected to two-dimensional bending (Fig. 17.5). Seeking the elastic curve of the beam in the form of a polynomial

$$w(x) = C_0 + C_1 x + C_2 x^2 + C_3 x^3$$

which is an integral of the equation of bending

$$EIw^{IV}(x) = 0$$

and using the boundary conditions in the form

$$w(0) = q_1 \qquad w'(0) = q_2 \qquad w(a) = q_3 \qquad w'(a) = q_4$$

we find

$$C_0 = q_1 \qquad C_1 = q_2 \qquad C_2 = \frac{1}{a^2}(-3q_1 - 2aq_2 + 3q_3 - aq_4)$$

$$C_3 = \frac{1}{a^3}(2q_1 + aq_2 - 2q_3 + aq_4)$$

so that

$$w(x) = \sum_{i=1}^{4} q_i H_i(x)$$

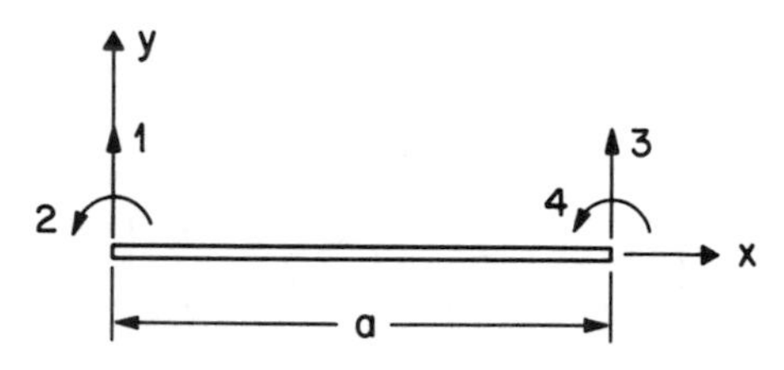

FIGURE 17.5 Beam element subjected to two-dimensional bending.

where

$$\left.\begin{aligned} H_1(x) &= 1 - 3\frac{x^2}{a^2} + 2\frac{x^3}{a^3} \\ H_2(x) &= x\left(1 - 2\frac{x}{a} + \frac{x^2}{a^2}\right) \\ H_3(x) &= \frac{x^2}{a^2}\left(3 - 2\frac{x}{a}\right) \\ H_4(x) &= -\frac{x^2}{a}\left(1 - \frac{x}{a}\right) \end{aligned}\right\} \tag{17.23}$$

The strain energy of a beam subjected to bending is

$$U = \frac{1}{2}EI\int_0^a [w''(x)]^2\,dx = \frac{1}{2}\sum_{i=1}^{4}\sum_{j=1}^{4} k_{ij}q_iq'_j$$

where

$$k_{ij} = EI\int_0^a H_i''(x)H_j''(x)\,dx$$

are the sought elements of the stiffness matrix. Using the expressions (17.23) for the Hermite functions, we obtain the stiffness matrix in the form

$$[K] = \frac{2EI}{a^3}\begin{vmatrix} 6 & 3a & -6 & 3a \\ & 2a^2 & -3a & a^2 \\ & & 6 & -3a \\ \text{Symmetrically} & & & 2a^2 \end{vmatrix}$$

17.3.3 Displacement Method

In Section 14.3, we examined two major methods used in structural analysis to solve statically indeterminate problems—the deformation method and the force method. In the deformation method, the redundant deformations are used as the principal unknowns. In the force method, it is the forces that are chosen as the principal unknowns. Similar approaches can be applied when the FEM is used. Modification of the deformation method, in this case, is such that nodal displacements are chosen as the principal unknowns.

In the preceding section, we showed how stiffness matrices of finite elements can be determined. The next step is to obtain the *global stiffness matrix*, i.e., the stiffness matrix for the whole structure. Such a matrix was obtained in Section 13.4 on the basis of energy considerations. Here we will use the Lagrange theorem to determine the global stiffness matrix. The stresses $\{\sigma\}_i$ within every ith finite element are equilibrated by its nodal forces $\{F\}_i$. These forces are related to the nodal displacements $\{q\}_i$ as follows:

$$\{F\}_i = [K]_i\{q\}_i \tag{17.24}$$

where $[K]_i$ is the stiffness matrix of the given element. Let us combine Eq. (17.24) written for a single finite element into a matrix equation for the whole structure. In accordance with the Lagrange theorem, the increment in the works of all the external forces acting on a structure through its virtual displacements is equal to the change in the strain energy of the system:

$$\delta W = \delta U \tag{17.25}$$

The increment in the works of the external forces is

$$\delta W = \sum_{s=1}^{n} \{\delta q^{(s)}\}^T\{P\}^{(s)} \tag{17.26}$$

where $\{P\}^{(s)} = \{P_1^{(s)}, P_2^{(s)}, \ldots, P_i^{(s)}, \ldots, P_t^{(s)}\}$ is the column vector of the external nodal forces, $\{\delta q^{(s)}\} = \{\delta q_1^{(s)}, \delta q_2^{(s)}, \ldots, \delta q_i^{(s)}, \ldots, \delta q_t^{(s)}\}$ is the column vector of the increments in the nodal displacements for the Sth node, and t is equal to the number of the unknown nodal displacements in the sth node. If

$$\{\bar{q}\} = \{\{q\}^{(1)} \quad \{q\}^{(2)}, \ldots, \{q\}^{(s)}, \ldots, \{q\}^{n}\}$$

is the vector of the nodal displacements for the entire structure, and

$$\{\bar{P}\} = \{\{P\}^{(1)} \quad \{P\}^{(2)}, \ldots, \{P\}^{(s)}, \ldots, \{P\}^{n}\}$$

is the vector of the external nodal loading for the entire structure, then the variation (17.25) can be presented as

$$\delta W = \{\delta\bar{q}\}^T\{\bar{P}\} \tag{17.27}$$

The increment in the strain energy of the structure δU is equal to the sum of the works of all the internal nodal forces moving through the corresponding virtual displacements:

$$\delta U = \sum_{i=1}^{m} \{\delta q\}_i^T \{F\}_i$$

Introducing a vector,

$$\{F\} = \{\{F\}_1, \{F\}_2, \ldots, \{F\}_m\}$$

of the internal nodal forces and a vector

$$\{q\} = \{\{q\}_1, \{q\}_2, \ldots, \{q\}_m\}$$

of the internal nodal displacements, we have

$$\delta U = \{\delta q\}^T \{F\} \tag{17.28}$$

After substituting Eqs. (17.27) and (17.28) in the condition (17.25), we obtain

$$\{\delta \bar{q}\}^T \{\bar{P}\} = \{\delta q\}^T \{F\} \tag{17.29}$$

For any specific structure, we can establish a connection between the vector $\{q\}$ of the internal nodal displacements for every finite element in its local coordinates and the vector $\{\bar{q}\}$ of the nodal displacements for the entire structure in "global" coordinates. This relationship can be written in the form

$$\{q\} = [H]\{\bar{q}\} \tag{17.30}$$

We illustrate the meaning of the matrix $[H]$ using an example of a frame structure in Fig. 17.6 (it is assumed that the elements in the "empty" boxes

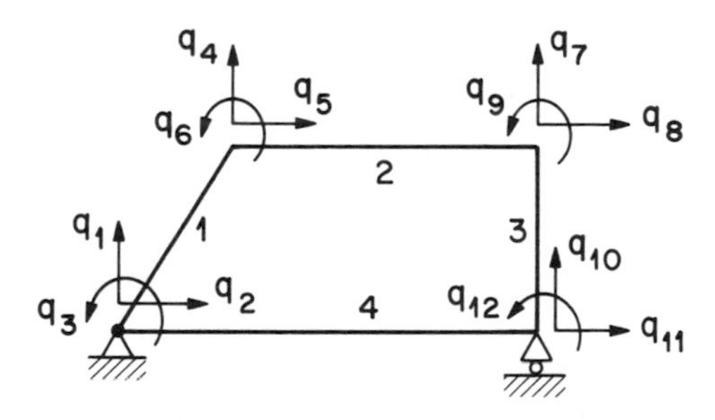

FIGURE 17.6 Frame structure subjected to node loads.

of the equation below are equal to zero):

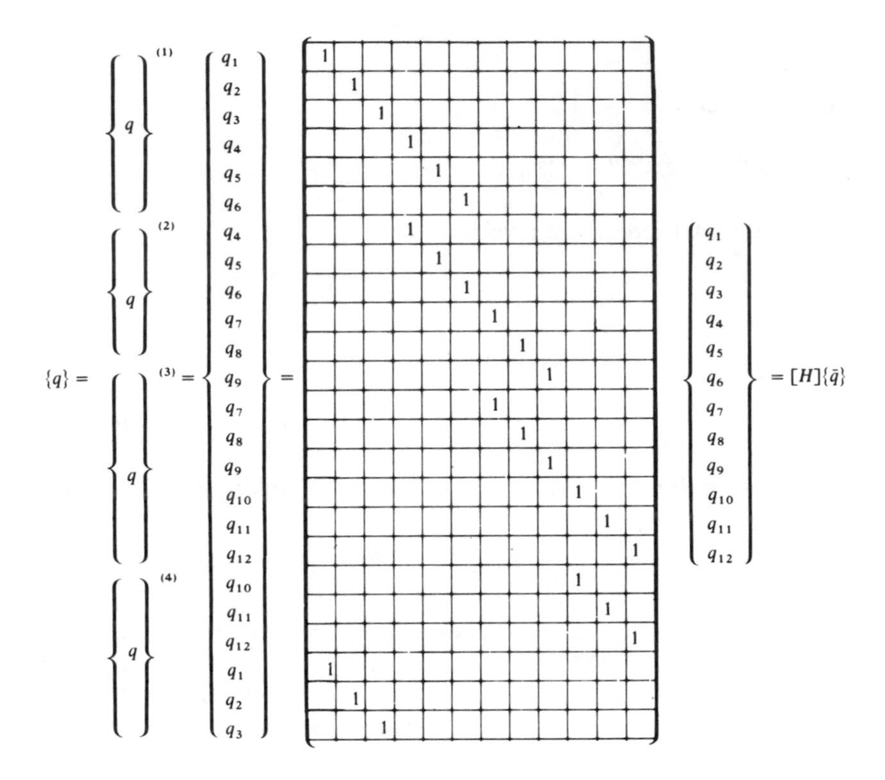

Since, in accordance with Eq. (17.30),

$$\{\delta q\} = [H]\{\delta \bar{q}\}$$

and $\{\delta \bar{q}\} \neq 0$, then Eq. (17.29) yields

$$\{\bar{P}\} = [H]^T\{F\} \tag{17.31}$$

This is the equation of equilibrium for the whole structure: it links the external forces $\{\bar{P}\}$ with the internal forces $\{F\}$. Considering Eq. (17.30), we can rewrite Eq. (17.31) as follows:

$$\mathbf{K\Delta} = \mathbf{F} \tag{17.32}$$

where $\mathbf{\Delta} = \{\bar{q}\}$ is the column matrix containing the collection of generalized displacements of the whole structure, $\mathbf{F} = \{\bar{P}\}$ is the global generalized force matrix, and

$$\mathbf{K} = [H]^T[K_g][H] \tag{17.33}$$

is the global stiffness matrix for the whole structure. In the latter formula, $[K_g]$ is the matrix establishing the connection between the forces $\{F\}$ and the displacements $\{q\}$ in global coordinates,

$$\{F\} = [K_g]\{q\}$$

Equation (17.32) coincides with Eq. (13.50), which was obtained on the basis of the energy considerations. It provides the basic FEM relationship in the global coordinate system and is, in effect, a matrix equilibrium equation. The obtained equilibrium equations contain, in the general case, six independent conditions of equilibrium, which correspond to the six degrees of freedom of the entire structure as a solid body. This results in the fact that the global stiffness matrix $\mathbf{K}$ is a singular one: its rank is smaller than its order by the number of the degrees of freedom of the structure as a solid body. The arbitrary rigid motion can be eliminated by imposing the appropriate number of boundary conditions restricting this motion. In the general case, the number of these conditions must be equal to six. In the case of a two-dimensional problem, it is sufficient to impose three restrictions on the motion of the nodal points. After excluding from the matrix $\mathbf{K}$ the rows and columns that correspond to the restricted displacements, we obtain a nonsingular stiffness matrix $\hat{\mathbf{K}}$. Then the equilibrium condition (17.32) can be rewritten as

$$\hat{\mathbf{K}}\hat{\mathbf{\Delta}} = \hat{\mathbf{F}}$$

i.e., in the form of the equation obtained in Section 13.4. Here $\hat{\mathbf{\Delta}}$ and $\hat{\mathbf{F}}$ are obtained from the matrices Δ and F by eliminating the rows and columns corresponding to the restricted degrees of freedom. Note that the elimination of the rows and columns in the stiffness matrix is a very complex operation since it is associated with the changes in the contents of all the matrices in the computer's memory. It should also be emphasized that obtaining the stiffness matrix on the basis of Eq. (17.33) is a very complex procedure as well since the size of the matrix $[H]$ can be very large. For instance, the FEM calculation of a relatively simple rod system consisting of 80 elements and 30 nodes results in a 960×180 matrix $[H]$. This requires approximately 20,000 memory cells for this matrix alone.

In conclusion, we would like to point out that although the force method can also be used in FEM analyses, it has several important shortcomings compared to the displacement method and therefore is not recommended.

17.3.4 Accuracy of the Finite-Element Method

In order to illustrate the accuracy that could be achieved by using FEM analysis, examine a rod of length a subjected to a uniform tensile load p (Fig. 17.7). Treating this rod as a single element with the end $x = 0$ fixed and the end $x = a$ free to extend, we have: $q_1 = 0$, $F_2 = pa$. Then Eq. (17.31) yields

$$q_2 = \frac{pa^2}{2EA}$$

Then, in accordance with Eq. (17.19), the axial displacements are distributed along the rod as follows:

$$q(x) = \frac{pa}{2EA} x$$

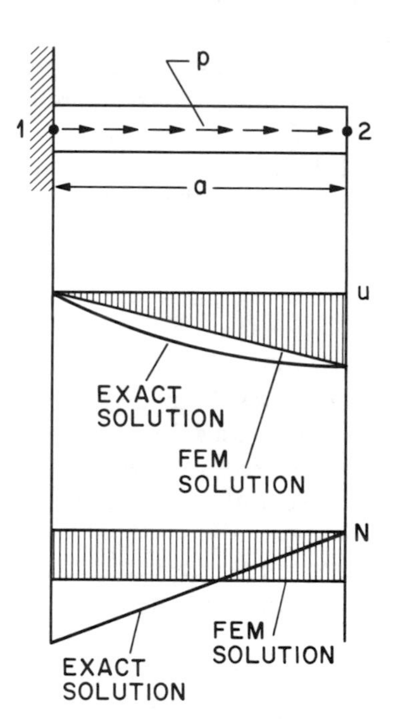

FIGURE 17.7 Rod subjected to uniform tensile load.

The exact solution can be obtained from the equation

$$EAq''(x) = -p$$

where the function $q(x)$ must satisfy the following boundary conditions:

$$q(0) = 0 \qquad q'(a) = 0$$

Then we obtain

$$q(x) = \frac{qx}{2EA}(2a - x)$$

so that

$$q_2 = q(a) = \frac{pa^2}{2EA}$$

Thus, the finite-element method results in the exact values of the displacements in the nodes but leads to approximate values of the displacements for the cross sections within the span of the beam.

As far as the axial force is concerned, the FEM predicts

$$N = \frac{pa}{2}$$

for the element by averaging the total force value, whereas the exact solution yields

$$N(x) = pa\left(1 - \frac{x}{a}\right)$$

In this connection, it should be emphasized that, by its very nature, the FEM cannot address problems associated with singularities; however, the predictions based on this method can be improved significantly if finer meshes are used.

18

Experimental Techniques

18.1 INTRODUCTORY REMARKS

As has been indicated in the introduction to this book, the overwhelming majority of studies dealing with mechanical behavior and performance of microelectronic components and systems are experimental. For many years, experimental stress analysis has been an important factor in the analysis and design of high-technology systems. Here are examples of problems in which experimental methods are indispensable:

- Evaluation of the mechanical characteristics of materials, with consideration of the environmental conditions (temperature, moisture, electromagnetic influence, etc.), the state of stress, stress levels, and other factors
- Search for, and evaluation of, various hypotheses that form, or may form, a foundation for a theory
- Provision of input data for theoretical analyses, and determination of the viability, accuracy, and limits of application of theoretical approaches
- Investigation of the type and magnitude of external forces, acting on a given component or structure
- Evaluation of the impact of various technological factors, deviations from ideal geometrical shapes, discrepancy of mechanical properties
- Evaluation of the internal forces, stresses, and strains in these cases for which theoretical techniques are unavailable or too complicated or for which the mechanical characteristics of materials are unknown, which is quite typical of microelectronic and fiber-optic materials

Unlike theoretical models, experimental research models are, as a rule, of the same physical nature as the actual component, although they reproduce the actual object in a simplified way and quite often on a different scale. In some cases, experimental models are based on rather frequently encountered similarities in the mathematical description of different objects. These are called analogous models. They are successfully used if one of the similar phenomena can be easier reproduced experimentally, is more visible, or is more accessible for measurements. For instance, the distribution of the shearing stresses over the cross-sectional area of a rod subjected to pure torsion is, under certain conditions, analogous to the distribution of the rotation angles of a deflected flexible membrane (*Prandtl's membrane analogy*). This enables us to replace the experimental investigation of stresses in a rod by measurement of the rotation angles of a membrane, which is much simpler.

The experimental stress analysis includes similitude theory, error theory, data processing, description of experimental equipment, and measurement techniques, as well as experimental design and statistical analysis for the given area of engineering. Since experimental stress analysis is beyond the scope of this book, our effort in this section will be concentrated on the following major experimental methods, most commonly used to evaluate the mechanical behavior of microelectronic and fiber-optic systems:

Method using bonded electrical strain gauges
Photoelastic stress and strain analysis
Moiré method

18.2 BONDED STRAIN GAUGES

In 1938, E. E. Simmons, Jr., at the California Institute of Technology, and A. Ruge, at Massachusetts Institute of Technology, independently developed techniques for bonding fine wire to structural elements. When these elements were loaded, the displaced surface transmitted all strains directly to the wire. This resulted in the bonded-wire strain gauge and later in the bonded-metal fail gauge.

The electrical resistance of a wire conductor is

$$R = \rho \frac{l}{A}$$

where ρ is the specific resistance of the wire material, l the length of the wire, and A its cross-sectional area. When the wire, bonded to a tested specimen, is subjected to tension and increases its length by the magnitude Δl, its cross-sectional area becomes smaller. In addition, the specific resistance of the material also changes. Similar phenomena, but in the opposite direction, take place when the wire experiences compression. Thus, the following relationship exists between the relative change in the electrical resistance of a wire conductor and its strain:

$$E = \frac{\Delta l}{l} = k\frac{\Delta R}{R} \tag{18.1}$$

If the axial deformation of the wire were accompanied only by the change in its length, while the ρ and A values remained unchanged, then the factor k would be equal to unity. In reality, however, because of the inevitable change in ρ and A, this factor is greater than unity and is usually within the range 2.0–3.5. The k value is called *strain sensitivity*, or *gauge factor*, and is determined experimentally. By measuring the change in the electrical resistance ΔR, we can calculate the strain E on the basis of the formula (18.1). This is the basic principle of the experimental method using bonded strain gauges. The electrical resistance of a gauge is determined by applying an electrical potential to its ends so that the gauge must be electrically isolated from the tested specimen if the latter is also a conductor. A typical foil strain gauge configuration is shown in Fig. 18.1. The length b is called the *gauge base*.

In practice, the gauge factor is lower than that of the strain sensitivity of the grid material. This is due to the "end loops" in a fail or wire grid pattern, resulting in larger ΔR. In addition, this pattern causes an undesirable *transverse sensitivity* of the gauge. The transverse sensitivity is defined as the ratio of gauge response perpendicular to the grid versus gauge response parallel to the grid;

$$K_T = \frac{F_T}{F_L}$$

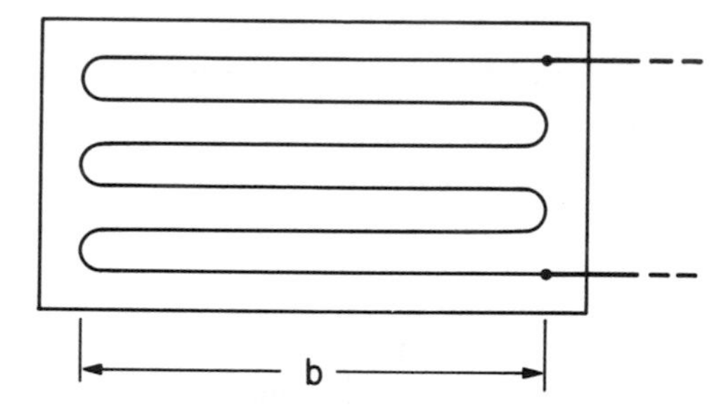

FIGURE 18.1 Typical foil strain gauge.

The transverse sensitivity value is measured by the manufacturer and is indicated in the gauge specifications. The arrangement of the wire in loops enables us to obtain a significant wire length for the given base value, thereby increasing its electrical resistance and the accuracy of the measurements. The base b should not be larger than the length of the more or less constant longitudinal deformation of the specimen at the location of the measurement point; otherwise, the gauge will record the mean value of the normal strain within the base b rather than the actual local strain. The base of modern gauges is between 3 and 40 mm.

When testing gauges, any change in temperature of the structure will result in the thermostatlike effect caused by the thermal expansion or contraction mismatch between the gauge and the specimen. This effect can be compensated for by employing an arrangement known as a *Wheatstone bridge* (Fig. 18.2). In this arrangement, a dummy gauge is used. It is applied to material similar to the test specimen and is placed in the adjacent lag of the wheatstone bridge in a location where the temperature of this material is approximately the same as the temperature of the test specimen in the area of the gauge location. In this case, the change in the resistance of the lags R_1 and R_4 due to thermally induced strains will be the same and will not result in any imbalance of the bridge.

Let the variable resistance R_r in Fig. 18.2 be in the zero location ($\Delta R = 0$) and the tested specimen be unstrained ($\Delta R_g = 0$) so that the resistance of the active gauge is $R_1 = R_g - \Delta R_g = R_g$ and the resistance of the dummy gauge $R_y = R_g + \Delta R = R_g$. From Fig. 18.2, the Wheatstone bridge is balanced when

$$\frac{R_1}{R_2} = \frac{R_4}{R_3}$$

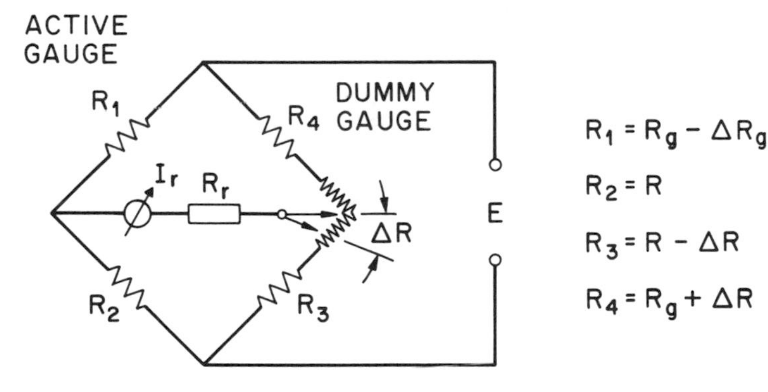

FIGURE 18.2 Wheatstone bridge.

This condition is fulfilled since $R_1 = R_4 = R_g$ and $R_2 = R_3 = R$. Hence, the bridge is balanced, and the measured current I_r is zero (there is no current in the variable resistance). If the resistance R_g of the gauge is changed as a results of the loading of the specimen under testing, so that its new resistance is $R_1 = R_g - \Delta R_g$ and the resistances R_3 and R_4 are changed by using the variable resistance by a small value ΔR, then we can obtain the following formula for the measured current I_r:

$$I_r = \frac{E}{2(R_g + R)}\left[\frac{\Delta R_g}{R_g} - \frac{\Delta R}{R_g}\left(1 + \frac{R_g}{R}\right)\right] \tag{18.2}$$

This formula gives sufficiently accurate results if the resistance R_r is significantly smaller than the resistances R and R_g.

The change ΔR_g in the resistance $R_1 = R_g$ of the active gauge can be measured on the basis of the formula (18.2) by one of the following two methods.

Let an initially balanced bridge become unbalanced because of the applied load, which results in the change ΔR_g in the resistance of the active gauge. Then, using the variable resistance R_r, we find a resistance ΔR that results in $I_r = 0$. Then Eq. (18.2) yields

$$\Delta R_g = \Delta R\left(1 + \frac{R_g}{R}\right)$$

This method is sometimes referred to as the *zero method.*

In the second method, the variable resistance is not used to bring the current I_r to a zero value. Instead, the actual I_r value is measured. With $\Delta R = 0$, the formula (18.2) yields

$$\frac{\Delta R_g}{R_g} = I_g \frac{2(R_g + R)}{E}$$

In a uniaxial stress-strain problem, the state of stress may be determined from the measured strain by Hooke's law $\sigma = EE$. In a two-dimensional stress field, the state of stress, if the direction of the principal stresses is known, can be obtained from the two principal strains E_1 and E_2 as

$$\sigma_1 = \frac{E}{1 - \nu^2}(E_1 + \nu E_2) \qquad \sigma_2 = \frac{E}{1 - \nu^2}(E_2 + \nu E_1)$$

When the principal stress directions are not readily apparent, strain gauges may be used in *rosette configurations* to determine the principal strain values and directions. A rosette is, as a rule, three or four gauges bonded in the same area (in the same "point") in different directions. A *rectangular rosette* (Fig. 18.3a) and *delta rosette* (Fig. 18.3b) are the most commonly used. Rectangular rosettes are employed when the principal directions at the given point are approximately known; then the rosette gauges are oriented in these directions. Then the extreme values of the strain and the angle α are calculated by the formulas

$$E_{\substack{\min \\ \max}} = \frac{1}{2}(E_1 + E_2) \pm \frac{1}{\sqrt{2}}\sqrt{(E_1 - E_2)^2 + (E_2 - E_3)^2}$$

$$\tan 2\alpha = \frac{2E_2 - E_1 - E_3}{E_1 - E_3}$$

The delta rosette enables us to find the principal strains and directions for an arbitrary orientation of the rosette. The calculation formulas in this case are

$$E_{\substack{\min \\ \max}} = \frac{1}{3}(E_1 + E_2 + E_3) \pm \frac{\sqrt{2}}{3}\sqrt{(E_1 - E_2)^2 + (E_2 - E_3)^2 + (E_3 - E_1)^2}$$

$$\tan 2\alpha = \sqrt{3}\,\frac{E_2 - E_3}{2E_1 - E_2 - E_3}$$

In addition to the direct determination of strain, bonded gauges are employed for force and motion measurements. When strain gauges are

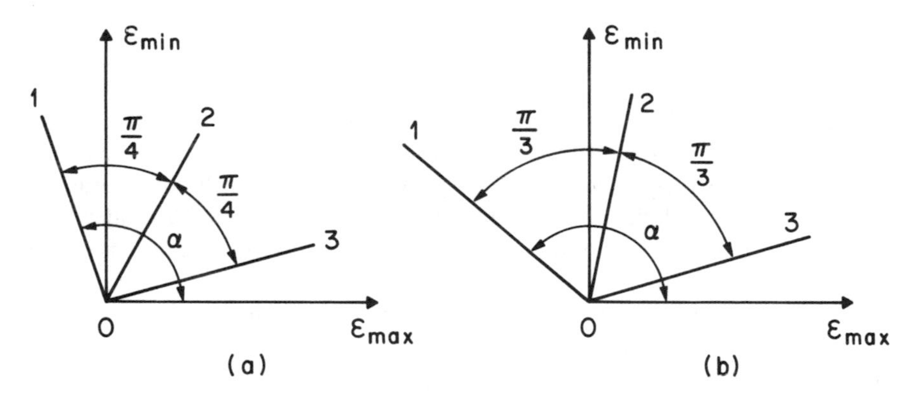

FIGURE 18.3 Rosettes: (a) rectangular rosette; (b) delta rosette.

designed as part of such devices, they are referred to as *transducers* and play a large part in industrial and laboratory applications. Such transducers are employed in weighing scales, displacement devices, torque meters, dynamometers, and pressure-measuring equipment.

18.3 PHOTOELASTIC ANALYSIS

In 1816, Sir David Brewster observed that glass exhibited color patterns when viewed under stress in polarized light. Since the early 1930s, the term *photoelasticity* has been employed to describe the experimental techniques for the use of transparent plastic materials to model a two-dimensional state of stress. As shown in Section 6.2, the two-dimensional state of stress for isotropic bodies is independent of the elastic constants and, therefore, the stresses in identically loaded bodies of the same configuration of the boundary will be the same for any material (the Maurice Lévy theorem). Hence, in order to model a two-dimensional stress condition of a single-connected body, it is sufficient to ensure the geometrical similarity of the model and the full-scale object, as well as the similarity of the loadings. For multiconnected bodies, such a modeling is possible only in the case in which the loads in each of the internal closed contours are self-equilibrated. Otherwise, the state of stress in a multiconnected body is elastic-constant–dependent, and there is no similarity in the states of stress and strain for bodies made of different materials.

Photoelastic stress analysis is based on the property of some transparent materials to become *doubly refracting* when stressed. Application of external loading makes such *optically active materials* optically anisotropic, and their principal optical axes are oriented in the directions of the principal stresses.

The photoelastic effect is achieved by placing an optical filter in the way of light and confining it to one plane—the plane of polarization (Fig. 18.4). A basic *plane polariscope* is an assembly consisting of a light source and two polarizing elements (Fig. 18.5). The element nearest to the light is called the *polarizer* and the second element the *analyzer*. When the planes of both elements are parallel, light in the plane of the page will be able to pass through to the camera or viewer. This arrangement is called a *parallel polariscope*. By slowly rotating the analyzer, the observer will note a gradual diminishing of the light intensity until the element is rotated 90°, at which time the light emission is completely extinguished. This arrangement is called a *crossed polariscope*.

For any relative orientation of the polarizer and the analyzer, when a specimen is placed between them and then loaded, the ray of light is split into two components propagating at right angles to each other, which are also the planes of principal stresses in the specimen. In addition to this, the

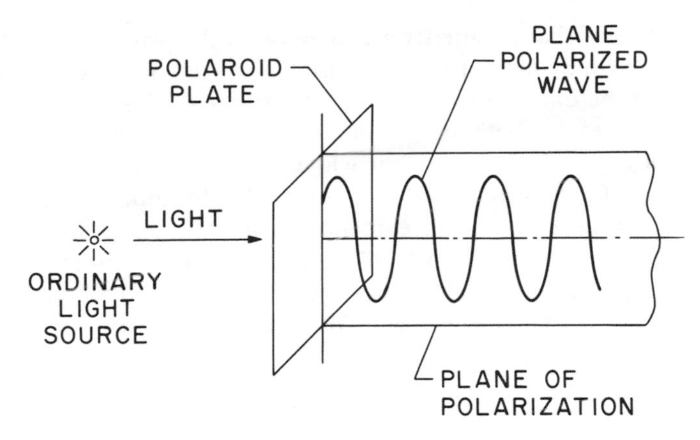

FIGURE 18.4 Plane-polarized light.

light travels along such mutually perpendicular paths at two different velocities, depending on the magnitude of the principal stresses. The lag of these velocities is proportional to the algebraic difference between the principal stresses. On the other hand, there is also a phase lag between the two light rays after they pass through the specimen. This phase lag is also proportional to the difference in the principal stresses and can be measured

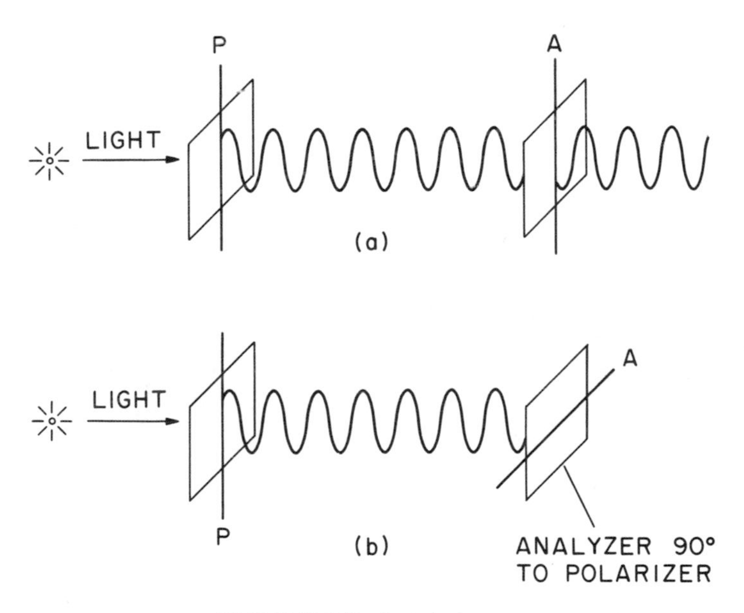

FIGURE 18.5 Basic polariscope.

by means of, say, the ray interference method. If the proportionality factors are known, then, by measuring the phase lag of the polarized light rays, we can evaluate the difference in the principal stresses. This is, in essence, the principal basis of the method. One can use either monochromatic light, i.e., a light with a given wavelength, or white light, which is a combination of numerous light rays with different wavelengths. In addition, the light can either be plane-polarized or have circular polarization.

In Fig. 18.6, we show how a plane-polarized monochromatic light is applied. Let the plane of polarization be located vertically and, hence, the light is also polarized in the vertical plane. Then the light ray can be presented

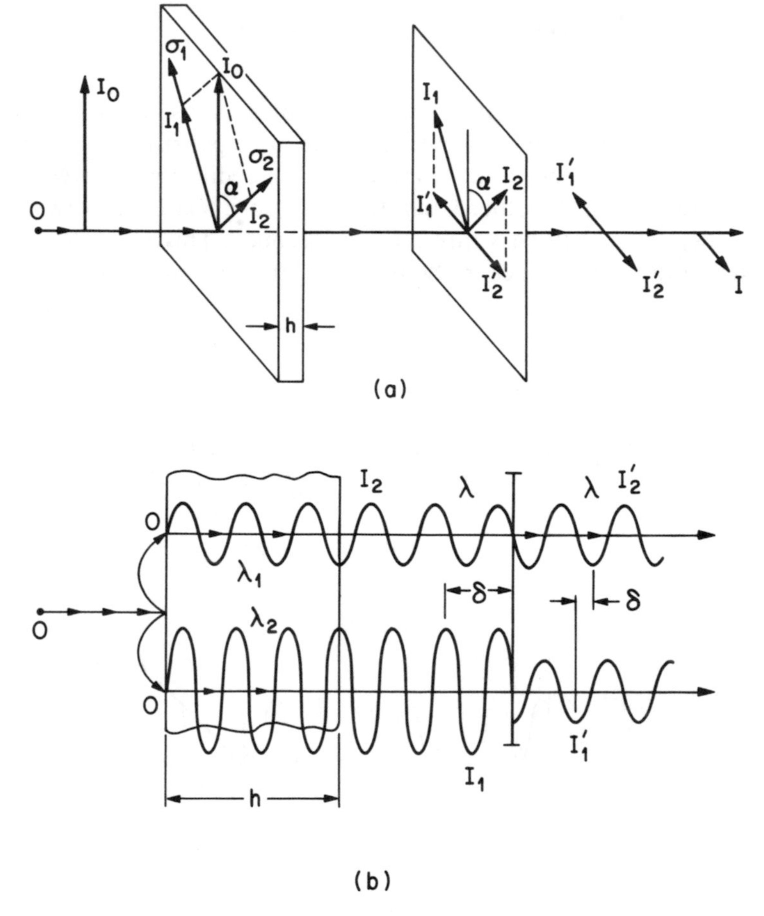

FIGURE 18.6 Plane-polarized monochromatic light.

as a wave as follows:

$$I_0 = A \sin \phi$$

where A is the amplitude (intensity) of light and ϕ is the phase angle. At the stressed specimen, the ray I_0 splits into two rays I_1 and I_2, with mutually perpendicular planes of polarization. If $\sigma_1 \neq \sigma_2$, then these rays propagate at different velocities when passing through the specimen's thickness. This results in a phase lag between the rays after they leave the specimen. Further, when traveling in the air, these rays have equal velocities while the obtained phase lag and the mutual perpendicularity of the planes of polarization remain. This can be presented as follows:

$$I_1 = A \sin \alpha \sin\left(\phi + 2\pi \frac{\delta}{\lambda}\right)$$

$$I_2 = A \cos \alpha \sin \phi$$

where δ is the phase lag and λ the wavelength. When the interference method is used to measure the phase lag, the two rays I_1 and I_2 must be superposed in one plane. This can be done by putting them through a polarizer whose plane of polarization is perpendicular to the plane of polarization of the ray I_0 before it passes through the specimen. As indicated earlier, the latter polarizer is called an analyzer and, together with the first element, the original polarizer forms a polariscope. After the rays I_1 and I_2 pass through the analyzer, they reduce to the following projections on the plane of polarization (Fig. 18.6):

$$I'_1 = -I_1 \cos \alpha = -A \sin \alpha \cos \alpha \sin\left(\phi + 2\pi \frac{\delta}{\lambda}\right)$$

$$I'_2 = I_2 \sin \alpha = A \sin \alpha \cos \alpha \sin \phi$$

These two rays, as a result of their interference behind the analyzer, give the following resulting ray:

$$I = I'_1 + I'_2 = \frac{1}{2} A \sin 2\alpha \left[\sin \phi - \sin\left(\phi + 2\pi \frac{\delta}{\lambda}\right)\right] \tag{18.3}$$

As mentioned earlier, the phase lag δ is proportional to the difference in the principal stresses, as well as to the thickness of the specimen plate:

$$\delta = hc(\sigma_1 - \sigma_2) \tag{18.4}$$

Here c is the stress optical coefficient, which is a material constant.

As we can see from Eq. (18.3), the brightness of the resulting ray I behind the analyzer depends on the angle α and the phase lag δ or, as follows from Eq. (18.4), on the difference in the principal stresses $\sigma_1 - \sigma_2$. When the angle α is zero, or $\pi/2$, $3\pi/2$, $5\pi/2$, etc., i.e., when the direction of one of the principal stresses at the given point of the specimen coincides with the plane of polarization, then the ray passing through this point will not get through and will not light the screen. Thus, there will be dark bands (*fringes*), called *isoclinics*, on the screen. In all the points of the isoclinics, one of the principal stresses is located in the plane of the polarization of the light, i.e., in the plane of the polarizer. By turning the polariscope in the plane perpendicular to the direction of the ray propagation, we can obtain isoclinics corresponding to different angles of rotation of the principal stresses, and then to the trajectories of the principal stresses.

The dependence of the brightness of the resulting ray I on the phase lag δ is exhibited in accordance with Eqs. (18.3) and (18.4) as follows. If $\delta = 0$, which corresponds to the stress-free situation $\sigma_1 = \sigma_2 = 0$ or to the condition $\sigma_1 = \sigma_2$, then there will be no light behind the analyzer. These points are called *isotropic*. The same effect can be observed when $\delta = \lambda, 2\lambda, \ldots, n\lambda$, i.e., when the difference in the principal stresses is such that the phase lag between the rays is equal to a whole number of the wavelengths. If the phase lag is $\delta = \lambda/2, 3\lambda/2, \ldots, [(n+1)/2]\lambda$, then the brightness I of the ray is maximum. When the phase angle δ is changed, the brightness of the ray is also changed from zero to the maximum value. Thus, the field of the specimen on the screen will be covered by dark and bright fringes, with a graduated change from shade to light. Every fringe of a constant brightness connects the points with equal difference of the principal stresses. These fringes are called *isochromes*. The dark isochrome corresponding to $\delta = 0$ ($\sigma_1 = \sigma_2 = 0$ or $\sigma_1 = \sigma_2$) is called the *zero isochrome*. The isochrome $\delta = \lambda$ is called an isochrome of the first order, and the isochrome $\delta = 2\lambda$ is called an isochrome of the second order, etc. Hence, the stress field of the specimen will be covered by both isoclinics and isochromes. When the polariscope is rotated, the isoclinics change their positions on the screen while the isochromes remain unchanged. This rotation can be used to obtain a picture of the isochromes that is not interfered with by the isoclinics. This is necessary for measuring the phase lag. If the polariscope is rotated fast enough, then the isoclinics, moving rapidly over the screen, will become invisible. The same effect can be obtained by using a light having circular polarization. A ray with circular polarization consists of two parallel polarized rays with two mutually perpendicular planes of polarization, equal amplitudes, and a phase lag equal to a quarter of the wavelength. The resulting vector of these two rays will change its orientation continuously, rotating around the axis of the ray

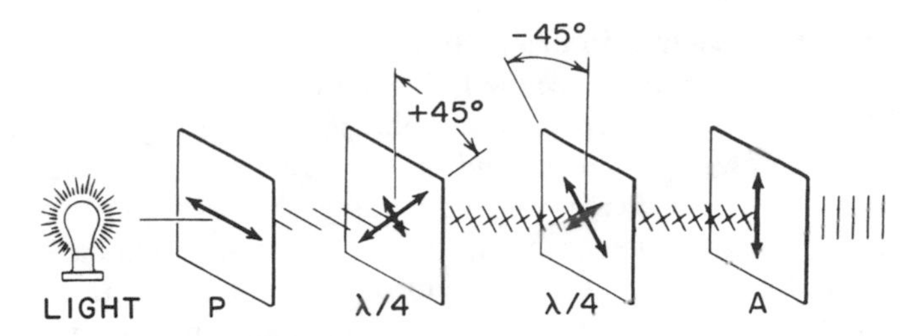

FIGURE 18.7 Circular polariscope.

propagation, which is similar to the rotation of the polariscope. In order to obtain circular polarization, it is necessary to put a plane-polarized ray through a plate whose optical axes are oriented 45° to the plane of the ray polarization and that creates a phase lag equal to a quarter of the wavelength (a so-called *quarter-wave plate disk*). Such circular polariscope is shown in Fig. 18.7. If we put in front of an analyzer a quarter-wave plate disk whose effect is opposite to the effect of the first disk, then the interference picture will be the same as it is in the absence of the quarter-wave plate disks and will depend only on the phase lag of the rays or on the difference in the principal stresses. In this case, the ray will not disappear on the isoclinics, and we can obtain a picture of the isochromes not interfered with by the isoclinics. In Fig. 18.8, the isochromes in a circular plate subjected to two diametric forces are shown.

FIGURE 18.8 Isochromes in a diametrically loaded disk.

If we observe the change in the isochromes with an increase in the loading of the specimen, then we will see the following picture. Before the loading is applied ($\sigma_1 = \sigma_2 = 0$), the polarized light passes through the specimen without undergoing any changes and is extinguished by the analyzer. The screen is dark. With an increase in the loading, the first light occurs in the points with the largest difference in the principal stresses, then darkness, then light again, and so on, until the very end of the load application. A similar change in light and dark occurs in all the points, depending on the change in the difference $\sigma_1 - \sigma_2$. The new isochromes appear in the points of maximum differences $\sigma_1 - \sigma_2$ and gradually shift to the areas of smaller differences. The higher the stress level, the denser the isochromes.

In order to determine the difference $\sigma_1 - \sigma_2$ on the basis of the isochromes, we first have to establish experimentally the so-called *fringe stress coefficient*, i.e., the value of the difference in the principal stresses that causes a phase lag of one wavelength for a specimen of unit thickness:

$$f = \frac{h}{N}(\sigma_1 - \sigma_2) \tag{18.5}$$

In this formula, N is the number of fringes, h the specimen thickness, and f the fringe stress coefficient. Indeed, with the known wavelength λ for the monochromatic light source, the relative retardation of light δ is related to the wavelength λ and the number of fringes N as follows:

$$\delta = N\lambda$$

On the other hand, the relative retardation is expressed by the formula (18.4). After excluding the δ value, we find

$$N\lambda = hc(\sigma_1 - \sigma_2)$$

Defining the fringe stress coefficient f as the ratio of the wavelength to the stress optical coefficient

$$f = \frac{\lambda}{c}$$

we obtain the formula (18.5). The fringe stress coefficient is determined by using a specimen of the same material, which is subjected to a known state of stress, say, to a uniaxial loading under the stress σ. In this case, $\sigma_1 = \sigma$, $\sigma_2 = 0$, $\sigma_1 - \sigma_2 = \sigma$. If the changes in the light/darkness occurred N times when the stress σ changed from zero to σ_1, then, with the thickness h of the plate, the fringe stress coefficient is

$$f = \sigma_1 \frac{h}{N}$$

With the known f value, the difference in the principal stresses can be found from Eq. (18.5) as

$$\sigma_1 - \sigma_2 = \frac{Nf}{h}$$

This equation is the basic solution employed in the photoelastic method of stress analysis. The smaller the fringe stress coefficient, the more optically active the material, and the denser the fringe pattern of the isochromes for the given state of stress. This enables us to evaluate with greater accuracy the difference in the principal stresses.

It is important to emphasize that photoelastic stress analysis enables us to determine the directions (isoclinics) and the algebraic differences (isochromes) of the principal stresses in any point of the specimen but does not contain any information that would provide a way to evaluate the stresses themselves. In order to evaluate each of the principal stresses, we can measure the change in the specimen's thickness Δh in the given points and evaluate the strain $\varepsilon = \Delta h/h$ in the through-thickness direction. Then the sum of the principal stresses can be determined by the formula

$$\sigma_1 + \sigma_2 = -\frac{\varepsilon}{\nu} E$$

known from the two-dimensional theory of elasticity. This approach, however, requires very sensitive equipment for measuring small Δh values.

18.4 MOIRÉ METHOD*

One of the oldest, but still very much alive, experimental techniques for full-field evaluation of the in-plane and/or out-of-plane displacement patterns is the *moiré method.* There are two main types of moiré: geometric and interferometric.

18.4.1 Geometric Moiré

Moiré is the French word for a fabric known as "watered silk," which exhibits patterns of light and dark bands. This moiré effect occurs whenever two

* This section was written by A. S. Voloshin, Lehigh University.

similar, but not quite identical, arrays of equally spaced lines or dots are arranged so that one array can be viewed through the other. Almost everyone has seen the effect in two parallel snow fences or in two layers of window screen placed one over the other.

The geometric interpretation of the moiré effect was first published by Tollenaar in 1945. In 1960, Morse, Durelli, and Sciammarella presented a complete analysis of the geometry of moiré fringes in strain analysis, in which the fundamental equations of the moiré method are derived and presented in the form of curves that can be used to give strains and rotations with a minimum of computation.

A second method of analysis of moiré fringe patterns, presented by Weller and Shepard in 1948, described an application in which moiré fringes were used to measure displacements. In 1954, Dantu followed the same approach, introducing the interpretation of moiré fringes as components of displacements for plane elasticity problems. Moiré fringes have also been used by Theocaris to measure out-of-plane displacements and by Lightenberg to measure slopes and moment distributions in flat slabs. The use of moiré fringes for measuring displacements in structural models has been outlined in a paper by Durelli and Daniel.

When two gratings are superimposed, one against another, the interference moiré fringes are the result of their having a difference either in pitch (defined as the distance between the center of any two neighboring lines) or in orientation. This is shown in Fig. 18.9. Either the dark or light bands may

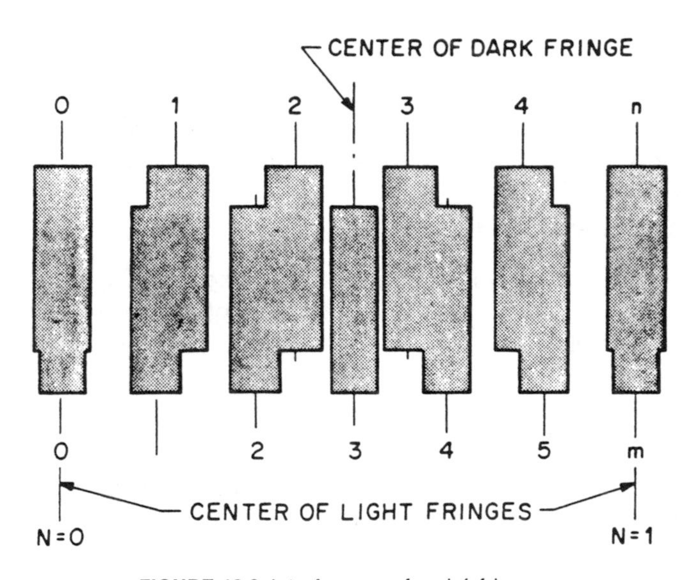

FIGURE 18.9 Interference of moiré fringes.

be called moiré fringes. If this pattern is placed at a far distance so that the eye can no longer resolve the individual lines, we would see only the moiré fringes. The centerlines of the dark gratings may be denoted with numbers 0 to m for one grating and 0 to n for the other; and the centerlines of light fringes are denoted by the numbers 0 to N, where N is an integer referred to as the fringe order. It is easy to see from Fig. 18.9 that N and the parameters m and n identifying the lines of the two interfering gratings are related by the relation

$$N = m - n \tag{18.6}$$

The formation of moiré fringes when two gratings are superimposed may be used to measure in-plane displacements. Here a reference coordinate system, from which all measurements are made, has to be defined. One of the gratings may be selected to serve that purpose; it is called the reference (or master) grating, and the direction normal to the grating lines is called the principal direction. The specimen grating is the one that is attached to a structure surface whose deformation is sought. The pitch p of the specimen grating is usually identical to that of the reference grating before deformation. To find the relation between the number of fringes N and the displacement U (in a direction normal to the grating lines), we may consider a specimen with gratings of pitch p that is in perfect alignment with a master grating of the same pitch. Since the two gratings are assumed to be in perfect alignment, no fringes are formed in the undeformed state. If, over a certain length, the number of undeformed grating lines is m and the specimen is made to deform by an amount equal to the pitch, one moiré fringe forms and the number of lines over the same length changes by one (decreases if the specimen is stretched and increases if compressed). It can be shown that, in the general case in which the number of lines in the deformed state becomes n, the corresponding displacement may be given by

$$U = (m - n)p \tag{18.7}$$

The number of fringes formed is related to m and n by Eq. (18.6) and, therefore, Eq. (18.7) may be rewritten as

$$U = Np \tag{18.8}$$

This is a basic equation in the moiré method for displacement measurements. All that is needed, in addition to knowledge of the pitch p, is the number of fringes formed, which may be obtained by simple fringe counting over the line of interest. A major limitation on the choice of such a line is that it has to start and end at fringe centers. Starting and ending fringes need not be of the same type. The equation is also applicable to a fringe pattern in which the first fringe is dark and the last one is bright, or vice versa. Here the displacement resolution of the method is brought to its maximum, which is half the pitch value. No further information is available about intermediate displacements at points not lying at fringe centers. Also, the determination of the fringe centers is a subjective process and is strongly affected by human judgment. Despite these limitations, the geometric moiré has received wide acceptance as a convenient full-field displacement-measuring technique.

18.4.2 Moiré Interferometry

Moiré interferometry combines the concepts and techniques of geometrical moiré and optical interferometry. In 1956, J. Guild showed that all moiré phenomena can be treated as cases of optical interference. In 1980, Post introduced moiré interferometry as a high-sensitivity displacement-measuring technique, along with procedures for producing high-frequency specimen gratings. Since its introduction, several improvements have been added to the technique. This includes simpler optical systems, setups for U and V measurements, optical separation of U and V fields, better methodologies for specimen-grating production, and measurements of out-of-plane displacements, as well as in-plane measurements. Moiré interferometry has increased the sensitivity of displacement measurements by at least an order of magnitude.

A schematic description of moiré interferometry is shown in Fig. 18.10. A diffraction grating (usually crossed-line grating) is produced on the specimen. This is a high-reflection symmetrical, phase-type grating. When loads are applied to the specimen, the grating moves and deforms, together with the specimen surface.

Two beams of coherent light of wavelength λ illuminate the specimen grating obliquely from angles $+\beta$ and $-\beta$. This two-beam interference creates walls of constructive and destructive interference (that is, a virtual grating) in the zone of their intersection; the virtual grating is cut by the plane of the specimen surface, where an array of parallel and very closely spaced fringes is formed. These fringes are essentially bright and dark bars, and they act like the reference grating of geometric moiré. This array of fringes and the walls of interference from which they are derived are called a virtual reference grating. Their frequency F is given by the governing equation for two-beam interference as

$$F = (2/\lambda) \sin \beta \tag{18.9}$$

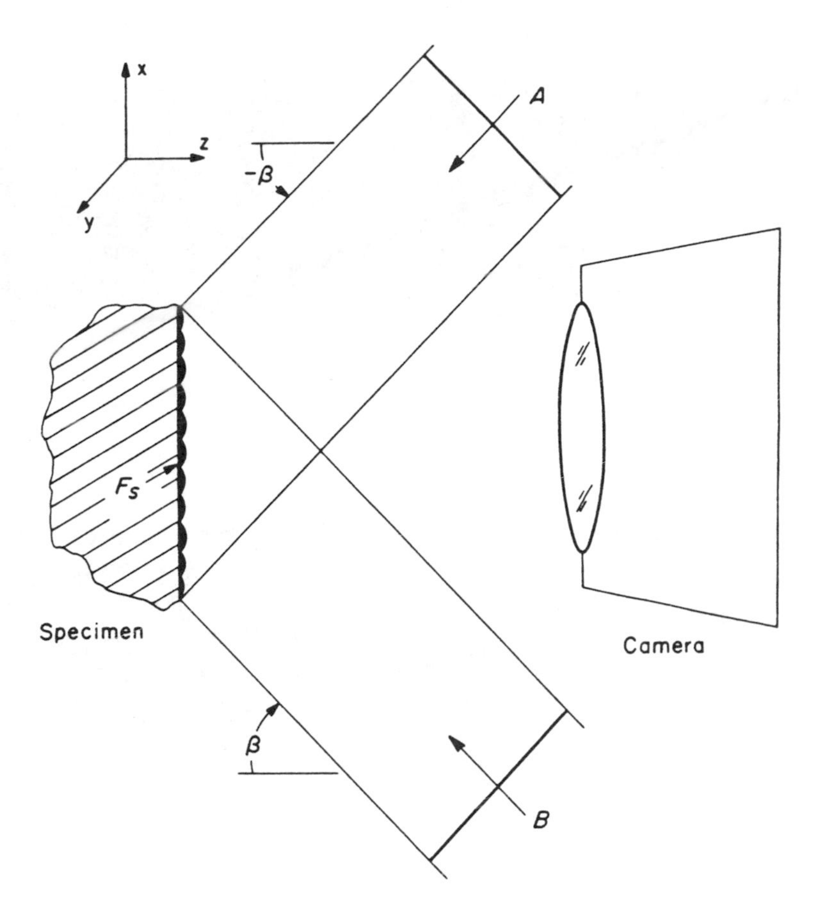

FIGURE 18.10 Schematics of moiré interferometry.

The specimen grating and virtual reference grating interact to form a moiré pattern, which is viewed and recorded by the camera.

Representative patterns of moiré interferometry fringes are shown in Fig. 18.11. They depict the in-plane displacements of every point on the specimen surface as contour maps of equal-displacement fringes. Quantitatively, for each point in the fringe pattern,

$$u = N_x/F$$
$$v = N_y/F \tag{18.10}$$

where u and v are components of displacement in the x and y directions, respectively; N_x and N_y are fringe orders when lines of the reference grating

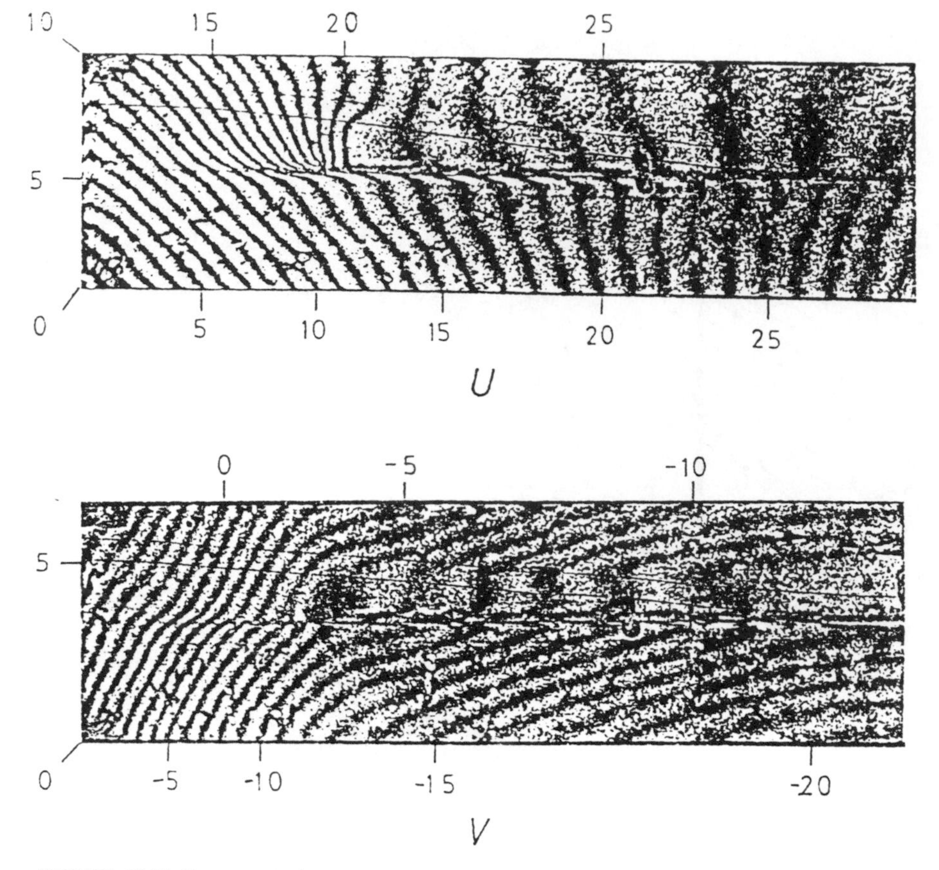

FIGURE 18.11 Representative patterns of moiré fringes: *u* and *v* fringe patterns for the cross section of a thermally loaded electronic package.

are perpendicular to the x and y directions, respectively; and F is the frequency of the reference grating.

To assign fringe orders, it is necessary to define a point of zero displacement as a reference point and to assign $N = 0$ to the fringe passing through that point and then number the adjacent fringes relative to that fringe. The reference point may be any point whose displacement is known. This could be a boundary point, a point at which load is applied, or a point on a symmetry axis for the specimen. For strain analysis, relative displacements rather than the displacements themselves are of interest. Knowledge of the location of a reference point is immaterial. Any arbitrary (and convenient) point can be considered a reference, and the fringe passing through it may be assigned zero order. All other fringes may then be numbered relative to that fringe.

Once the displacement map has been constructed, strains may be determined from strain-displacement relations as follows:

$$\left.\begin{aligned} \varepsilon_x &= \frac{\partial u}{\partial x} \\ \varepsilon_y &= \frac{\partial v}{\partial y} \\ \gamma_{xy} &= \frac{\partial u}{\partial y} + \frac{\partial v}{\partial x} \end{aligned}\right\} \tag{18.11}$$

The derivatives on the right-hand side of Eq. (18.11) may be determined from measured displacements, and the equation may be rewritten as

$$\left.\begin{aligned} \varepsilon_x &\approx \frac{\Delta u}{\Delta x} \\ \varepsilon_y &\approx \frac{\Delta v}{\Delta y} \\ \gamma_{xy} &\approx \frac{\Delta u}{\Delta y} + \frac{\Delta v}{\Delta x} \end{aligned}\right\} \tag{18.12}$$

Use of Eq. (18.12) requires that Δx and Δy be sufficiently small in order to have faithful representations of the derivatives. In order to have small increments Δx and Δy, it is necessary to collect displacement data from the fringe pattern at points that are very closely spaced. This is possible only if the fringe density of the pattern is very high. High fringe densities in moiré patterns are easily acquired when the displacements are high, such as in cases of high loads and compliant materials. However, cases of low fringe densities are often encountered, usually when stiff structures are considered and when applied loads are limited to small values that are inadequate for inducing high displacement levels. A more complicated situation arises when both high and low fringe densities exist at different regions of the same field. Low fringe density does not yield sufficient displacement data for reliable strain analysis when simple fringe counting is used to arrive at the displacement data.

Another issue that has to be taken into account when using the moiré method for displacement and strain analysis is the existence of a null field.

In general, the specimen grating will not have lines that are perfectly straight and uniformly spaced. Also, the optical elements used to form the reference grating will not be so accurate that a perfect reference grating is formed. The result is that a few fringes will usually appear in the field of view before the specimen is loaded. These must be subtracted from the pattern obtained after loading in order to determine the load-induced displacements. This can be done manually by subtracting the fringe orders at corresponding points in the load and no-load pattern. Doing so further reduces the amount of displacement data since, for each point considered in the load pattern, there should exist a corresponding point that lies on a fringe center in the no-load pattern. This condition is nearly impossible to satisfy. Thus, the commonly adopted approach to coping with the null field is to ignore it after assuming that the initial pattern is sparse while the full-load pattern exhibits a great number of fringes. This will not work for cases in which the full-load pattern has a low fringe density or when transient analysis is desired.

To relax the high fringe density requirement and extend the useful utilization of moiré interferometry to patterns of low fringe density, with the initial patterns properly taken into account without limiting the number of displacement data points extracted from the field, a new technique was recently introduced utilizing modern digital image analysis.

The approach employs fractional fringe analysis to compute the displacements at any point in the field (the point need not be a fringe center). Digital image processing is used to accurately collect light intensity information needed for the computations. The introduced approach also eliminates ambiguities associated with the determination of the locations of fringe centers for fringe counting, thus reducing random errors in the collected data. Moreover, this approach enables fast and automated analysis of moiré fringe patterns while increasing the sensitivity of the method by up to two orders of magnitude.

The optical law for moiré interferometry may then be expressed in terms of the continuous fringe order $\Psi(x)$ rather than the displacement as

$$I(x) = I_0 + I_1 \cos 2\pi\Psi(x) \tag{18.13}$$

It can be seen that each time $\Psi(x) = n$, where n is an integer, the intensity will be at a maximum (at the center of a bright fringe). When $\Psi(x) = 1/2\,(2n + 1)$, the intensity will be at a minimum (at the center of a dark fringe). The bright fringes are the loci of points at which the displacements, in the x direction, of the specimen grating with respect to the reference grating are equal to an integer number multiplied by the pitch of the master grating. The dark fringes have similar interpretation but in terms of half-pitches. When $\Psi(x)$ is an integer or an odd multiple of one half, the displacement information is basically the same as would be obtained by simple fringe counting as evident from Eq. (18.8). This is where fringe counting stops. No displacement information is available about other points in the field between the bright fringe at $\Psi = 0$ and the dark fringe at $\Psi = 1/2$. This is where Eq. (18.13) picks up. Knowledge of the light intensity, $I(x)$, at the point x, together with the amplitudes I_0 and I_1, results in the fact that $\Psi(x)$ is a fraction between 0 and 1/2. This function determines the displacement at point x when multiplied by the pitch of the reference grating. The same statement is true for any intermediate point between any two fringes. This is where the name "fractional fringe analysis" comes from. The fringe orders need not be integers (or odd multiples of 1/2); they may be any fractional number.

Equation (18.13) may be rewritten to yield the displacements directly as follows:

$$U(x) = (1/2\pi F) \operatorname{arc\,cos} [(I(x) - I_0)/I_1] \tag{18.14}$$

The preceding equation may be applied over each half-fringe separately, thus determining the displacements relative to the starting point. The actual cumulative displacements are found by adding the displacement at the starting points to all values determined. Mathematically, this is expressed as follows:

$$U(x) = U_0 + (1/2\pi F) \operatorname{arc\,cos} [(I(x) - I_0)/I_1] \tag{18.15}$$

where U_0 is the displacement at the starting point, and x is zero at that point. U_0 is the reference grating pitch multiplied by either an integer or an odd multiple of 1/2, depending on whether the starting point is the center of a bright or a dark fringe. The amplitudes I_0 and I_1 are determined from light intensity.

The task of determining the displacement field has become that of determining the light intensity field. A convenient way to record light intensity involves video cameras in conjunction with digital image–processing systems. Recent advances in the field of digital image analysis have led to significant progress in automated measurements of light intensity in photoelasticity over fairly large fields and with high accuracy, as well as in geometric moiré for displacement measurements (A. S. Voloshin, C. P. Burger, R. E. Rowlands, T. S. Richard, and A. F. Bastawros).

The capabilities of enhanced moiré interferometry have recently been applied in two active areas—microelectronics and fracture mechanics. Steady-state, as well as transient, thermal strain analysis in electronic packages and thermal expansion studies in composite sheets used in circuit boards have been investigated by enhanced moiré interferometry. Full-field strain maps were developed for electronic packages at elevated temperatures. Details and levels of total and net strains in the different components of the devices were visible. Transient thermal strains at very small regions of prime interest on the chip were also monitored. Valuable design data have been rendered by this technique.

Enhanced moiré interferometry has also been used to develop an accurate calibration procedure for stress chips by simply applying some external load to the package and measuring the strains at the strain gauge locations while monitoring the resistance changes. Accurate gauge factors were determined, and crude classical calibration methodologies have been eliminated.

The other application considered was the determination of stress intensity factors in Mode I, Mode II, or mixed mode. It has been shown that enhanced moiré interferometry is capable of delivering displacement information, free from null field errors and with minimum random errors, in the density required for accurate and full utilization of available power series solutions for the crack problem.

Enhanced moiré interferometry possesses excellent potential in many areas. An accurate, convenient, computerized full-field displacement-measuring technique of remarkable sensitivity, it can be applied to almost any structure undergoing small deformations. Its potential is evident in microelectronics, where no other approach has ever provided the amount or the quality of data it can provide. Concerns about reliability and increasing complexity in microelectronic devices necessitate reliable techniques such as enhanced moiré interferometry for proper assessment of the mechanical performance of devices. Some of the areas in which this approach has immediate application include delamination problems in plastic IC packages, monitoring of power dissipation in high-power modules, strain analysis of circuit boards, and strain analysis in solder bumps.

Enhanced moiré interferometry may also be used for basic characterization of new, advanced materials such as composites, polymers, and ceramics. For many of these materials, conventional testing methodologies are insufficient to detect details of behavior. Moiré interferometry, with its extensive range and high resolution, will be an appropriate tool to assess these materials.

QUESTIONS AND PROBLEMS

1. How are the notions of concentrated forces and concentrated moments introduced? When can these abstractions be used?
2. What are the major types of beam supports? How are they characterized?
3. What beams are called statically determinate and statically indeterminate? How is the degree of the static indeterminacy defined?
4. Formulate the definitions of the lateral force and the bending moment.
5. What are the hypotheses underlying the engineering theory of beam bending?
6. Write the formulas for the normal stresses acting over the cross sections of a beam subjected to bending. Explain these formulas.
7. What is the neutral axis of a beam cross section? What is the neutral surface?
8. Write and explain the differential relationship in the bending of beams.
9. Write the formula for the shearing stress acting over the cross sections of a beam subjected to bending. Explain this formula.
10. Write the differential equation of bending of beams, as well as the boundary conditions. Explain both the basic equation and the boundary conditions.
11. Derive the expressions for all the elements of bending for a simply supported beam subjected to uniformly distributed lateral load q. Use the solution based on direct integration of the equation of bending. How are the constants of integration determined?
12. Solve a problem similar to problem 11 for the case of three-point bending.
13. What is the essence of the method of initial parameters? What are its advantages over the method of direct integration of the equation of bending?
14. Let the elastic curve of a beam be given by an analytical expression. How can all the other elements of bending be evaluated? Determine the external loading and the forces at the supports.
15. Obtain analytical solutions and draw diagrams for the lateral forces, bending moments, rotation angles, and deflections for a beam subjected to four-point bending. Use the differential relationships between the elements of bending to make sure that there are no errors in the diagrams.
16. Describe the superposition principle. How can it be used to solve statically indeterminate problems of beam bending?
17. Describe the reference tables of beam deflections. How can they be used in engineering practice?
18. Formulate the hypotheses underlying an approximate method of evaluating shear deformation during beam bending. In what cases might the consideration of shear be necessary?
19. What are the properties of Winkler's elastic foundation?
20. Write the equation of bending for a beam on elastic foundation. What is the difference between this equation and the equation of bending of a regular beam?
21. Write and explain the solution of the equation of bending of a beam on elastic foundation in the case of a simply supported beam subjected to a uniformly distributed load.

22. What properties of the functions $V_i(\alpha x)$ facilitate obtaining solutions to the problems of bending of beams on elastic foundation?
23. What parameter determines the effect of the elastic foundation on the elements of bending of a prismatic beam?
24. How is the effect of shear on the theory of beams on elastic foundation considered? When is it important?
25. What are the peculiarities of bending of beams under combined action of lateral and axial loads? Indicate the differential equations of bending in the case of tensile and compressive forces.
26. Can the superposition principle be used in the theory of beams subjected to simultaneous action of lateral and axial forces? How does the axial force affect the elements of bending of such beams?
27. Obtain a solution to the equation of bending of a beam experiencing bending under a uniform lateral load and tension caused by forces applied to its end cross sections. The beam is simply supported at the ends.
28. Obtain the solution to the preceding problem in the presence of an elastic foundation.
29. What parameter characterizes the effect of axial loading on the bending of beams subjected to the combined action of lateral and axial loading?
30. What causes the nonlinear bending of beams subjected to lateral forces if the axial displacements of their end cross sections is restricted? How does such nonlinearity affect the deflections and the stresses in the beam?
31. What causes thermal bending of a single-material beam? What does the maximum normal stress of a beam subjected to nonuniform temperature distribution depend on?
32. What might be the advantages of applying variational methods?
33. How are generalized coordinates, generalized displacements, and generalized forces defined?
34. What are the coordinate functions?
35. Formulate the principle of virtual displacements. From what law does it follow?
36. Formulate the Lagrange theorem.
37. Formulate the reciprocal theorem for forces. Illustrate this theorem using an example of a single-span beam.
38. Formulate Klapeiron's theorem.
39. For which elastic systems is Castigliano's theorem valid? What does it mean?
40. How is the theorem of least work obtained? How can it be used to solve statically indeterminate problems in the theory of rod systems?
41. Formulate the reciprocal property of displacements and the reciprocal theorem for works.
42. On what energy theorems is the Rayleigh–Ritz method based?
43. How should the coordinate functions be chosen in the Rayleigh–Ritz method?
44. Describe the Bubnov–Galerkin procedure in the theory of bending of beams. What does this method have in common with the Rayleigh–Ritz method? How do these methods differ from one another?
45. When is the utilization of the Rayleigh–Ritz and Bubnov–Galerkin methods advisable?

46. Using the Rayleigh–Ritz method, determine the maximum deflection and the maximum bending moment for a beam subjected to a uniform lateral load. Compare the obtained results with the exact solutions, retaining just one term in the obtained series.
47. How is the finite-element method related to the Rayleigh–Ritz method? What are the major advantages of the former method?
48. What is the stiffness matrix of a finite element? What is the global stiffness matrix?
49. What is a library-of-elements type? What is it needed for?
50. What structures are called frames? What is the difference between trusses and frames?
51. What are the major assumptions underlying the theory and calculations of frames?
52. What are complex frames? Why is it advisable to use the deformation method in the theory of such frames?
53. Describe the force method and the deformation method.
54. Define a plate structure. What is a membrane?
55. On what hypotheses is the technical theory of the bending of plates based? What is the physical meaning and significance of these hypotheses?
56. Write the basic differential equation for deflection of plates. Describe its physical meaning.
57. Examine the following special cases of the general linear equation of the bending of plates: plates of unit width, cylindrical bending, spherical bending.
58. What boundary conditions should be given at the contour of the plate?
59. How should we write the boundary conditions during plate bending: (1) for the deflection at the simply supported edge; (2) for the deflection at the clamped edge; (3) for the deflection at the free edge?
60. What boundary conditions are called geometric? Kinematic? Static? Mixed? Why?
61. Describe the essence of the Navier solution for simply supported plates.
62. Describe the essence of the M. Lévy solution in the theory of plates. What are its advantages over the Navier solution?
63. Obtain the M. Lévy solution for a plate clamped along two opposite edges.
64. Describe the use of the superposition method in the theory of plates having no simply supported opposite edges.
65. Obtain a solution to a problem of bending of a square plate simply supported along the contour and subjected to a uniform load using Navier's method.
66. Solve the preceding problem using M. Lévy's method.
67. Solve problem 65 using the Rayleigh–Ritz method.
68. Explain the physical meaning of von Kármán's equations. When should these equations be used?
69. What is the effect of nonlinearity on the deflections and stresses for rectangular plates supported by a nondeformable contour?
70. Provide the classification of thin plates from the viewpoint of their analysis.
71. What is the essence of Euler's approach in the theory of buckling of bars?
72. Write the Euler formula for the critical force in the buckling of single-span bars. How do the boundary conditions affect the critical force?

73. How does the elastic foundation affect the critical force of a bar subjected to compression?
74. How does initial curvature affect the prebuckling behavior of a beam subjected to compression? Does this curvature affect the critical force?
75. What is the buckling mode for a curved beam subjected to lateral loading? What is the physics of buckling? Write the formula for the critical pressure. What does it depend on?
76. What are the major differences between buckling of bars and buckling of rectangular plates uniformly compressed in one direction?
77. Describe the finite-difference method and the area of its application.
78. Obtain a solution for bending of a beam subjected to bending due to a uniform load on the basis of a finite-difference method. The beam is simply supported at the ends.
79. Describe the major contents of the collocation method. Illustrate its application, using an example of a simply supported beam experiencing uniform lateral loading.
80. Describe the major contents of the mesh method and how it is related to the finite-difference method.
81. What physical laws are used when strain gauges are applied in experimental stress analysis?
82. What are rosettes used for?
83. What are dummy resistances used for in the strain gauge method?
84. What are the physical fundamentals for applying the photoelastic method for stress measurements?
85. What are isoclinics and isochromes? What data can be obtained using isoclinics and isochromes?
86. How are the differences in the principal stresses and their directions determined in photoelastic analysis?
87. Is photoelastic analysis sufficient to determine the principal stresses? If not, how should it be supplemented to evaluate these stresses?
88. In what cases and on what basis can we consider two states of stress similar when photoelastic analysis is applied?
89. What is the essence of the moiré method?

BIBLIOGRAPHY

Baumeister, T., and Marks, L. S., *Mechanical Engineer's Handbook*, McGraw-Hill, New York, 1958.

Blake, A., *Practical Stress Analysis in Engineering Design*, Marcel Dekker, New York, 1982.

Blake, A., *Handbook of Mechanics, Materials, and Structures*, Wiley, New York, 1985.

Bleich, F., *Buckling Strength of Metal Structures*, McGraw-Hill, New York, 1952.

Boresi, A. P., *Advanced Mechanics of Materials*, Wiley, New York, 1985.

Budynas, R. G., *Advanced Strength and Applied Stress Analysis*, McGraw-Hill, New York, 1977.

Bulson, P. S., *The Stability of Flat Plates*, American Elsevier, New York, 1969.

Chia, C. Y., *Nonlinear Analysis of Plates*, McGraw-Hill, New York, 1980.

Crandall, S. H., *Engineering Analysis*, McGraw-Hill, New York, 1956.

Dally, J. W., and Riley, W. F., *Experimental Stress Analysis*, McGraw-Hill, New York, 1965.

Donnell, L. H., *Beams, Plates, and Shells*, McGraw-Hill, New York, 1976.

Faupel, J. H., and Fischer, F. E., *Engineering Design*, Wiley, New York, 1981.

Frocht, M. M., *Photoelasticity*, Wiley, New York, 1957.

Fung, Y. C., *Foundations of Solid Mechanics*, Prentice-Hall, Englewood Cliffs, N. J., 1965.

Gerard, G., *Introduction to Structural Stability Theory*, McGraw-Hill, New York, 1962.

Gere, J. M., and Weaver, W., *Analysis of Framed Structures*, Van Nostrand, Princeton, N. J., 1965.

Gjelsvik, A., *The Theory of Thin Walled Bars*, Wiley-Interscience, New York, 1981.

Griffel, W., *Handbook of Formulas for Stress and Strain*, Ungar, New York, 1966.

Higdon, A. *et al.*, *Mechanics of Materials*, Wiley, New York, 1976.

Hoff, N. J., *The Analysis of Structures*, Wiley, New York, 1956.

Keyser, C. A., *Materials of Engineering*, Prentice-Hall, Englewood Cliffs, N. J., 1956.

Lanczos, C., *The Variational Principles of Mechanics*, Univ. of Toronto Press, Toronto, Ont., 1970.

Langhaar, H. L., *Energy Methods in Applied Mechanics*, Wiley, New York, 1962.

Lipson, C., and Juvinall, R. C., *Handbook of Stress and Strength*, Macmillan, New York, 1963.

McKinley, J. W., *Fundamentals of Stress Analysis*, Matrix, Portland, Ore., 1979.

Norris, C. H., and Wither, J. B., *Elementary Structural Analysis*, McGraw-Hill, New York, 1960.

Panc, V., *Theories of Elastic Plates*, Noordhoff, Leiden, The Netherlands, 1975.

Popov, E. P., *Introduction to Mechanics of Solids*, Prentice-Hall, Englewood Cliffs, N. J., 1968.

Popov, E. P., Nagarajan, S., and Fu, Z. A., *Mechanics of Materials*, Prentice-Hall, N. J., 1976.

Richards, T. H., *Energy Methods in Stress Analysis*, Ellis Horwood, London, England, 1977.

Roark, R. J., *Formulas for Stress and Strain*, McGraw-Hill, New York, 1965.

Roark, R. J., and Young, W. C., *Formulas for Stress and Strain*, McGraw-Hill, New York, 1975.

Sechler, E. E., *Elasticity in Engineering*, Wiley, New York, 1952.

Seeley, F. B., and Smith, J. O., *Advanced Mechanics of Materials*, Wiley, New York, 1952.

Shanley, F. R., *Mechanics of Materials*, McGraw-Hill, New York, 1967.

Shigley, J. E., *Mechanical Engineering Design*, McGraw-Hill, New York, 1963.

Shigley, J. E., *Applied Mechanics of Materials*, McGraw-Hill, New York, 1976.

Szilard, R., *Theory and Analysis of Plates*, Prentice-Hall, Englewood Cliffs, N. J., 1974.

Timoshenko, S. P., and Woinowski-Krieger, S., *Theory of Plates and Shells*, McGraw-Hill, New York, 1959.

Timoshenko, S. P., and Gere, J. M., *Theory of Elastic Stability*, McGraw-Hill, New York, 1961.

Timoshenko, S. P., *Strength of Materials*, Van Nostrand, New York, 1968.

Ugural, A. C., *Stresses in Plates and Shells*, McGraw-Hill, New York, 1981.

Zienkiewicz, O. C., *The Finite Element Method in Engineering and Science*, McGraw-Hill, London, 1971.

Appendix

Tables of Beam Deflections

Statically Determinate Beams 404–409

Statically Indeterminate Beams 410–412

TABLES OF DEFLECTIONS OF STATICALLY DETERMINATE BEAMS

	Beam, loading, and diagrams of moments and shear forces	Elastic curve	Angles of rotation	Bending moments	Shear forces	Support reactions
			1. Simply Supported Beams			
1	Concentrated load in the mid-cross-section	$w=\frac{Pl^3}{16EI}\left[\frac{x}{l}-\frac{4}{3}\frac{x^3}{l^3}+\Big\|_{\frac{l}{2}}\frac{8}{3}\left(\frac{x}{l}-\frac{1}{2}\right)^3\right]$ $w_{max}=\frac{Pl^3}{48EI}$ for $x=\frac{1}{2}$	$w'=\frac{Pl^2}{16EI}\left[1-4\frac{x^2}{l^2}+\Big\|_{\frac{l}{2}}8\left(\frac{x}{l}-\frac{1}{2}\right)^2\right]$ $w'_0=\frac{Pl^2}{16EI}$ for $x=0$ $w'_l=-\frac{Pl^2}{16EI}$ for $x=l$	$M=-Pl\left[\frac{x}{2l}-\Big\|_{\frac{l}{2}}\left(\frac{x}{l}-\frac{1}{2}\right)\right]$ $M_{max}=-\frac{Pl}{4}$ for $x=\frac{l}{2}$	$N=-P\left(\frac{1}{2}-\Big\|_{\frac{l}{2}}1\right)$	$R_1=-\frac{P}{2}$ $R_2=-\frac{P}{2}$
2	Concentrated load in the mid-cross-section	$w=\frac{Pl^3}{6EI}\left[\frac{c'}{l}\cdot\frac{x}{l}\left(1-\frac{c'^2}{l^2}-\frac{x^2}{l^2}\right)+\Big\|_c\left(\frac{x-c}{l}\right)^3\right]$ $w_c=\frac{Pc^2c'^2}{3EIl}$ for $x=c$ $w_{max}=0.0641\frac{Pc'l^2}{EI}\times\left(1-\frac{c'^2}{l^2}\right)^{\frac{3}{2}}$ for $x=\sqrt{\frac{l^2-c'^2}{3}}$, if $c>c'$	$w'=\frac{Pl^2}{6EI}\left[\frac{c'}{l}\left(1-\frac{c'^2}{l^2}-3\frac{x^2}{l^2}\right)+\Big\|_c 3\left(\frac{x-c}{l}\right)^2\right]$ $w'_0=\frac{Pcc'}{6EI}\left(1+\frac{c'}{l}\right)$ for $x=0$ $w'_l=\frac{Pcc'}{6EI}\left(1+\frac{c}{l}\right)$ for $x=l$	$M=-Pl\left[\frac{c'x}{l^2}-\Big\|_c\left(\frac{x-c}{l}\right)\right]$ $M_{max}=-\frac{Pcc'}{l}$ for $x=c$	$N=-P\left(\frac{c'}{l}-\Big\|_c 1\right)$	$R_1=-P\frac{c'}{l}$ $R_2=-P\frac{c}{l}$

In all the tables, the terms after the sign $\Big|\Big|_c$ should be considered only for cross-sections $x > c$.

Cont'd

	Beam, loading, and diagrams of moments and shear forces	Elastic curve	Angles of rotation	Bending moments	Shear forces	Support reactions
3	Two symmetric forces in the span	$w=\frac{Pl^3}{6EI}\left[\frac{x}{l}\left(3\frac{cc'}{l^2}-\frac{x^2}{l^2}\right)+\right.$ $+\Vert_c\left(\frac{x-c}{l}\right)^3+$ $\left.+\Vert_{l-c}\left(\frac{x-c'}{l}\right)^3\right]$ $w_c=\frac{Plc^2}{6EI}\left(3\frac{c'}{l}-\frac{c}{l}\right)$ for $x=c$ $w_{max}=\frac{Pl^2c}{6EI}\left(\frac{3}{4}-\frac{c^2}{l^2}\right)$ for $x=\frac{l}{2}$	$w'=\frac{Pl^2}{2EI}\left[\frac{cc'}{l^2}-\frac{x^2}{l^2}+\right.$ $+\Vert_c\left(\frac{x-c}{l}\right)^2+$ $\left.+\Vert_{l-c}\left(\frac{x-c'}{l}\right)^2\right]$ $w'_0=-w'_l=\frac{Pcc'}{2EI}$	$M=-Pl\left[\frac{x}{l}-\Vert_c\frac{x-c}{l}-\right.$ $\left.-\Vert_{c'}\frac{x-c'}{l}\right]$ $M_{max}=-Pc$ for $c\le x\le c'$	$N=-P(1-\Vert_c 1-\Vert_{c'}1)$	$R_1=R_2=-P$
4	Concentrated moment at a support cross-section	$w=-\frac{\boldsymbol{M}l^2}{6EI}\left(\frac{x}{l}-\frac{x^3}{l^3}\right)$ $w_{max}=-0.0778\frac{\boldsymbol{M}l^2}{EI}$ for $x=0.577l$	$w'=-\frac{\boldsymbol{M}l}{6EI}\left(1-3\frac{x^2}{l^2}\right)$ $w'_0=-\frac{\boldsymbol{M}l}{6EI}$ $w'_l=\frac{\boldsymbol{M}l}{3EI}$	$M=\boldsymbol{M}\frac{x}{l}$ $M_{max}=\boldsymbol{M}$ for $x=l$	$N=\frac{\boldsymbol{M}}{l}$	$R_1=+\frac{\boldsymbol{M}}{l}$ $R_2=-\frac{\boldsymbol{M}}{l}$
5	Two concentrated moments at support cross-sections	$w=-\frac{l^2}{6EI}\frac{x}{l}\left(1-\frac{x}{l}\right)\times$ $\times\left[\boldsymbol{M}'\left(2-\frac{x}{l}\right)+\right.$ $\left.+\boldsymbol{M}''\left(1+\frac{x}{l}\right)\right]$	$w'=-\frac{l}{2EI}\left[\boldsymbol{M}'\left(\frac{2}{3}-\right.\right.$ $\left.-2\frac{x}{l}+\frac{x^2}{l^2}\right)+$ $\left.+\boldsymbol{M}''\left(\frac{1}{3}-\frac{x^2}{l^2}\right)\right]$ $w'_0=\frac{2\boldsymbol{M}'+\boldsymbol{M}''}{6EI}l$ $w'_l=\frac{\boldsymbol{M}'+2\boldsymbol{M}''}{6EI}l$	$M=\boldsymbol{M}'\left(1-\frac{x}{l}\right)+\boldsymbol{M}''\frac{x}{l}$ $M=\boldsymbol{M}'$ for $x=0$ $M=\boldsymbol{M}''$ for $x=l$	$N=-\frac{\boldsymbol{M}'-\boldsymbol{M}''}{l}$	$R_1=-\frac{\boldsymbol{M}'-\boldsymbol{M}''}{l}$ $R_2=\frac{\boldsymbol{M}'-\boldsymbol{M}''}{l}$

Cont'd

	Beam, loading, and diagrams of moments and shear forces	Elastic curve	Angles of rotation	Bending moments	Shear forces	Support reactions
6	Concentrated moment in the span	$w=\frac{\boldsymbol{M}l^2}{6EI}\left[\frac{x}{l}\left(1-3\frac{c'^2}{l^2}-\frac{x^2}{l^2}\right)+\Vert_c 3\left(\frac{x-c}{l}\right)^2\right]$ $w_c=\frac{\boldsymbol{M}cc'}{3EI}\left(\frac{c-c'}{l}\right)$	$w'=\frac{\boldsymbol{M}l}{6EI}\left[1-3\frac{c'^2}{l^2}-3\frac{x^2}{l^2}+\Vert_c 6\left(\frac{x-c}{l}\right)\right]$ $w_0'=\frac{\boldsymbol{M}l}{6EI}\left(1-3\frac{c'^2}{l^2}\right)$ $w_l'=\frac{\boldsymbol{M}l}{6EI}\left(1-3\frac{c^2}{l^2}\right)$ $w_c'=\frac{\boldsymbol{M}l}{3EI}\left(3\frac{cc'}{l^2}-1\right)$	$M=-\boldsymbol{M}\left[\frac{x}{l}-\Vert_c 1\right]$	$N=-\frac{\boldsymbol{M}}{l}$	$R_1=-\frac{\boldsymbol{M}}{l}$ $R_2=\frac{\boldsymbol{M}}{l}$
7	Uniformly distributed load over entire span $Q=ql$	$w=\frac{Ql^3}{24EI}\left(\frac{x}{l}-2\frac{x^3}{l^3}+\frac{x^4}{l^4}\right)$ $w_{max}=\frac{5}{384}\frac{Ql^3}{EI}$ for $x=\frac{l}{2}$	$w'=\frac{Ql^2}{24EI}\left(1-6\frac{x^2}{l^2}+4\frac{x^3}{l^3}\right)$ $w_0'=-w_l'=\frac{Ql^2}{24EI}$	$M=-\frac{Ql}{2}\left(\frac{x}{l}-\frac{x^2}{l^2}\right)$ $M_{max}=-\frac{Ql}{8}=-0.125Ql$ for $x=\frac{l}{2}$	$N=-\frac{Q}{2}\left(1-\frac{2x}{l}\right)$	$R_1=R_2=-\frac{Q}{2}$
8	Uniformly distributed load over a portion of a span $Q=qc'$	$w=\frac{Ql^2c'}{24EI}\left[\frac{x}{l}\left(1+2\frac{c}{l}-\frac{c^2}{l^2}-2\frac{x^2}{l^2}\right)+\Vert_c \frac{(x-c)^4}{l^2c'^2}\right]$	$w'=\frac{Qlc'}{24EI}\left[1+2\frac{c}{l}-\frac{c^2}{l^2}-6\frac{x^2}{l^2}+\Vert_c 4\frac{(x-c)^3}{lc'^2}\right]$ $w_0'=\frac{Qlc'}{24EI}\times$ $\times\left(1+2\frac{c}{l}-\frac{c^2}{l^2}\right)$ $w_l'=-\frac{Qlc'}{24EI}\left(1+\frac{c}{l}\right)^2$	$M=-\frac{Ql}{2}\left[\frac{c'x}{l^2}-\Vert_c \frac{(x-c)^2}{lc'}\right]$ $M_{max}=-\frac{Qc'}{8}\left(1+\frac{c}{l}\right)^2$	$N=-Q\left(\frac{c'}{2l}-\Vert_c \frac{x-c}{c'}\right)$	$R_1=-\frac{Qc'}{2l}$ $R_2=-Q\left(1-\frac{c'}{2l}\right)$

Cont'd

	Beam, loading, and diagrams of moments and shear forces	Elastic curve	Angles of rotation	Bending moments	Shear forces	Support reactions
9	Linearly distributed load over entire span R₁ Q R₂ x l y M N	$w=\frac{Ql^3}{180EI}\left(7\frac{x}{l}-10\frac{x^3}{l^3}+3\frac{x^5}{l^5}\right)$ $w_{max}=0.01304\frac{Ql^3}{EI}\approx\frac{5}{384}\frac{Ql^3}{EI}$ for $x=0.519l$	$w'=\frac{Ql^2}{180EI}\times\left(7-30\frac{x^2}{l^2}+15\frac{x^4}{l^4}\right)$ $w'_0=\frac{7}{180}\frac{Ql^2}{EI}\approx 0.0389\frac{Ql^2}{EI}$ $w'_l=-\frac{2}{45}\frac{Ql^2}{EI}\approx -0.04444\frac{Ql^2}{EI}$	$M=-\frac{Ql}{3}\left[\frac{x}{l}-\left(\frac{x}{l}\right)^3\right]$ $M_{max}=-0.1283Ql$ for $x=0.577l$	$N=-\frac{Q}{3}\left(1-3\frac{x^2}{l^2}\right)$	$R_1=-\frac{Q}{3}$ $R_2=-\frac{2}{3}Q$
10	Linearly distributed load over a portion of a span R₁ Q R₂ x c c′ y M N	$w=\frac{Ql^3}{180EI}\left[\frac{c'}{l}\left(7+6\frac{c}{l}-3\frac{c^2}{l^2}-10\frac{x^2}{l^2}\right)\frac{x}{l}+\Vert_c 3\frac{(x-c)^5}{l^3c'^2}\right]$	$w'=\frac{Ql^2}{180EI}\left[\frac{c'}{l}\left(7+6\frac{c}{l}-3\frac{c^2}{l^2}-30\frac{x^2}{l^2}\right)+\Vert_c 15\frac{(x-c)^4}{l^2c'^2}\right]$ $w'_0=\frac{Qlc'}{180EI}\times\left(7+6\frac{c}{l}-3\frac{c^2}{l^2}\right)$ $w'_l=-\frac{Qlc'}{180EI}\times\left(8+9\frac{c}{l}+3\frac{c^2}{l^2}\right)$	$M=-\frac{Ql}{3}\left[c'\frac{x}{l^2}-\Vert_c\frac{(x-c)^3}{lc'^2}\right]$	$N=-\frac{Q}{3}\left[\frac{c'}{l}-\Vert_c 3\frac{(x-c)^2}{c'^2}\right]$	$R_1=-\frac{Q}{3}\frac{c'}{l}$ $R_2=-Q\left(1-\frac{1}{3}\frac{c'}{l}\right)$

Cont'd

	Beam, loading, and diagrams of moments and shear forces	Elastic curve	Angles of rotation	Bending moments	Shear forces	Support reactions
	2. Cantilever Beams					
11	Concentrated load at the end 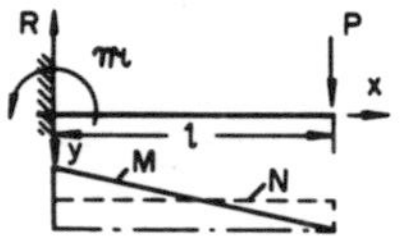	$w=\frac{Pl^3}{3EI}\left[\frac{x^2}{l^2}\left(\frac{3}{2}-\frac{x}{2l}\right)\right]$ $w_{max}=\frac{Pl^3}{3EI}$ for $x=l$	$w'=\frac{Pl^2}{EI}\left[\frac{x}{l}\left(1-\frac{x}{2l}\right)\right]$ $w'_0=0$ $w'_l=\frac{Pl^2}{2EI}$	$M=P(l-x)$ $M_{max}=Pl$ for $x=0$	$N=-P$	$R=-P$ $\boldsymbol{M}=Pl$
12	Concentrated moment at the end 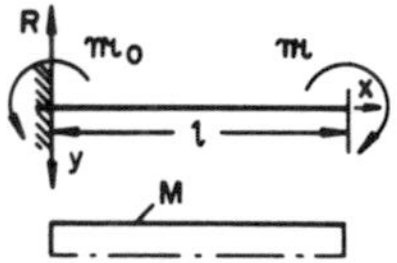	$w=\frac{\boldsymbol{M}x^2}{2EI}$ $w_{max}=\frac{\boldsymbol{M}l^2}{2EI}$ for $x=l$	$w'=\frac{\boldsymbol{M}x}{EI}$ $w'_0=0$ $w'_l=\frac{\boldsymbol{M}l}{EI}$	$M=M_{max}=\boldsymbol{M}$	$N=0$	$R=0$ $\boldsymbol{M}=\boldsymbol{M}$
13	Uniformly distributed load over the entire span 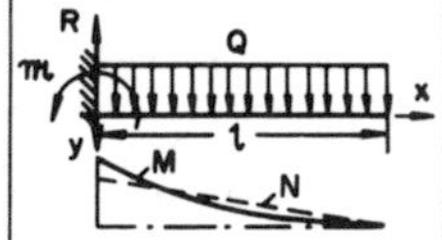	$w=\frac{Ql^3}{24EI}\frac{x^2}{l^2}\left[6-4\frac{x}{l}+\frac{x^2}{l^2}\right]$ $w_{max}=\frac{Ql^3}{8EI}$ for $x=l$	$w'=\frac{Ql^2}{6EI}\frac{x}{l}\times$ $\times\left[3-3\frac{x}{l}+\frac{x^2}{l^2}\right]$ $w'_0=0$ $w'_l=\frac{Ql^2}{6EI}$	$M=\frac{Ql}{2}\left(1-\frac{x}{l}\right)^2$ $M_{max}=\frac{Ql}{2}$ for $x=0$	$N=-Q\left(1-\frac{x}{l}\right)$	$R=-Q$ $\boldsymbol{M}=\frac{Ql}{2}$

Cont'd

	Beam, loading, and diagrams of moments and shear forces	Elastic curve	Angles of rotation	Bending moments	Shear forces	Support reactions
14	Uniformly distributed load over a portion of a span	$w=\frac{Ql^3}{24EI}\left[6\left(1+\frac{c}{l}\right)\frac{x^2}{l^2}-4\frac{x^3}{l^3}+\|_c\frac{(x-c)^4}{c'l^3}\right]$ $w_l=\frac{Ql^3}{24EI}\left(2+6\frac{c}{l}+\frac{c'^3}{l^3}\right)$ for $x=l$ $w_c=\frac{Qlc^2}{4EI}\left(1+\frac{c}{3l}\right)$ for $x=c$	$w'=\frac{Ql^2}{2EI}\left[\left(1+\frac{c}{l}\right)\frac{x}{l}-\frac{x^2}{l^2}+\|_c\frac{(x-c)^3}{3c'l^2}\right]$ $w'_0=0$ $w'_l=\frac{Ql^2}{6EI}\left(1+\frac{c}{l}+\frac{c^2}{l^2}\right)$ $w'_c=\frac{Qlc}{2EI}$	$M=\frac{Ql}{2}\left[1+\frac{c}{l}-2\frac{x}{l}+\|_c\frac{(x-c)^2}{c'l}\right]$ $M_{max}=\frac{Q(c+l)}{2}$ for $x=0$	$N=-Q\left(1-\|_c\frac{x-c}{c'}\right)$	$R=-Q$ $\mathbf{M}=\frac{Q(c+l)}{2}$
15	Linearly distributed load	$w=\frac{Ql^3}{60EI}\frac{x^2}{l^2}\left[10-10\frac{x}{l}+5\frac{x^2}{l^2}-\frac{x^3}{l^3}\right]$ $w_{max}=\frac{Ql^3}{15EI}$ for $x=l$	$w'=\frac{Ql^2}{12EI}\frac{x}{l}\left(4-6\frac{x}{l}+4\frac{x^2}{l^2}-\frac{x^3}{l^3}\right)$ $w'_0=0$ $w'_l=\frac{Ql^3}{12EI}$ for $x=l$	$M=\frac{1}{3}Ql\left(1-\frac{x}{l}\right)^3$ $M_{max}=\frac{Ql}{3}$ for $x=0$	$N=-Q\left(1-\frac{x}{l}\right)^2$	$R=-Q$ $\mathbf{M}=\frac{Ql}{2}$
16	Concentrated moments at the ends*	$w=\frac{Pl^3}{12EI}\left[3\left(\frac{x}{l}\right)^2-2\left(\frac{x}{l}\right)^3\right]$ $w_0=0$ for $x=0$; $w_l=\frac{Pl^3}{12EI}$ for $x=l$	$w'=\frac{Pl^2}{2EI}\left[\frac{1}{l}-\left(\frac{x}{l}\right)^2\right]$ $w'_0=w'_l=0$ $w'_{max}=\frac{Pl^2}{8EI}$ for $x=\frac{l}{2}$	$M=\frac{Pl}{2}\left[1-2\frac{x}{l}\right]$ $M=\frac{Pl}{2}$ for $x=0$ $M=-\frac{Pl}{2}$ for $x=l$	$N=-P$	$R=-P$ $\mathbf{M}_1=-\mathbf{M}_2=\frac{Pl}{2}$

* The rotation angle at the right end is zero

TABLES OF DEFLECTIONS OF SOME STATICALLY INDETERMINATE BEAMS (BEAMS CLAMPED AT BOTH ENDS)

	Beam, loading, and diagrams of moments and shear forces	Elastic curve	Bending moments	Shear forces	Support moments	Support reactions
1	Concentrated force in the span	$w=\frac{PB}{6}\frac{c'^2}{l^2}\frac{x^2}{l^2}\left[3\frac{c}{l}-\frac{3c+c'}{l}\frac{x}{l}\right]+\Vert_c\frac{PB}{6}\left[\frac{x-c}{l}\right]^3$ $w_c=\frac{PB}{3}\frac{c^3}{l^3}\frac{c'^3}{l^3}$ for $x=c$	$M=P\frac{c'^2}{l}\left[\frac{c}{l}-\frac{3c+c'}{l}\frac{x}{l}\right]+\Vert_c P(x-c)$	$N=-P\frac{c'^2}{l^2}\frac{3c+c'}{l}+\Vert_c P$	$\mathcal{M}_1=\frac{cc'^2}{l^2}P$ $\mathcal{M}_2=-\frac{c^2c'}{l^2}P$	$R_1=-\frac{c'^2}{l^3}(3c+c')P$ $R_2=-\frac{c^2}{l^3}(c+3c')P$
2	Two symmetric concentrated forces	$w=\frac{PB}{6}\frac{x^2}{l^2}\left[3\frac{cc'}{l^2}-\frac{x}{l}\right]+\Vert_c\frac{PB}{6}\left[\frac{x-c}{l}\right]^3+\Vert_{c'}\frac{PB}{6}\left[\frac{x-c'}{l}\right]^3$ $w_c=\frac{PB}{6}\frac{c^3}{l^3}\frac{2c'-c}{l}$ for $x=c$	$M=P\left[\frac{cc'}{l}-x\right]+\Vert_c P(x-c)+\Vert_{c'}P(x-c')$	$N=-P+\Vert_c P+\Vert_{c'}P$	$\mathcal{M}_1=\frac{cc'}{l}P$ $\mathcal{M}_2=-\frac{cc'}{l}P$	$R_1=-P$ $R_2=-P$

Notes: 1) The origin is at the left end. 2) In the expressions for the elastic curves the notation $B=\frac{l^3}{EI}$ is used. 3) The support moments are positive if they are directed counterclockwise. 4) Terms after the sign $\Vert_c$ refer only to cross-sections $x>c$.

Cont'd

	Beam, loading, and diagrams of moments and shear forces	Elastic curve	Bending moments	Shear forces	Support moments	Support reactions
3	Uniformly distributed load over the entire cross-section	$w=\frac{QB}{24}\frac{x^2}{l^2}\left(1-2\frac{x}{l}+\frac{x^2}{l^2}\right)$ $w_{\max}=\frac{1}{384}QB$ for $x=\frac{l}{2}$	$M=\frac{Ql}{12}\left(1-6\frac{x}{l}+\right.$ $\left.+6\frac{x^2}{l^2}\right)$ $M\frac{l}{2}=-\frac{Ql}{24}$	$N=\frac{Q}{2}\left(-1+2\frac{x}{l}\right)$	$M_1=\frac{Ql}{12}$ $M_2=-\frac{Ql}{12}$	$R_1=-\frac{Q}{2}$ $R_2=-\frac{Q}{2}$
4	Uniformly distributed load over a portion of the span on the lift	$w=\frac{qB}{24}\left\{\frac{x^2}{l^2}\left[\frac{x}{l}x-2\left(2-\right.\right.\right.$ $\left.-2\frac{c^2}{l^2}+\frac{c^3}{l^3}\right)c\frac{x}{l}+\frac{c^2}{l}\left(6-\right.$ $\left.\left.-8\frac{c}{l}+3\frac{c^2}{l^2}\right)\right]-$ $\left.-\|_c\frac{(x-c)^4}{l^3}\right\}$	$M=\frac{ql^2}{12}\left\{6\frac{x^2}{l^2}-6\frac{c}{l}\times\right.$ $\times\left(2-2\frac{c^2}{l^2}+\frac{c^3}{l^3}\right)\frac{x}{l}+$ $+\frac{c^2}{l^2}\left(6-8\frac{c}{l}+\right.$ $\left.\left.+3\frac{c^2}{l^2}\right)\right\}-\|_c\frac{q}{2}(x-c)^2$	$N=\frac{ql}{2}\left[2\frac{x}{l}-\frac{c}{l}\left(2-\frac{c^2}{l^2}+\right.\right.$ $\left.\left.+\frac{c^3}{l^3}\right)\right]-\|_c q(x-c)$	$M_1=\frac{qc^2}{12}\left(6-8\frac{c}{l}+3\frac{c^2}{l^2}\right)$ $M_2=-\frac{qc^2}{12}\left(4-3\frac{c}{l}\right)\frac{c}{l}$	$R_1=-\frac{qc}{2}\left(2-2\frac{c^2}{l^2}+\right.$ $\left.+\frac{c^3}{l^3}\right)$ $R_2=-\frac{qc}{2}\left(2-\frac{c}{l}\right)\frac{c^2}{l^2}$
5	Same, with the load on the right	$w=\frac{qB}{24}\left\{\frac{x^2}{l}\left[\frac{x}{l}\left(2\frac{c^4}{l^4}-\right.\right.\right.$ $\left.-4\frac{c^3}{l^3}+4\frac{c}{l}-2\right)+1-$ $\left.-6\frac{c^2}{l^2}+8\frac{c^3}{l^3}-3\frac{c^4}{l^4}\right]+$ $\left.+\|_c\frac{(x-c)^4}{l^3}\right\}$	$M=\frac{ql^2}{12}\left[6\left(\frac{c^4}{l^4}-2\frac{c^3}{l^3}+\right.\right.$ $\left.+2\frac{c}{l}-1\right)\frac{x}{l}+\left(1-\right.$ $\left.\left.-6\frac{c^2}{l^2}+8\frac{c^3}{l^3}-3\frac{c^4}{l^4}\right)\right]+$ $+\|_c\frac{q}{2}(x-c)^2$	$N=\frac{ql}{2}\left(\frac{c^4}{l^4}-2\frac{c^3}{l^3}+\right.$ $\left.+2\frac{c}{l}-1\right)+\|_c q(x-c)$	$M_1=\frac{ql^2}{12}\left(1-6\frac{c^2}{l^2}+8\frac{c^3}{l^3}-\right.$ $\left.-3\frac{c^4}{l^4}\right)$ $M_2=-\frac{ql^2}{12}\left(1-4\frac{c^3}{l^3}+3\frac{c^4}{l^4}\right)$	$R_1=-\frac{ql}{2}\left(1-2\frac{c}{l}+\right.$ $\left.+2\frac{c^3}{l^3}-\frac{c^4}{l^4}\right)$ $R_2=-\frac{ql}{2}\left(1-2\frac{c^3}{l^3}+\right.$ $\left.+\frac{c^4}{l^4}\right)$

Cont'd

	Beam, loading, and diagrams of moments and shear forces	Elastic curve	Bending moments	Shear forces	Support moments	Support reactions
6	Linearly distributed load over the entire span $\mathfrak{M}_1$ Q l $\mathfrak{M}_2$	$w=\frac{QB}{60}\frac{x^2}{l^2}\left[\frac{x^3}{l^3}-3\frac{x}{l}+2\right]$ $w\frac{e}{2}=\frac{1}{384}QB$ for $x=\frac{l}{2}$ $w_{max}\approx\frac{1}{382}QB$ for $x=0.525l$	$M=\frac{Ql}{30}\left[10\frac{x^3}{l^3}-\right.$ $\left.-9\frac{x}{l}+2\right]$ $M_{max}=-\frac{Ql}{23.3}$ for $x=0.548l$	$N=\frac{Q}{10}\left[10\frac{x^2}{l^2}-3\right]$	$\boldsymbol{M}_1=\frac{Ql}{15}$ $\boldsymbol{M}_2=-\frac{Ql}{10}$	$R_1=-0.3Q$ $R_2=-0.7Q$
7	Linearly distributed load over a portion of the span $\mathfrak{M}_1$ Q c c' l $\mathfrak{M}_2$	$w=\frac{QB}{60}\left\{\frac{c'}{l}\frac{x^2}{l^2}\left[2+\frac{c}{l}-\right.\right.$ $-3\frac{c^2}{l^2}+\frac{x}{l}\left[2\frac{c^2}{l^2}+\right.$ $\left.\left.\left.+\frac{c}{l}-3\right]\right]+\Vert_c\frac{(x-c)^5}{l^3c'^2}\right\}$	$M=\frac{Qc'}{30}\left[2+\frac{c}{l}-\right.$ $-3\frac{c^2}{l^2}+3\left[2\frac{c^2}{l^2}+\right.$ $\left.\left.+\frac{c}{l}-3\right]\frac{x}{l}\right]+$ $+\Vert_c\frac{Q(x-c)^3}{3c'^2}$	$N=\frac{Qc'}{10l}\left[\frac{c^2}{l^2}+\frac{c}{l}-3\right]+$ $+\Vert_c Q\frac{(x-c)^2}{c'^2}$	$\boldsymbol{M}_1=\frac{Q}{30}\frac{c'^2}{l^2}(2l+3c)$ $\boldsymbol{M}_2=-\frac{Qc'}{30}\left[10\frac{c}{l}+3\frac{c'^2}{l^2}\right]$	$R_1=-\frac{Q}{10}\frac{c'^2}{l^2}\times$ $\times\left[3+2\frac{c}{l}\right]$ $R_2=-\frac{Q}{10}\left[10-\right.$ $\left.-3\frac{c'^2}{l^2}-2\frac{cc'^2}{l^2}\right]$
8	Concentrated moment within the span $\mathfrak{M}_1$ $\mathfrak{M}$ $\mathfrak{M}_2$ c c' l	$w=\frac{\boldsymbol{M}B}{l}\left[\frac{c'}{l}\frac{x^2}{l^2}\left[\frac{3}{2}\frac{c}{l}-\right.\right.$ $\left.-\frac{1}{2}-\frac{x}{l}\frac{c}{l}\right]+$ $\left.+\Vert_c\left[\frac{x-c}{l}\right]^2\frac{1}{2}\right]$	$M=\boldsymbol{M}\frac{c'}{l^2}\left[(2c-c')-\right.$ $\left.-6\frac{c}{l}\right]+$ $+\Vert_c\boldsymbol{M}$	$N=-6\frac{cc'}{l^3}\boldsymbol{M}$	$\boldsymbol{M}_1=\frac{c'(2c-c')}{l^2}\boldsymbol{M}$ $\boldsymbol{M}_2=\frac{c(2c'-c)}{l^3}\boldsymbol{M}$	$R_1=-\frac{6cc'}{l^3}\boldsymbol{M}$ $R_2=\frac{6cc'}{l^3}\boldsymbol{M}$

Index

Absolutely stiff plates, 305
Adiabatic modulus of elasticity, 159
Adiabatic process, 156
Airy, G. B., 69
Airy (stress) function, 69
Albrey, T., 150
All-round compression (*see* Hydrostatic pressure)
Apparent modulus, 153, 154
Argyris, J. H., 12
Aron, H., 11
Axisymmetric deformation, 91

Backward differences, 355
Bars, 56, 170
  See also Beams
Beams, 169
  See also Bending of beams
Beltrami, E., 47
Beltrami–Michell's equations, 47, 161
Beltrami's strength theory, 63
Bending of beams, 169
  application of Castigliano's theorem, 219
  application of Rayleigh–Ritz and Bubnov–Galerkin methods, 226–234
  application of theorem of least work, 223
  beams on elastic foundation, 190, 191
  bending moment and shearing force, 173
  boundary conditions, 179
  combined action of lateral and axial loads, 198
  differential equations of bending, 179
  effect of shear (beams on elastic foundation), 194
  elastoplastic bending, 130
  large deflections (elastica), 264
  nonlinear bending, 201
  reference tables of deflections of beams, 186, 403
  shearing stress and shear deformation, 187
  supports, 170–172
  thermal bending, 209
Bending of frames, 243
  complex frames, 248
  simplest frames, 243
Bending of plates, 256
  circular plates
    axisymmetric bending of absolutely rigid plates, 317
    large deflections under uniform load, 312
    von Kármán's equations, 308
  classification of plates from the viewpoint of their analysis, 305
  Kirchhoff–Love hypotheses, 257–258
  rectangular plates
    application of Rayleigh–Ritz and Bubnov–Galerkin method, 285
    application of superposition method, 283
    boundary conditions, 265
    cylindrical bending, 264
    differential equation of bending, 262
    large deflections, 292
    large deflections under uniform load, 298
    M. Lévy solution, 272
    "membrane" stresses, 292
    Navier's solution, 268
    plate clamped along two opposite edges, 276
    plate under distributed edge moments, 279
    pure bending, 57
    spherical bending, 265
    von Kármán's equations, 298
  thin and thick plates, 256
Bernoulli, J., 11
Bessel, F. W., 349
Bessel equation, 349
Bessel function, 349
Betti, F., 225
Biharmonic equation, 70

Biharmonic operator, 262, 308, 311
Bimaterial assemblies, 81
Bland, D. R., 150
Body forces, 22, 25, 91
Boltzmann, L., 12
Boundary conditions
  in bending of beams, 179
  in bending of plates, 265
  in elasticity theory, 23, 162, 164
  geometric (kinematic), 214, 268
  mixed, 268
  static, 23, 268
  in torsion, 100, 101
  in two-dimensional problems, 70
Bowing (*see* Bending)
Bredt, R., 109
Bredt's theorem in torsion, 109, 164
Bryan, G. H., 340, 346
Bubnov, I. G., 279
Bubnov–Galerkin method, 227, 233, 285, 287, 298, 313, 351
Buckling of beams (bars), 324
  bars on elastic foundation, 327
  curved bars, 335
  effect of initial curvature, 329
  Euler's formulas, 324
  large deflections (elastica), 331
Buckling of plates, 340
  circular plates, 348
  rectangular plates
    plate compressed in two opposite directions, 346
    plate simply supported along opposite edges, 344
    plate uniformly compressed in one direction, 340
Bulk modulus, 43, 126, 161

Canonical equations, 229
Canonical form of stress surface, 29
Cantilever, 52, 71, 172, 173, 188, 231
Castigliano, A., 219
Castigliano's theorem, 219–221
Cauchy, A. L., 11, 29, 35
Cauchy's formulas, 35, 45, 50, 53, 66
Central differences, 355
Circular cylinder, 92
Circular plate
  bending, 308–317
  buckling, 348
  See also Bending of plates and Buckling of plates
Circular ring, 86, 89, 92, 317
Circular shaft, 53, 163
Circulation of a vector, 109
Circulation of the shearing stress, 109

Claugh, R. W., 234
Cleavage plane, 121
Coating, 97
Coaxial cylinders, 94
Coefficient of compliance, 79, 80, 81
Coefficient of thermal expansion, 48, 94, 145
Collocation method, 358
Compatibility conditions, 35, 36, 67
Complementarity theorem, 26
Components of strain, 33, 37
Components of stress, 19, 20, 21
Compression, 20, 55
Concentrated forces and moments, 170
Conditions of compatibility (*see* Compatibility conditions)
Constants, elastic (*see* Hooke's law)
Coordinate functions, 214
Coulomb, C. A., 11, 62
Courant, R., 12
Creep, 144, 146, 147, 154
Curved bar, 334
Cylinder
  symmetrical deformation, 94
  thermal stresses, 94
  thick, under pressure, 93

Darrow, K. K., 3
Deflection of beams (*see* Bending of beams)
Deflection of foundations, 254
Deformation method, 235, 367
Dilatation, 38, 126, 161
Direction cosines, 22, 23, 26
Displacement method, 235, 367
Displacements, 33
Drucker, D. C., 12
Duhamel, J. M. C., 44, 47, 48, 161, 162
Duhamel–Neumann's equations, 47, 48, 162
Duhamel–Neumann's hypothesis, 44, 161

Elastic limit, 42
Elasticity theory
  direct and inverse problems in, 45, 161
  elementary problems in, 49, 163
  hypotheses in, 15, 16
  semi-inverse method in, 46, 161
Ellipsoid, stress, 29, 30, 55
Elliptic bar, 104, 105, 165
Energy
  functional, 227, 286
  law of conservation of, 215
  strain, 60, 228, 235, 239
Enthalpy, 156
Entropy, 155
Equilibrium conditions, 22, 25, 26, 67, 160, 176
Euler, L., 11

Euler's force, 326
Euler formulas, 191

Filon, L. N. G., 74
Filon's solution, 74, 164
Finite differences, 353
Finite-element method, 6, 7, 360, 372
First-order method, 355
Flexible plates of small deflection, 306
Flexural rigidity of plates, 59, 256
Force method, 254
Forward differences, 354
Fourier, J. B., 73, 269, 341
Fourier series, 73, 269, 341
Four-point bending, 174
Fracture, 115
Fracture mechanics, 121
Fracture modes, 117, 118
Fracture tests, 121
Fracture toughness, 121
Frames, 243
  See also Bending of frames
Free energy, 156
Free enthalpy, 156

Galerkin, B. G., 5
Galilei, G., 62
Generalized coordinates, 214
Generalized displacements, 214
Generalized forces, 215
Generalized Prandtl's formula, 104
Germain, S., 11, 256, 262
Gibbs, J. W., 156
Gibbs' potential, 156
Gilbert, D., 5
Grave, D., 1
Green, G., 103, 109
Green's formula, 103, 109
Griffith, A. A., 115
Griffith theory, 115–117, 165
Gross, B., 150

Hardening, 42
Harmonic function, 47
Harmonic operator, 47, 69, 263
Hencky, H., 12
Hermit functions (polynomials), 365
Hill, R., 12
Hollow shaft, torsion of, 110
Homogeneous deformation, 16
Hooke, R., 11
Hooke's law, 16, 32, 38–44, 45, 53, 67, 91, 161, 239
  with consideration of thermoelastic strains, 43, 44
Huber, M. T., 64
Huber's theory, 64
Huber–von Mises–Hencky theory, 64, 65, 126, 163, 164
Hydrostatic pressure, 31, 32, 55, 125, 138
Hypothesis of plane cross sections, 175

Ideal plasticity, 132
Ilyushin, A. A., 128
Ilyushin's theorem on unloading, 130
Influence coefficients, 251
Interpolation polynomials, 354
Invariants of stress (*See* Stress invariants)
Irwin, G. R., 117, 165
Irwin's stress intensity approach, 117–119, 165
Isochrometric lines, 384, 385
Isoclinic lines, 384
Isothermal process, 156
Isotropy, 16

Kármán, T. von, 292, 308
Kelsey, S., 234
Kelvin, Lord (W. Thomson), 4, 12
Kirchhoff, G. R., 11, 256
Klapeiron B. P. E., 219
Klapeiron's theorem, 219
Kron, G., 12

Lagrange, J. L., 11, 262
Lagrange theorem, 216, 217, 227
Lamé, G., 30, 39, 47
Lamé constants, 39, 40
Lamé equations, 47, 48, 162
Lamé problem, 93
Laplace, P. S., 47, 309
Laplacean operator, 47, 309
Lateral force, 173
  (*See also* Shearing force)
Least work, principle of, 222
Leibnitz, G. W., 62
Lévy, M., 256
Lévy theorem, 68, 85, 164
Light, polarized (*see* Polarized light)
Linear viscoelasticity theory, 150
Lode, W., 12
Love, A., E. H., 11

Mariotte, F., 62
Martin, H. C., 234
Maxwell, J. C., 12
Maxwell model, 149
Maxwell's theorem of reciprocity of displacements, 225
Membrane forces, 292, 311
Membranes, 256, 306
Mesh method, 353

Method of initial parameters, 183
Michell, A. G. M., 47
Mises, R. von, 12
Mohr, O., 62
Mohr circles, 62, 63
Mohr strength theory, 62, 65
Moiré method, 387
Moment of inertia, 176, 210
Multiply connected region, 37, 103, 112, 113, 164
Muskhelishvili, N., 91

Nádái, A., 12, 31, 65
Nádái hypothesis, 65
Navier, H., 11, 257
Navier's solution, 268–272
Neutral axis (surface), 176
Newton's interpolation formula, 354
Nonlinearity, 16

Orawan, E., 117
Orthogonal functions, 76

Photoelasticity, 69, 380
Plane strain, 66, 164
Plane stress, 66, 164
Plastic deformation, 16, 125
Plastic flow, 124, 125, 138, 165
Plasticity, 123
Plastics, 143
Plates (*see* Bending of plates and Buckling of plates)
Point-slope formula, 355
Poisson, S. D., 11
Poisson's equation, 100
Poisson's ratio, 40, 43, 161
Polar coordinates, 82, 164
Polariscope, 380, 381
Polarized light, 381
Polarizer, 380
Polymers, 143
Polynomials, 70, 164
Prager, W., 12
Prandtl, L., 101
Prandtl–Reuss equations, 141
Prandtl's formula, 101, 104
Prandtl's membrane analogy, 375
Prandtl's stress function, 100, 164
Prerupture deformation of plastics, 149
Pressure, 92
Principal axes, 26
Principal directions, 26, 37
Principal plane, 27
Principal strain, 37
Principal stress, 27, 30
Principle
  of least work, 222
  of Saint-Venant, 46
  of superposition, 16
  of virtual displacements, 215, 216
Prismatic bar, 49, 164
Proportional limit, 41
Pure bending, 49, 57

Rayleigh, Lord (J. W. Strutt), 226
Rayleigh–Betti theorem, 225
Rayleigh–Ritz method, 226, 229, 286
Recurrence formulas, 356
Reciprocal property of displacements, 224
Reciprocal theorem for forces, 217
Reciprocal theorem for works, 225
Rectangular plate (*see* Bending of plates and Buckling of plates)
Redundant unknowns, 172
Relaxation function, 151
Relaxation of stress, 144, 146
Relaxation time, 150, 152
Research models, 1
Residual deformations, 15
Retardation function, 151
Retardation time, 151
Ribière solution, 73, 164
Rigid-body displacements, 35
Ring (*see* Circular ring)
Ritz, W., 226
Ros, M., 12
Rotation, 33
Ruge, A., 375
Runge–Kutta method, 356
Rupture, 148

Saint-Venant, B. de, 11, 36, 306
Saint-Venant–Lévy–von Mises equations, 141
Saint-Venant's conditions of compatibility, 36, 67
Saint-Venant's hypothesis, 98
Saint-Venant's principle, 46, 161
Saint-Venant's semi-inverse method, 46
Saint-Venant's torsion function, 98, 164
Schleicher, F., 12
Second-order methods, 356
Section modulus, 72, 73, 106, 130
Semi-infinite beam on elastic foundation, 193
Shear modulus, 41, 43, 126, 161
Simmons, E. E., 375
Simple loading, 142
Simply connected region, 37, 103, 161
Small elastoplastic strains, 125
Solution procedures, 45
Specific heat, 157

Spherical tensor, 31, 160
Spring constant, 191
Statically indeterminate problems (systems), 26, 160, 172
Stiffness matrix, 240, 363, 368
Strain
 at point, 33
 compatibility of, 35, 36, 67, 161
 components of, 33, 37
 effective, 127
 invariants, 37, 161
 nominal, 124
 normal and shearing, 33
 plane, 27
 principal axes of, 37, 160
 principal planes of, 37, 160
 rotation, 33
 surface, 37
 tensor, 37, 161
 thermoelastic, 43
 "true," 124
Strain energy, 60, 61, 163
Strain-hardening coefficient, 124
Strain gauges, 375
Strain rosette, 379
Strength
 coefficient, 124
 theories (hypotheses), 60–65, 163
 ultimate, 42
Stress
 at point, 19
 components of, 19, 86, 160
 concentration, 164
 deviatoric, 31, 32, 160
 effective, 126
 invariants of, 26, 29, 32, 126, 160
 mean, 18
 nominal, 123
 normal and shearing, 20, 21
 octahedral, 31, 126, 160
 plane, 27
 principal, 26, 31, 160
 radial, 82–85
 relaxation, 144, 146
 shearing, 20, 26, 30, 82–85, 112, 160, 178, 179
 signs of, 20
 surface, 29
 tensor, 20, 21, 28, 31, 39, 160
 "true," 123
Stress ellipsoid, 30, 160
Stress functions, 69, 100, 164
Stress-strain relation, 42, 124, 156, 158, 161
 See also Hooke's law
Stress surface, 29
Structural analysis, 8–10
Superposition, principle of, 16, 162, 279, 283
Synge, J. L., 12

Temperature, stress due to nonuniform distribution of, 209
Tension, 20, 56
Thermal expansion, 48, 94, 145
Thermal stress, 94, 209
Thermodynamic potential, 155
Thermodynamics of viscoelastic deformation, 155
Thermoplastic materials, 143
Thermosetting materials, 143
Three-point bending, 174
Time effects in plastics, 144–148
Timoshenko, S. P., 279
Topp, J. L., 234
Torque, 55
Torsion
 boundary conditions in, 54
 circular shaft in, 53
 displacements in, 53, 54
 elliptic shaft in, 104
 multiconnected sections, 112
 of prismatic bars, 53
 of rectangular shaft, 107
 section modulus in, 106
 stress function for, 98
 of thin-walled sections, 100, 108, 110
Torsional rigidity, 55, 112, 165
Trapezoidal formula, 356
Tubes
 subjected to internal and external pressure, 93
 torsion of (*see* Torsion)
Turner, M. J., 234
Twist angle, 55, 109, 112, 114

Ultimate axial force, 130
Ultimate bending moments, 130

Variational methods, 213
Virtual displacements, 215, 216
Viscoelasticity, 143
Viscoelastic deformation, 155
Voight, W., 12
Voight-Kelvin model, 149

Wheatstone bridge, 377
Winkler, E., 190
Winkler's hypothesis, 190

Yield stress, 42, 164
Young, T., 11
Young's modulus, 40, 41, 145, 161